Environmental and Microbial Biotechnology

Series Editor

Ram Prasad, Department of Botany, Mahatma Gandhi Central University, Motihari, Bihar, India

Innovative and novel advances in microbial biotechnology are providing great understandings in to the machineries of nature, presenting fascinating prospects to apply principles of biology to different arenas of science. Sustainable elucidations are emerging to address the concerns on improving crop productivity through microbes, depleting natural resources, environmental pollution, microbial degradation of pollutants, nanomaterials, nanotoxicity & safety issues, safety of food & agricultural products etc. Simultaneously, there is an increasing demand for natural bio-products of therapeutic and industrial significance (in the areas of healthcare, environmental remediation, microbial biotechnology). Growing awareness and an increased attention on environmental issues such as climate change, energy use, and loss of non-renewable resources have carried out a superior quality for research that provides potential solutions to these problems. Emerging microbiome approaches potentially can significantly increase agriculture productivity & human healthcare and henceforth can contribute to meet several sustainable development goals.

The main objectives have provided an impetus for research on plants and microorganisms that produce novel bio-products with variable properties and understanding their mechanisms of action at cellular and molecular level. Hence, research activities of the environmental and microbial Biotechnology are comprehensively focused up on major sectors viz., bioresources, biorefining, bioremediation of organic and inorganic pollutants, environmental risk analysis of microorganisms, environmental assessment using microbiological indicators, enzymes for environment, food & industrial applications, nanomaterials & nanotoxicity, sustainable ecobiotechnology, biofertilizer, biocontrol agents for agriculture improvement and natural products for healthcare applications.

This book series is a state-of-the-art for a wide range of scientists, researchers, students, policy makers and academician involve in understanding and implementing the knowledge on environmental and microbial biotechnology to develop biologics for proper health care to continue life in smooth and sustainable strategy without any adverse effect.

More information about this series at http://www.springer.com/series/16324

Inamuddin • Mohd Imran Ahamed •
Ram Prasad
Editors

Microbial Biosurfactants

Preparation, Properties and Applications

Springer

Editors
Inamuddin
Department of Applied Chemistry, Zakir Husain College of Engineering and Technology, Faculty of Engineering and Technology
Aligarh Muslim University
Aligarh, India

Mohd Imran Ahamed
Department of Chemistry, Facualty of Science
Aligarh Muslim University
Aligarh, Uttar Pradesh, India

Ram Prasad
Department of Botany
Mahatma Gandhi Central University
Motihari, Bihar, India

ISSN 2662-1681 ISSN 2662-169X (electronic)
Environmental and Microbial Biotechnology
ISBN 978-981-15-6606-6 ISBN 978-981-15-6607-3 (eBook)
https://doi.org/10.1007/978-981-15-6607-3

This Springer imprint is published by the registered company Springer Nature Singapore Pte Ltd.
The registered company address is: 152 Beach Road, #21-01/04 Gateway East, Singapore 189721, Singapore

Preface

Microbial biosurfactants are biomolecules obtained from bacteria, yeast, fungi, as well as animals and plants. Biosurfactants owing to their biocompatible nature can be used as emulsifying, defoaming, anti-adhesive, antioxidant, antimicrobial, antibiofilm, and antioxidant agents. Microbial biosurfactants have a wide range of applications starting from household detergents to cosmetics, environmental biotechnology to agriculture, and food processing to biomedical industries. This book reviews the applications of microbial biosurfactants in the food industry, such as antioxidant, environmental biotechnology, biomedicine, energy, and household detergents. It is written by experts having considerable experience in the area of preparation and characterization of biosurfactants. It aims to cater to the needs of people from a wide variety of disciplines from food industry to household applications as well as from industries to biomedicine, energy, and environmental biotechnology. It is an in-depth resource for graduate and postgraduate students, researchers, biotechnologists, industrialist, material scientists, and R&D professionals of food industries working in the area of biosurfactants. It contains 14 chapters. The summaries of the chapters are given below.

Chapter 1 discusses the characteristics of biosurfactants advantageous for the food industry. Also, it addresses the concept of food additive and how the multiple functions of biosurfactants can be used, such as emulsifier, antimicrobial, antibiofilm, and antioxidant agent.

Chapter 2 presents a brief description of the main contaminants in food processing and the biosurfactants applied to avoid them. It focuses on the application of glycolipids and lipopeptides as the major microbial biosurfactants, such as preservatives, and antimicrobial, antioxidant, and antibiofilm agents in food processing.

Chapter 3 discusses several aspects of biosurfactants such as sources, structure, isolation, and potential role and applications of biosurfactants. The microbial-derived surfactants can replace synthetic surfactants in a variety of industrial applications as detergents, emulsifiers, solubilizers, and foaming and wetting agents. This chapter discusses the antioxidant property of biosurfactants with examples.

Chapter 4 discusses the classification and physical and chemical properties of biosurfactants. Furthermore, it highlights the factors affecting the production of biosurfactants and the methods of cultivation in the laboratory and industrial scale.

Chapter 5 focuses on various classes of microbial biosurfactants, their basic chemical properties, and the details of genes encoding various types of biosurfactants. It also elaborates the potential industrial and pharmaceutical applications of biosurfactants.

Chapter 6 reviews the biodegradation pathway of PAHs involving enzymes and microorganisms. The production of biosurfactants by microorganisms and their contribution toward the degradation of insoluble PAHs are properly discussed.

Chapter 7 discusses the different types of biosurfactants and the structure of surfactin, along with its membrane interaction, synthesis, and regulation. Breast cancer is a global issue. Surfactin, which is a biosurfactant, can act against breast cancer.

Chapter 8 describes the current research and knowledge of microbial biosurfactants with anticancer potential. Information on the structure and production of biosurfactants is detailed. The main emphasis is given to the anticancer activity in the treatment of breast cancer, lung cancer, leukemia, melanoma, colon cancer, and drug delivery systems.

Chapter 9 discusses the various sources of oil and petroleum pollutants and technologies for their remediation using biosurfactants. A major focus is given on the mode of action of biosurfactant and biosurfactant producing microorganisms for the removal of oil pollutants from soil and water.

Chapter 10 deals with the main classes of biosurfactants with anti-adhesive action. It also discusses the process of microbial adhesion for the formation of biofilms and the studies involving the applications of microbial biosurfactants as disruptive agents on different surfaces.

Chapter 11 discusses the applications of surfactants, especially biosurfactants, on the treatment of waste-activated sludge. Recent developments on value-added biometabolite production, bioenergy recovery, dewatering, decontamination of organic contaminants, and heavy metal removal are covered. Besides, state-of-the-art processes to promote biotransformation of organics from sludge are presented.

Chapter 12 emphasizes the role of biosurfactants in the medical and pharmaceutical industries. Important physicochemical properties of biosurfactants are included. Potential applications in cancer treatment, drug delivery, wound healing, and antimicrobial therapy are described in detail. Moreover, future perspectives are also included.

Chapter 13 discusses different types of biosurfactants and their production. Various applications of biosurfactants are reported especially emphasizing their antibacterial property.

Chapter 14 describes the chemical nature of biosurfactants and media composition required for microbial growth. Even genetic regulation and biosynthesis of surfactants are also discussed with a diverse group of genes. Additionally, the applications of biosurfactants in different industries like textile, leather, petroleum, cosmetic, household detergents, and washing soaps are also discussed.

Aligarh, India Inamuddin
Aligarh, India Mohd Imran Ahamed
Motihari, Bihar, India Ram Prasad

Contents

Application of Microbial Biosurfactants in the Food Industry

1

Italo José Batista Durval, Ivison Amaro da Silva,
and Leonie Asfora Sarubbo

Abstract

The food industry has evolved over the centuries, accompanying changes in dietary habits. Technologies have emerged to improve both the flavor and useful life of food products. The quest for efficient additives that do not affect the health of consumers, increase durability, offer nutraceutical advantages, and satisfy market niches has led to increasing research into natural alternatives for the replacement of synthetic additives. Biosurfactants emerge as a biocompatible solution with multiple functions that can be used as emulsifying, antimicrobial, antibiofilm, and antioxidant agents.

Keywords

Food additives · Food preservatives · Emulsifier · Antimicrobial · Antibiofilm · Antioxidant

1.1 Surfactants in the Food Industry

Technological advances in the nineteenth century enabled the manipulation of food products through the use of additives, favoring the mass production of foods with a pleasant flavor. This led to the further development and ever-increasing use of such

I. J. B. Durval (✉) · I. A. da Silva
Northeast Biotechnology Network (RENORBIO), Federal Rural University of Pernambuco, Recife, Pernambuco, Brazil

Advanced Institute of Technology and Innovation (IATI), Recife, Pernambuco, Brazil

L. A. Sarubbo
Advanced Institute of Technology and Innovation (IATI), Recife, Pernambuco, Brazil

Centre of Science and Technology, Catholic University of Pernambuco, Recife, Pernambuco, Brazil

Inamuddin et al. (eds.), *Microbial Biosurfactants*, Environmental and Microbial Biotechnology, https://doi.org/10.1007/978-981-15-6607-3_1

additives. Moreover, changes in lifestyle in recent decades have transformed eating habits, with the increased incorporation of added ingredients to food products (Onaolapo and Onaolapo 2018).

Chemically synthesized surfactants are used in numerous food formulations. Biosurfactants have also been used for this purpose, such as lecithin and some proteins used in salad dressings and cake frosting. However, chemically synthesized surfactants are toxic, and, therefore, biosurfactants have gained ground due to their biodegradable nature and low toxicity, making these natural compounds more attractive as novel functional additives for use in the food industry (Sharma 2016).

There is a growing demand on the part of consumers for the replacement of more harmful synthetic products with less harmful natural products that perform the same functions. In this context, biosurfactants emerge as a biocompatible solution that can be used as emulsifying, antimicrobial, antibiofilm, and antioxidant agents (Table 1.1) with applications in the formulation of food products (Ranasalva et al. 2014).

Biosurfactants can be used in baked goods and ice creams during the cooking of fats and oils and for the control of consistency, extending the useful life of the product and solubilizing aromatic oils. Rhamnolipids improve the stability of dough as well as the volume, texture, and conservation of baked goods (Vijayakumar and Saravanan 2015).

1.1.1 Food Additives

According to the World Health Organization (2018), food additives are substances added to maintain or improve the safety, freshness, flavor, texture, and/or appearance of foods. Numerous additives have been developed over the years to fulfill the needs of food products, as large-scale food production is very different from making products on a small scale at home. Additives are needed to ensure that processed foods remain in a good state and safe to consume throughout their entire journey from factories or industrial kitchens and during transportation to warehouses and stores until finally reaching the consumer. Food additives are derived from plants, animals, or minerals or may be synthetic. There are thousands of additives—all of which are designed for a specific purpose, making foods safer or more attractive.

Among the fundamental principles of the use of additives, safety is paramount, and the adoption of the procedures necessary for the acquisition of innocuous, healthy foods is indispensable. Therefore, previous to authorization usage, an additive must be submitted to a satisfactory toxicological assessment, considering possible secondary effects such as accumulation, synergisms, or even protectiveness (Brazil 1997, 2002).

Authorization for the use of additives in the food industry and the inspection of these foods follow different standards depending on the country. This responsibility falls to the Food and Drug Administration (FDA) in the United States and the *Agência Nacional de Vigilância Sanitária* (Anvisa [National Health Surveillance

Table 1.1 Biosurfactants with uses in the food industry

Microorganism	Biosurfactant type	Function	Target	References
Bacillus sp. MTCC 5877	Glycolipid	Antiadhesive, antimicrobial	*E. coli*	Anjun et al. (2016)
Candida utilis	Carbohydrate-lipid-protein complex	Emulsifier	–	Campos et al. (2019)
A. piechaudii CC-ESB2	–	Emulsifier, antioxidant	–	Chen et al. (2015)
Pseudomonas aeruginosa ATCC-10145	Rhamnolipid	Antimicrobial	*S. lutea*, *M. luteus*, *B. pumilus*, *P. chrysogenum*, *C. albicans*	El-Sheshtawy and Doheim (2019)
Candida albicans SC5314	Sophorolipid	Emulsifier, antimicrobial	*Pseudomonas aeruginosa* (MTCC 424), *Escherichia coli* (MTCC 723), *Bacillus subtilis* (MTCC 441), and *Staphylococcus aureus* (MTCC 9886)	Gaur et al. (2019)
Candida glabrata CBS138	Sophorolipid	Emulsifier, antimicrobial	*Pseudomonas aeruginosa* (MTCC 424), *Escherichia coli* (MTCC 723), *Bacillus subtilis* (MTCC 441), and *Staphylococcus aureus* (MTCC 9886)	Gaur et al. (2019)
Bacillus licheniformis VS-16	Phospho-lipopeptide	Antibiofilm	*E. coli*	Giri et al. (2017)
Bacillus spp.	Surfactin	Antimicrobial	*Bacillus cereus*, *Listeria monocytogenes*, *Staphylococcus aureus*, *Streptococcus pneumoniae*, *Salmonella typhimurium*, *Serratia marcescens*, and *Klebsiella pneumoniae*	Isa et al. (2020)
Acinetobacter indicus M6	Glycolipoprotein	Antibiofilm, antimicrobial	*P. aeruginosa* ATCC 9027, *Staphylococcus aureus* ATCC 6538	Karlapudi et al. (2020)

(continued)

Table 1.1 (continued)

Microorganism	Biosurfactant type	Function	Target	References
Nesterenkonia sp.	Lipopeptide	Emulsifier, antioxidant, antibiofilm, antimicrobial	*Staphylococcus aureus*	Kiran et al. (2017)
Lactobacillus casei ATCC 393	–	Antioxidant, antibiofilm, antimicrobial	*S. aureus ATCC 6538*, *S. aureus* 9P, *S. aureus* 29P	Merghni et al. (2017)
Candida lipolytica UCP 0988	Rufisan	Antiadhesive, antimicrobial	*S. agalactiae*, *S. mutans*, *S. mutans* NS, *S. mutans* HG, *S. sanguis* 12, *S. oralis* J22	Rufino et al. (2011)
Pseudomonas spp.	Glycolipid	Antibiofilm	*Staphylococcus aureus*	Silva et al. (2017)
Lactobacillus pentosus	Glycolipopeptide	Emulsifier	–	Vecino et al. (2015)
Bacillus subtilis C19	Lipopeptide	Antimicrobial	*C. albicans*	Yuliani et al. (2018)
Bacillus subtilis SPB1	Lipopeptide	Antioxidant	–	Zouari et al. (2016)

Agency]) in Brazil. Countries belonging to economic blocks, such as the European Union, adopt norms determined by the union health board.

1.1.2 Biosurfactants as Food Preservatives

1.1.2.1 Emulsifying Agents

An emulsion is a mixture of different systems consisting of one or more immiscible liquids, which are spread in another in the form of droplets (Santos et al. 2016). These types of systems are characterized by low stability, which can be magnified by surfactants, thus reducing interfacial tension, consequently lessening the surface energy between the two phases, and forestalling the union of the particles by the formation of hysteric and electrostatic barriers (McClements and Gumus 2016).

In food, the emulsifier acts promoting the stability of the formed emulsion. The reduction of surface tension at the oil-water interface is the key point and results in the control of developed droplets, as well as in the stabilization of aerated systems. As a result, the emulsifiers work by improving the consistency and texture of the formulated food, promoting the solubilization of aromas as well. Another function is the increasing of shelf life. Therefore, emulsifiers are essential for food industry, in which water-oil foams and emulsions are often used (Satpute et al. 2018; Radhakrishnan et al. 2011; Patino et al. 2008).

Natural food emulsifying agents derived from plants, such as lecithin and gum arabic, already enjoy considerable participation and acceptance in the market. However, lecithin has functional limitations when employed in products submitted to modern processing conditions, such as microwave cooking and irradiation. Cream, butter, margarine, and mayonnaise are examples of emulsions.

The production of emulsifiers from microbial cultures is an alternative to existing additives, enabling the acquisition of more resistant products that meet the requirements of modern food processing technologies (Nitschke and Costa 2007). There are reports of the use of biosurfactants as emulsifiers for the processing of raw materials with applications in baked goods (affecting the rheological characteristics of dough) and processed meats (emulsification of fat).

Biosurfactants can be used as emulsifiers to control the clustering of fat globules, stabilize aerated systems, and improve the consistency of fatty products. Studies report the use of rhamnolipids to improve the properties of butter, croissants, and frozen pastries (Muthusamy et al. 2008). A bioemulsifier produced by *Candida utilis* was used in salad dressings (Campos et al. 2014, 2015), and a biosurfactant produced by *Bacillus subtilis* was used in the formulation of cookies (Zouari et al. 2016).

Microorganisms such as *Candida utilis*, *Candida valida*, *Hansenula anomala*, *Rhodospiridium diobovatum*, *Rhodotorula graminis*, *Klebsiella* sp., and *Acinetobacter calcoaceticus* and the alga *Porphyridium cruentum* were identified as robust producers of bioemulsifiers, presenting stability superior to commercial emulsifiers (Barros et al. 2007).

1.1.2.2 Antibiofilm Agents

A biofilm is a set of entangled microorganisms that reside within an extracellular polymer matrix adhered to a surface. About 5 to 35% of the biofilm is comprised of microorganisms, and the rest is extracellular matrix (Jamal et al. 2018).

In the food industry, bacterial biofilms are a potential source of contamination, the transmission of disease and the deterioration of food products. Thus, reducing the formation of biofilm on the surface of foods is of extreme importance to providing quality products to consumers (Campos et al. 2013). Methods for the prevention or eradication of biofilm encompass physical, chemical, or biological processes as well as the development of novel or modified packaging materials (Silva et al. 2017). Due to their considerable surface activity, biosurfactants are effective at avoiding the formation of biofilm (Sharma 2016).

Biosurfactants are suggested to reduce hydrophobic interactions, which diminish the hydrophobicity of the surface and impede the adherence of microbes (Mnif and Ghribi 2015). Therefore, biosurfactants have the ability to interrupt the formation of biofilm by controlling microbial interactions with interfaces and altering the chemical and physical conditions of the environment of the developing biofilm (Kiran et al. 2010).

A biosurfactant isolated from the bacterium *Lactobacillus paracasei* exhibited antibiofilm activity against *Candida albicans*, *Staphylococcus aureus*, *Staphylococcus epidermidis*, and *Streptococcus agalactiae*, which are all well-known food pathogens (Gudiña et al. 2010). A biosurfactant derived from *Bacillus licheniformis* reduced formation of *E. coli* biofilm by 54% (Giri et al. 2017). A biosurfactant produced by *Nocardiopsis* sp. MSA13 significantly interrupted the formation of biofilm by *Vibrio alginolyticus* (Kiran et al. 2014). A glycolipid produced by *Brevibacterium casei* significantly inhibited the production of biofilm by *Vibrio* spp., *E. coli*, and *Pseudomonas* spp. (Kiran et al. 2010).

1.1.2.3 Antimicrobial Agents

Numerous secondary metabolites derived from microorganisms have been described as antimicrobial agents. The quest for compounds with action against pathogens emerges from the need to inhibit the activity of these microorganisms, especially in the food industry. Conventional antibiotics no longer serve this purpose due to the occurrence of increasingly resistant pathogens (Sharma 2016).

Biosurfactants have been successfully used to inhibit or retard the development of common microorganisms in food products. The mechanism of action depends on the physicochemical characteristics of the bioactive compound. Some of the mechanisms described include a change in permeability, destabilization and rupture of the cell membrane, or destruction of protein conformations, with the alteration of vital functions, including the generation and transport of energy (Fracchia et al. 2015).

Anjun et al. (2016) performed tests with effective results using a biosurfactant produced by *Bacillus* sp. to inhibit the growth of *E. coli.* Yuliani et al. (2018) investigated a biosurfactant produced by *Bacillus subtilis* C19 and found activity against five pathogens (*Staphylococcus aureus*, *Escherichia coli*, *Pseudomonas aeruginosa*, *Salmonella enterica typhi* and *Listeria monocytogenes*). Biosurfactants isolated from *Pediococcus acidilactici* and *Lactobacillus plantarum* demonstrated antimicrobial activity against *Staphylococcus aureus* CMCC26003 (Yan et al. 2019). Previously, some food-related pathogens, such as *Bacillus cereus*, *Staphylococcus aureus*, and *Micrococcus luteus* decreased their proliferation when rhamnolipids derived from *P. aeruginosa* were added in the culture (Costa et al. 2010).

1.1.2.4 Antioxidant Agents

Antioxidants are a class of food additive used to avoid lipid oxidation, thereby increasing the useful life of food products. The generation of toxic compounds and the development of rancidness and undesirable flavors are the negative effects of lipid peroxidation, leading to a reduction in the quality and safety of the product (Nitschke and Silva 2017).

The necessity of synthetic antioxidants replacement in food industry led the search of natural compounds with antioxidant potential. Biosurfactants have significant antioxidant activity and therefore have the potential to fulfill this purpose (Sharma 2016). Biosurfactants isolated from strains of *Lactobacillus casei* demonstrated satisfactory activity regarding the sequestering of DPPH free radicals, with a greater effect achieved when the concentration of the biosurfactant was increased (Merghni et al. 2017).

Yalçin and Çavuşoğlu (2010) suggest that a lipopeptide produced by *Bacillus subtilis* RWI could be used as a natural alternative antioxidant. The authors evaluated its antioxidant activity based on its redox power, the sequestration of DPPH, and the chelation of iron ions, concluding that the biocompound has good antioxidant capacity for the elimination of free radicals.

1.1.3 Industrial Prospects

The properties of versatility, biocompatibility, and sustainability have led to growing interest in biosurfactants, especially in the food industry, where there is a quest to discover novel compounds for use as gelling, emulsifying, or dispersing agents, such as xanthan gum and emulsan (McClements and Gumus 2016). In addition, increasing of shelf life and additional nutraceutical benefits can be achieved by the use of biosurfactants.

High production costs result primarily from inefficient bioprocessing methods as well as the use of expensive substrates, which account for up to 50% of final cost of the product. In order for biosurfactants to gain a significant portion of the market, there is a need for the use of inexpensive substrates that provide high yields, improvements in processing technologies to facilitate the recovery of the product, greater knowledge in manipulating the metabolism of biosurfactant-producing microorganisms, and the selection of biosurfactants for specific applications (Campos et al. 2013). The development of fermentation technologies will also increase the possibility of modifying the structure and function of biopolymers in a controllable manner, enabling the development of "designer biopolymers."

New ingredients will be developed that can tolerate modern food processing techniques, such as ultrahigh temperatures, extrusion by microwave heating, etc., and can function adequately in new formulations with low salt, fat, and calorie contents. However, the success of food products and ingredients produced by biotechnology also depends on consumer acceptance. There is no doubt that new discoveries in biotechnology will offer solutions to the challenges faced by the food industry (Campos et al. 2013).

Despite the potential applications, the food industry does not yet employ biosurfactants as additives on a large scale. Moreover, the use of biosurfactants as novel ingredients in foods requires the approval of regulatory agencies.

References

Anjun F, Gautam G, Edgard G, Negi S (2016) Biosurfactant production through *Bacillus sp.* MTCC 5877 and its multifarious applications in food industry. Bioresour Technol 212:262–269. https://doi.org/10.1016/j.biortech.2016.02.091.

Barros FFC, Quadros CP, Maróstica Júnior MR, Pastore GM (2007) Surfactina: propriedades químicas, tecnológicas e funcionais para aplicações em alimentos. Quim Nova 30:409–414. https://doi.org/10.1590/S0100-40422007000200031

Brazil (1997) Ministry of Health by National Health Surveillance Agency. Resolution RDC no. 540. Accessed 27 Oct 1997

Brazil (2002) Ministry of Health by National Health Surveillance Agency. Resolution RDC no. 234. Accessed 19 Aug 2002

Campos JM, Stamford TLM, Sarubbo LA, Luna JM, Rufino RD, Banat IM (2013) Microbial biosurfactants as additives for food industries. Biotechnol Prog 29:1097–1108. https://doi.org/10.1002/btpr.1796.

Campos JM, Stamford TLM, Sarubbo LA (2014) Production of a bioemulsifier with potential application in the food industry. Appl Biochem Biotechnol 172:3234–3252. https://doi.org/10.1007/s12010-014-0761-1

Campos JM, Stamford TLM, Rufino RD, Luna JM, Stamford TCM, Sarubbo LA (2015) Formulation of mayonnaise with the addition of a bioemulsifier isolated from Candida utilis. Toxicol Rep 2:1164–1170. https://doi.org/10.1016/j.toxrep.2015.08.009

Campos JM, Stamford TLM, Sarubbo LA (2019) Characterization and application of a biosurfactant isolated from *Candida utilis* in salad dressings. Biodegradation 30:313–324. https://doi.org/10.1007/s10532-019-09877-8

Chen C, Lin T, Shieh Y (2015) Emulsification and antioxidation of biosurfactant extracts from Chinese medicinal herbs fermentation in vitro. J Biosci Bioeng 120:387–395. https://doi.org/10.1016/j.jbiosc.2015.02.010.

Costa SGVA, Nitschke M, Lépine F, Déziel E, Contiero J (2010) Structure, properties and applications of rhamnolipids produced by *Pseudomonas aeruginosa* L2-1 from cassava wastewater. Process Biochem 45:1511–1516. https://doi.org/10.1016/j.procbio.2010.05.033

El-Sheshtawy HS, Doheim MM (2019) Selection of *Pseudomonas aeruginosa* for biosurfactant production and studies of its antimicrobial activity. Egypt J Petrol 23:1–6. https://doi.org/10.1016/j.ejpe.2014.02.001

Fracchia L, Banat JJ, Cavallo M, Ceresa C, Banat IM (2015) Potential therapeutic applications of microbial surface-active compounds. AIMS Bioeng 2:144–162. https://doi.org/10.3934/bioeng.2015.3.144

Gaur VK, Regar RK, Dhiman N, Gautam K, Srivastava JK, Patnaik S, Kamthan M, Manickam N (2019) Biosynthesis and characterization of sophorolipid biosurfactant by *Candida* spp.: application as food emulsifier and antibacterial agent. Bioresour Technol 285:121314. https://doi.org/10.1016/j.biortech.2019.121314

Giri SS, Sen SS, Jun JW, Sukumaran V, Park SC (2017) Role of Bacillus licheniformis VS16-derived biosurfactant in mediating immune responses in carp rohu and its application to the food industry. Front Microbiol 8:2–13. https://doi.org/10.3389/fmicb.2017.00514

Gudiña EJ, Teixeira JA, Rodrigues LR (2010) Isolation and functional characterization of a biosurfactant produced by *Lactobacillus paracasei*. Colloids Surf 76:298–304. https://doi.org/10.1016/j.colsurfb.2009.11.008

Isa MHM, Shamsudin NH, Al-Shorgani NKN, Alsharjabi FA, Sl Kalil M (2020) Evaluation of antibacterial potential of biosurfactant produced by surfactin-producing *Bacillus* isolated from selected Malaysian fermented foods. Food Biotechnol 34:1–24. https://doi.org/10.1080/08905436.2019.1710843

Jamal M, Ahmad W, Andleeb S, Jalil F, Imran M, Nawaz MA, Hussain T, Ali M, Rafiq M, Kamil MA (2018) Bacterial biofilm and associated infections. J Chin Med Assoc 81:7–11. https://doi.org/10.1016/j.jcma.2017.07.012

Karlapudi AP, Venkateswarulu TC, Srirama K, Kota RK, Mikkili I, Kodali VP (2020) Evaluation of anti-cancer, anti-microbial and anti-biofilm potential of biosurfactant extracted from an A*cinetobacter* M6 strain. J King Saud Univ Sci 32:223–227. https://doi.org/10.1016/j.jksus.2018.04.007

Kiran GS, Sabarathnam B, Selvin J (2010) Biofilm disruption potential of a glycolipid biosurfactant from marine *Brevibacterium casei*. FEMS Immunol Med Microbiol 59:43–438. https://doi.org/10.1111/j.1574-695X.2010.00698.x

Kiran GS, Nishanth LA, Priyadharshini S, Anitha K, Selvin J (2014) Effect of Fe nanoparticle on growth and glycolipid biosurfactant production under solid state culture by marine *Nocardiopsis sp*. MSA13A. BMC Biotechnol 14:48. https://doi.org/10.1186/1472-6750-14-48

Kiran GS, Priyadharshini S, Sajayan A, Priyadharsini GB, Poulose N, Selvin J (2017) Production of lipopeptide biosurfactant by a marine *Nesterenkonia* sp. and its application in food industry. Front Microbiol 8:1138. https://doi.org/10.3389/fmicb.2017.01138.

McClements DJ, Gumus CE (2016) Natural emulsifiers—biosurfactants, phospholipids, biopolymers, and colloidal particles: molecular and physicochemical basis of functional performance. Adv Colloid Interface 234:3–26. https://doi.org/10.1016/j.cis.2016.03.002

Merghni A, Dallel I, Noumi E, Kadmi Y, Hentati H, Tobji S, Amor A, Mastouri M (2017) Antioxidant and antiproliferative potential of biosurfactants isolated from *Lactobacillus casei* and their anti-biofilm effect in oral *Staphylococcus aureus* strains. Microb Pathog 104:84–89. https://doi.org/10.1016/j.micpath.2017.01.017

Mnif I, Ghribi D (2015) Review lipopeptides biosurfactants: mean classes and new insights for industrial, biomedical, and environmental applications. Biopolymers 104:129–147. https://doi.org/10.1002/bip.22630

Muthusamy K, Gopalakrishnan S, Ravi TK, Sivachidambaram P (2008) Biosurfactants: properties, commercial production, and application. Curr Sci 94:736–747

Nitschke M, Costa S (2007) Biosurfactants in food industry. Trends Food Sci Technol 18:252–259. https://doi.org/10.1016/j.tifs.2007.01.002.

Nitschke M, Silva SS (2017) Recent food applications of microbial surfactants. Crit Rev Food Sci Nutr 58:631–638. https://doi.org/10.1080/10408398.2016.1208635

Onaolapo AY, Onaolapo OJ (2018) Food additives, food and the concept of ‘food addiction’: is stimulation of the brain reward circuit by food sufficient to trigger addiction? Pathophysiology 25:263–276. https://doi.org/10.1016/j.pathophys.2018.04.002

Patino JMR, Sánchez CC, Niño MMR (2008) Implications of interfacial characteristics of food foaming agents in foam formulations. Adv Colloid Interface 140:95–113. https://doi.org/10.1016/j.cis.2007.12.007

Radhakrishnan N, Kavitha V, Madhavacharyulu E, Gnanamani A, Mandal AB (2011) Isolation, production and characterization of bioemulsifiers of marine bacteria of coastal Tamil Nadu. Indian J Mar Sci 40:76–82

Ranasalva N, Sunil R, Poovarasan G (2014) Importance of biosurfactant in food industry. IOSR JAVS 7(5):6–9. e-ISSN: 2319-2380

Rufino RD, Luna JM, Sarubbo LA, Rodrigues LRM, Teixeira JAC, Campos-Takaki GM (2011) Antimicrobial and anti-adhesive potential of a biosurfactant Rufisan produced by *Candida lipolytica* UCP 0988. Colloids Surf B Biointerfaces 84:1–5. https://doi.org/10.1016/j.colsurfb.2010.10.045

Santos DKF, Luna JM, Rufino RD, Santos VA, Sarubbo LA (2016) Biosurfactants: multifunctional biomolecules of the 21st century. Int J Mol 17:401–430. https://doi.org/10.3390/ijms17030401

Satpute SK, Zinjarde SS, Banat IM (2018) Recent updates on biosurfactants in the food industry. In: Microbial cell factories. CRC, Boca Raton, FL, pp 1–20. https://doi.org/10.1201/b22219-1

Sharma D (2016) Applications of biosurfactants in food. In: Sharma D (ed) Biosurfactants in food. Springerbriefs in food, health, and nutrition. Springer, Berlin, pp 43–80. https://doi.org/10.1007/978-3-319-39415-2_3

Silva SSE, Carvalho JWP, Aires CP, Nitschke M (2017) Disruption of *Staphylococcus aureus* biofilms using rhamnolipid biosurfactants. J Dairy Sci 100:7864–7873. https://doi.org/10.3168/jds.2017-13012

Vecino X, Barbosa-Pereira L, Devesa-Rey R, Cruz JM, Moldes AB (2015) Optimization of extraction conditions and fatty acid characterization of *Lactobacillus pentosus* cell-bound biosurfactant/bioemulsifier. J Sci Food Agric 95:313–320. https://doi.org/10.1002/jsfa.6720

Vijayakumar S, Saravanan V (2015) Biosurfactants-types, sources and applications. Res J Microbiol 10:181–192. https://doi.org/10.3923/jm.2015.181.192

World Health Organization (WHO) (2018) Food additives. https://www.who.int/news-room/fact-sheets/detail/food-additives. Accessed 27 Nov 2018

Yalçin E, Çavuşoğlu K (2010) Structural analysis and antioxidant activity of a biosurfactant obtained from *Bacillus subtilis* RW-I. Turk J Biochem 35:243–247

Yan X, Gu S, Cui X, Shi Y, Wen S, Chen H, Ge J (2019) Antimicrobial, anti-adhesive and anti-biofilm potential of biosurfactants isolated from *Pediococcus acidilactici* and *Lactobacillus* plantarum against *Staphylococcus aureus* CMCC26003. Microb Pathog 127:12–20. https://doi.org/10.1016/j.micpath.2018.11.039

Yuliani H, Perdani MS, Savitri I, Manurung M, Sahlan M, Wijanarko A, Hermansyah H (2018) Antimicrobial activity of biosurfactant derived from *Bacillus subtilis* C19. Energy Procedia 153:274–278. https://doi.org/10.1016/j.egypro.2018.10.043

Zouari R, Besbes S, Ellouze-Chaabouni S, Ghribi-Aydi D (2016) Cookies from composite wheat–sesame peels flours: dough quality and effect of Bacillus subtilis SPB1 biosurfactant addition. Food Chem 194:758–769. https://doi.org/10.1016/j.foodchem.2015.08.064

Microbial Biosurfactants for Contamination of Food Processing

2

Isadora Ferreira da Silva, Nathália Andrade Neves, Bianca Marise Pereira e Silveira, Kele Aparecida Costa Vespermann, and Maria Emília Rodrigues Valente

Abstract

Food security is one of the biggest concerns in food processing. In recent decades, the effects of contaminants in food crops are currently compromising food security and human health. Food contamination can be microbiological, chemical and physical. It can occur on different steps of food processing, such as transport, storage and packaging of raw or processed food, as well as during heating processes. Therefore, substances including the additives can help to maintain the food security during food processing until it reaches the consumers. They can provide food preservation, maintaining freshness and preventing bacterial contamination, among many others. The additives from natural sources have been receiving more attention from food manufacturers when compared to the synthetic ones, due to their higher quality and safety. An example of natural additives are biosurfactants, which are derived from microorganisms. Interests in the use of biosurfactants have been increasing in the food market, as a result of their capacity to replace synthetic additives in the food industry. The objective of this chapter is to present the main microbial biosurfactants used to avoid contamination during food processing. We briefly discuss their potential applications as food preservatives, presenting antimicrobial, antioxidant and antibiofilm activities against food pathogens.

I. Ferreira da Silva (✉)
Department of Applied Biochemistry, Federal University of Viçosa—UFV, Viçosa, MG, Brazil

N. Andrade Neves · B. M. Pereira e Silveira · K. A. Costa Vespermann
Institute of Science and Technology, Federal University of Valleys of Jequitinhonha and Mucuri—UFVJM, Diamantina, MG, Brazil

M. E. Rodrigues Valente
Department of Food Engineering, Federal University of Espírito Santo—UFES, Alegre, ES, Brazil

Inamuddin et al. (eds.), *Microbial Biosurfactants*, Environmental and Microbial Biotechnology, https://doi.org/10.1007/978-981-15-6607-3_2

Keywords

Biosurfactants · Microorganisms · Contamination · Additives · Food processing · Biofilms

Summary This chapter presents a brief description of the main contaminants in food processing and the biosurfactants applied in avoiding them. It is focused on the application of glycolipids and lipopeptides as the main classes of microbial biosurfactants, as preservatives, antimicrobial, antioxidant and antibiofilm agents during food processing.

2.1 Introduction

2.1.1 Food Contamination

The origins of food contaminants can be microbiological, physical or chemical. Data from the World Health Organization informs around 600 million people get sick after eating contaminated food, causing 420,000 deaths per year (World Health Organization 2019). Understanding the origin and when it happens during food processing can contribute to a more effective control and avoid contamination.

Chemical contaminants are one of the serious sources of food contamination. They include the environmental contaminants (organic pollutants, heavy metals, pesticides, disinfectants, detergents, deodorants, veterinary products); food processing contaminants that are formed during heating, cooking or packaging (such as furanes, polycyclic aromatic hydrocarbons and acrylamide) and processing contaminants that are released during cooking, processing or packaging (McKay and Scharman 2015).

In a microbiological contamination, bacteria, virus and parasites can spread in food and cause harm to humans. The most common sources of microbiological contamination include *Salmonella* spp., *Bacillus cereus*, *C. perfringens*, *E. coli*, *Shigella* spp., *Listeria monocytogenes*, *Clostridium botulinum*, *Yersinia enterocolitica* and *Vibrio cholera* (Hussain 2016). There is also the physical contamination, which happens when any foreign matter enters food and causes illness or injury when consumed. Examples of physical hazards could include glass, metal and plastic fragments, hair, fingernails, dirt and insects (Echavarri-Bravo et al. 2017; Aguiar et al. 2018).

The contamination could occur naturally or be introduced by humans. A significant amount of food contamination can occur during food processing steps, packaging, transportation and storage of raw or processed foods (Rather et al. 2017).

2.1.2 Contamination in Food Processing

Food processing involves activities for the conversion of raw food into more stable final products. Its goal is to increase the quality, nutritive value, taste and shelf life. The main stages of food processing include heating, packaging, storage, distribution and transport (Sethu and Ananth Viramuthu 2008). In order to avoid contamination in food processing, it is important to identify the most likely contaminants from each of these stages.

The environment of food processing inside the industry is usually one of the main sources of contamination. The exposure to contaminated surfaces plays a major role in possible food contamination in the industries. Equipment, utensils, hallways, workbenches and pipes are the main sources of contamination in industries. These contaminants can be in contact with the food or can be transferred to the food by air, other materials and people. Employees can contaminate the food during its processing through the direct transfer of microorganisms from their body to the food or by carrying them from a contaminated area to another (Masotti et al. 2019). The equipment design is also important in avoiding contamination as it influences the cleaning and rinsing operations and therefore the removal of contaminants (Faille et al. 2018). Good hygiene practices can limit the microbial contamination, decreasing cross-contamination and formation of biofilms. Once it is formed, the biofilm becomes a source of contamination, and it is considered one of the main concerns for the food industry (Carrasco et al. 2012).

In addition to the processing of food in industries, food contamination can also take place during transportation, including the transportation of raw food to the industries and the processed food from the industries to the stores. Around 200 billion metric tons of food is transported in the world every year; 35% is transported by land, 60% by sea and 5% by air. The possibility of contamination during transport and storage of food is highlighted by the large amount of food transported, along with the handling requirements for each food product (Ackerley et al. 2010). Food contamination during transport can occur from vehicle exhausts of fuel or due to cross-contamination. This contamination can often happen in long-distance ships, through the use of chemicals present in cleaning products or other sources (Nerín et al. 2016). The packages used in food are often not tested to resist and protect the food for transportation across long distances, leading to food contamination (Rather et al. 2017). A good example of cross-contamination happened in 1994 when around a quarter of a million people got gastroenteritis after eating Schwan's ice cream. The ice cream contained *Salmonella* that came from eggs that were earlier transported using the same trucks (Hennessy et al. 1996).

Heating treatment is the most recognized and used approach for food process in the industry or at home. The heating combined with external factors during cooking causes the release of toxic substances (Bordin et al. 2013). These compounds including acrylamide, furanes, nitrosamines and chloropropanols are generated during heating, cooking, fermentation, canning and backing (Nerín et al. 2016). For example, when using a microwave, the packaging material used such as plastics, paperboard and composites can have their components migrated to the food during heating. Fasano et al. (2015) detected significant amounts of plastic components (phthalates, bisphenol A (BPA), polybrominated diphenyl ethers (PBDE) and tetrabromobisphenol A (TBBPA)) after using a microwave to heat food in a polyethylene packaging during 3 min at 800 W.

The contamination during food storage is mainly due to changes in the storage conditions (high temperature and humidity) that can affect the packaging material properties (Nerín et al. 2016). Direct sunlight and packaging can accelerate the deterioration of food, causing an adsorption of unwanted off odours (Rather et al. 2017). Another contamination factor is the moisture, which can increase the susceptibility of the food to microbial contamination, leading to modifications of its texture and decreasing shelf life (Gaikwad et al. 2019). The contamination also depends on the type of food. For example, dry and canned foods usually present a long shelf life; however, they may deteriorate in colour, flavour and nutritive value over time (McCurdy et al. 2009). Fresh meat, such as seafood, beef and poultry should be maintained at low temperatures in a freezer and refrigerator. Choi et al. (2020) demonstrated that storing beef at 4 °C was essential for the maintenance of food quality, reducing significantly the growth of pathogens such as *E. coli* when compared to storage at 25 °C.

These changes during storage also depend on the packaging material, as it should exhibit very good barrier properties. The packaging needs to provide physical protection and helps with the increase in the shelf life of the product. The use of stabilizers, plasticizers, antioxidants and shipping agents are common in packaging processes to enhance the characteristics of the packaging material (Conte et al. 2013). However, some substances from the packaging material can be transferred to the food and can cause health risks to the consumers if they have toxic effects (Lau and Wong 2000). Therefore, strict legislation is applied worldwide for the use of substances in packaging materials. Adverse effects can be also caused by substances that were not intentionally added to the food packaging and are present in food. They can come from the packaging under a heating process as mentioned before and be generated from reactions of substances present in the packaging or by the reaction of substances with foodstuffs. Bauer et al. (2019) identified 50, in which 8 were NIAS, in baby food from contact of the food with the polyurethane layer of the plastic multilayer packaging.

The use of preservatives in foodstuff is crucial to prevent contamination and deterioration of food. These substances can be used during transportation, processing, storage and packaging of food. Benzoates, sulphites, sorbates, propionates, nitrites and parabens are the most used antimicrobials used in food. Regarding the synthetic antioxidants, *tert*-butylhydroquinone (TBHQ), butylated hydroxytoluene (BHT), propyl gallate and butylated hydroxyanisole (BHA) are considered as the most common (Carocho et al. 2015).

Although studied for decades, the use of these synthetic preservatives to avoid food contamination and deterioration can cause health problems and negatively affect the environment (Botterweck et al. 2000; Iammarino et al. 2013; Vandghanooni et al. 2013). For this reason, the interest in natural, safe and environmentally friendly preservatives has increased. Among these, the microbial surfactants have been extensively studied during the last few years. From this point on, we will discuss the use and relevance of the microbial biosurfactants in preventing contamination during food processing.

2.2 Microbial Biosurfactants Use in Food Processing

Surfactants are compounds that create micelles in a solution and adsorb to the interfaces between a solution and a different phase (gases-liquids, liquid-liquid) leading to reductions in the tension surface. This is possible because of the two different functional groups with different affinity within their molecule. The surfactants are composed of amphiphilic molecules, with a chemical structure consisting of a hydrophilic group (carbohydrates or amino acids) and a hydrophobic one (fatty acids). In food processing, the microbial surfactants can be used as emulsifying, antimicrobial, anti-adhesive, antibiofilm and antioxidant agents (Sharma et al. 2018).

The microbial surfactants present a broad variety of compounds produced by bacteria, yeast or fungi. They are favoured over the synthetic ones in this industry due to the non-toxicity nature, excellent biodegradability, high surface/interfacial activity, biocompatibility, stability under extreme conditions of pH, temperature and sanity (Nitschke and Silva 2018; Parthasarathi and Subha 2018).

The production of biosurfactants has been extensively studied. Inexpensive raw substrates available in high amounts can be used to produce biosurfactants. Agricultural and industrial wastes, by-products including hydrocarbons and oil waste can be used for the production of these compounds, enhancing the cost-effectiveness of the biosurfactant production process. The biosurfactants are generally extracellularly produced (Inès and Dhouha 2015a). Downstream and recovery process has also been widely investigated, since the most methods used increase significantly the costs of biosurfactant production process (Jimoh and Lin 2019). Improvement of microbial strain by using genetic strategies, optimization of media composition and development of scaled-up methods, among others, can also help in achieving a viable commercial production of high quantities of biosurfactants.

In food processing compounds like phospholipids, fatty acids, lipoheteropolysaccharides and protein-sugar-lipid complex molecules are used as biosurfactants (Nitschke and Silva 2018). They can be divided in classes according to their chemical composition, molecular weight, physicochemical properties and microbial source (Naughton et al. 2019). The high molecular weights are amphipathic polysaccharides, proteins, lipopolysaccharides and lipoproteins. The main group is comprised of compounds with lower molecular weights, including fatty acids, glycolipids, lipopeptides and phospholipids. Among these, glycolipids and lipopeptides are the most widely applied to avoid contamination in food processing.

2.2.1 Glycolipids

Glycolipids are the most popular group of biosurfactants. It presents carbohydrate and a fatty acid as the hydrolytic and hydrophobic portions, respectively (Abdel-Mawgoud and Stephanopoulos 2018). When compared with synthetic surfactants, the natural glycolipids present better surfactant activity. According to Liu et al. (2020), the natural glycolipids present a higher molecular richness than the synthetic ones. They present a distinctive distribution of the polarity groups over the glycolipid molecule and more branched structures in comparison with the synthetic glycolipids (Abdel-Mawgoud and Stephanopoulos 2018). However, the industrial production and application in large scale of these natural surfactants are not viable yet. As it is hard to separate and purify these natural surfactants, it is still not possible to fully understand their structure-activity relationships (Liu et al. 2020).

The glycolipids can be categorized into subclasses according to the carbohydrates and lipid portions, as follows: rhamnose lipids, trehalose lipids, sophorose lipids, cellobiose lipids, mannosylerythritol lipids (MEL), lipomannosyl-mannitols, lipoarabinomannans, lipomannans, diglycosyl diglycerides, monoacylglycerol and galactosyl diglyceride (Mnif and Ghribi 2016). Among these, rhamnolipids, sophorolipids and trehalolipids are the best known subclasses of glycolipids.

Rhamnolipids are an extensively studied biosurfactant. They are characterized by rhamnose molecules linked to β-hydroxydecanoic acid molecules (Inès and Dhouha 2015a). Their production was first described using *P. aeruginosa* in 1949. The fatty acids present in the rhamnolipid molecules ranges from 8 to 16 carbons. The β-hydroxydecanoic acid is predominant in the rhamnolipid produced by *P. aeruginosa*, whereas β-hydroxytetradecanoic acid is mostly found in the molecules produced by *Burkholderia* sp. (Henkel et al. 2012).

Due to the several advantages of the biosurfactants such as rhamnolipids, many companies have shown interests in exploiting the production of this biosurfactant as follows: Jeneil biosurfactant (USA), Paradigm Biomedical Inc. (USA), GlycoSurf LLC (USA), AGAE Technologies LLC (USA) Logos Technologies LLC (USA), Rhamnolipids Companies Inc. (USA), TeeGene Biotech Ltd. (United Kingdom) and Urumqi Unite Bio-Technology Co., Ltd. (China).

Regarding the substrates used for rhamnolipids, the most commonly ones are plant oils, sugars and glycerol. It has been found on literature, the utilization of oil wastes such as vegetable oil, palm oil, mango kernel oil, glycerol and glycerin, among others (Table 2.1).

Investigations to optimize the costs and the production of these compounds are important to enable bioprocesses, considering the potential associated with rhamnolipid. Genetic engineering has been used in order to improve its synthesis. Boles et al. (2005) used *Pseudomonas aeruginosa*, and strains with the insertion/exclusion of rhlAB, for the production of rhamnolipid using a complex medium (Na_2HPO_4; KH_2PO_4; NaCl; $CaCl_2$; $MgSO_4$; glucose, glutamate), for 48 h at 37 °C. The rhamnolipid yield improved from 0.1 mg/mL to 0.5 mg/mL when using a genetically modified strain in comparison with the wild strain. Zheng et al. (2020) also achieved an increase of 32.63% on the rhamnolipid yield by *Pseudomonas*

Table 2.1 Conditions of fermentation for production of glycolipids by microorganisms

Biosurfactant	Microorganism	Process conditions	Yield	Reference
Sophorolipids	*Rhodotorula babjevae* YS3	10% glucose, 19 °C, 200 rpm, 72 h, 5% inoculum	19.0 g/L	Sen et al. (2017)
Sophorolipids	*Candida albicans*	30 °C, 72 h, 2% glucose, 150 rpm	1320 mg/L	Gaur et al. (2019)
Trehalose	*Rhodococcus fascians*	28 °C, 24 h, Davis minimal media	0.14 mg/mL	Janek et al. (2018)
Mannosylerythritol lipid	*Pseudozyma aphidis*	28 °C, 180 rpm, soybean oil, 240 h	61.50 g/L	Niu et al. (2019)
Sophorolipids	*C. Bombicola*	20 g/L residual oleic acid, 30 °C, 450 rpm, 288 h, 10% inoculum	69.83 g/L, 0.24 g/L/h	Silveira et al. (2019)
Sophorolipid	*C. Bombicola*	200–800 rpm, 50 g/L rapeseed oil, 25 °C, 1023 h	20.22 g/g 1.07 g/L/h	Dolman et al. (2017)
Sophorolipids	*C. Bombicola*	25 °C, 200 rpm, 48 h, 168 h, 1–5 g/L rapeseed oil, 20–60 g/L glucose	11.4 g/g 1.45 g/L/h	Liu et al. (2019)
Sophorolipids	*C. Bombicola*	5 mL food waste hydrolysate, 100–120 g/L glucose 24 h, 30 °C, 150 rpm, 9% inoculum	0.26 g/g 0.39 g/L/h	Kaur et al. (2019)
Rhamnolipid	*Pseudomonas stutzeri*	46.55 g/1 glycerol, 150 rpm, 30 °C, 144 h	4.78 g/L	Sheikh et al. (2019)
Rhamnolipid	*P. aeruginosa*	25.4–116.0 g/L glycerin, 30 °C, 170 rpm, 288 h	17.6 g/L	Dobler et al. (2020)
Rhamnolipid	*P. aeruginosa*	12% rice-based distillers' dried grains with solubles, 48 h, 35 °C, 180 rpm	14.87 g/L	Borah et al. (2019)
Rhamnolipid	*P. guguanensis*	Vegetable oil, peptone + yeast extract 150 rpm, 30 °C, 168 h,	12.5 mg/mL	Ramya Devi et al. (2018)
Rhamnolipid	*P. aeruginosa*	10 g/L palm oil waste, glucose, 37 °C, 36 h	3.4 g/L	Radzuan et al. (2018)
Rhamnolipid	*P. aeruginosa*	Mango kernel oil, glucose, 120 h, 30 °C	2.8 g/L	Reddy et al. (2016)

aeruginosa after performing the fermentation in a 1.5 L bioreactor through a new continuous production process, based on a cyclic fermentation coupled with the fractionation of foam when compared to the process performed in a decoupled system.

Sophorolipid structure comprises a hydroxyl fatty acid and a disaccharide sophorose linked by β-1, 2 bond (Varjani and Upasani 2017). This biosurfactant can present an acidic structure, when the fatty acids' carboxylic end is free, or a lactonic ring structure, when this end is esterified (Sen et al. 2017). The composition of the sophorolipids such as the length of carbon chain and the fatty acid structure and the proportion of the acidic and lactonic forms depend on factors such as medium composition, environment conditions (pH, temperature, aeration) and strain used for their production (Díaz De Rienzo et al. 2015; Oliveira et al. 2015).

Different from rhamnolipids, sophorolipids are synthesized by non-pathogen yeast strains. The most common ones used are *Centrolene petrophilum*, *Candida bombicola*, *Rhodotorula bogoriensis* and *Candida apicola* (Banat et al. 2010). Díaz De Rienzo et al. (2015) obtained a mixture of 45% (v/v) of acidic and lactonic congeners of sophorolipids using the strain *Candida bombicola* ATCC 22214, in a process conducted at 30 °C, using 10% (w/v) glucose, 1% (w/v) yeast extract and 0.1% (w/v) urea (GYU medium). The same strain was also used in a process conducted with a working volume of 10 L, using *Candida* growth media for 7 days at 26 °C, achieving around 90% of acidic and lactonic congeners of sophorolipids (Zhang et al. 2016).

The sophorolipids usually present antimicrobial and antibiofilm activities important to avoid contamination in food processing (Sharma et al. 2018; Jimoh and Lin 2019). The production of this surfactant has shown greater interest from the companies when compared to the rhamnolipids (Sen et al. 2017). The current main companies exploring the production of this biosurfactant are Saraya (Japan),

Ecover Eco-Surfactant (Belgium), Groupe Soliance (France), MG Intobio Co., Ltd. (South Korea) and SyntheZyme LLC (USA).

Trehalolipids is another glycolipid biosurfactant, which is composed of a disac charide trehalose connected at C_6 to long-chain α-spread and β-hydroxy unsaturated fats (Varjani and Upasani 2017). This subclass of biosurfactants is considered chemically stable. Changes on pH values, salt concentrations and temperature usually do not cause any modification in their surface activities (Kitamoto et al. 2009). The first biosurfactant of this subclass was a trehalose dimycolate (TDM) described in 1930s and obtained by *Mycobacterium tuberculosis*. It plays a major role in infections caused by this pathogen (Kuyukina et al. 2015). The TDM was later found to be also produced by the genus *Nocardia* and *Corynebacterium*.

Similar to sophorolipids, the structure of the trehalolipids such as the size and structure of fatty acids, the quantity of carbon molecules and the degree of unsaturation depend on the strain and growth conditions used (Roy 2017). This subclass presents a high variety of structure. In addition to trehalose dimycolate (TDM), the trehalose trimycolates, mono-, tetra- and octa-acylated derivatives also represent anionic trehalose-type molecules (Niescher et al. 2006).

Trehalolipids can be produced by different microorganisms, including *Nocardia, Williamsia, Mycobacterium, Corynebacterium, Dietzia, Gordonia, Tsukamurella, Skermania* and *Rhodococcus.* The most widely studied subclass is trehalose dimycolates obtained from *Rhodococcus erythropolis* (Banat et al. 2010). The trehalolipids obtained from *Rhodococcus* are commonly involved in adhesion to surfaces and increase solubility of hydrophobic compounds. They assist the cells in accessing hydrophobic substrates, by promoting a contact between the cells and the substrates or by indirect contact through the adhesion to the emulsified oil (Bages-estopa et al. 2018).

Table 2.1 presents the microorganisms used, cultivation conditions and fermentation strategies in the production of glycolipids.

2.2.2 Lipopeptides

Lipopeptides are consisted of a fatty acid portion connected to a peptide chain. They can be divided into three different subclasses, depending on the sequence of amino acids presented: surfactins (cyclic lipoheptapeptide linked to a β-hydroxy-fatty acid group), iturins (heptapeptides cyclized by amide bond formed between the α-COO group of the seventh amino acid and the β-NH_2 group of the β-fatty acid) and fengycins (β-hydroxy fatty acid connected to a peptide domain, composing of 10 amino acids. 8 of them are presented in a cyclic structure) (Hentati et al. 2019).

Surfactin is the most widely studied natural lipopeptides. It is characterized as highly surface active and water soluble, and it is consisted of four isomers (surfactin A–D). This biosurfactant is considered as cyclic lipopeptides, comprising of a cyclic heptapeptide structure connected with a fatty acid containing 13–15 carbons. The type of microorganism and culture conditions during its production will influence on the amino acids composition and fatty acids presented in its molecule. The Asp and Glu residues are generally placed in the heptapeptide ring. The ring then presents a saddle shape containing two negative residues on each top end (Liu et al. 2020). Surfactin can be produced by a variety of gram-positive strains of endospore-producing, *Bacillus subtilis* (de Araujo et al. 2011).

A concentration of 20 μM of surfactin can cause a decrease in surface tension of water from 72 to 27 mN/m, which is significantly lower when compared to the reductions in surface tensions of most biosurfactants found in literature. Surfactin also presents a critical micelle concentration (CMC) of 23 mg/L in water, which is significant below the CMC of other biosurfactants (Chen et al. 2015). Surfactin has been considered as antiviral, antibacterial, antifungal, anti-mycoplasma and antibiofilm on metallic and polypropylene surfaces and also has haemolytic properties, presenting cation-carrier and pore-forming effects (Banat et al. 2010; Chen et al. 2015; Sharma et al. 2018; Meena et al. 2020).

The commercial surfactin produced by Kaneka Corporation (Japan) and Soft Chemical Laboratories (South Africa) are used for contamination in food processing.

Table 2.2 Conditions of fermentation for production of lipopeptides by microorganisms

Biosurfactant	Microorganism	Process conditions	Yield	Reference
Surfactin	*Bacillus subtilis*	30 °C, Luria-Bertani broth, 200 rpm, 72 h	547.0 mg/L	Meena et al. (2020)
Surfactin	*B. tequilensis*	30 g/L sucrose, 30 °C, 200 rpm, 48 h	1879.43 ± 30.4 mg/L	Singh and Sharma (2020)
Fengycin	*B. subtilis*	30 °C, 200 rpm, 16 h, 26.2 g/L mannitol, 21.9 g/L soybean meal	3.5 g/L	Wei et al. (2010)
Iturin A	*B. amyloliquefaciens*	30 g/L corn starch, 70 g/L soybean meal, 28 °C, 230 rpm, 72 h	2013.43 ± 32.86 mg/L	Xu et al. (2020)
Surfactin	*B. amyloliquefaciens and B. subtilis*	200 g/L distillers' grains, 30 °C, 160 rpm, 96 h	3.4 g/L	Zhi et al. (2017)
Iturin	*B. amyloliquefaciens*	4% sunflower oil cake, 1% inoculum, 37 °C, 48 h, 180 rpm	819 mg/L	Kumar et al. (2017)
Surfactin	*B. subtilis*	2 g/L mg-Al-layered double hydride, 10 g/L sucrose, 200 rpm, 30 °C 120 h	3.8 g/L	Kan et al. (2017)
Surfactin	*B. subtilis*	Trypticase soy broth 32 °C, 170 rpm, 96 h	99.6 ± 1.38 mg/L	Alvarez et al. (2020)

The lipopeptides are generally produced by bacteria, moulds or yeast (Inès and Dhouha 2015b). *Bacillus* species are the most known microorganism used in the production of surfactin, iturin and fengycin (Table 2.2).

2.3 Application of Microbial Surfactants in Food Processing

Food security is one of the biggest concerns in food processing. Therefore, the main goal of food processing is to obtain products that are safe and have good organoleptic properties. The use of natural additives instead of synthetic ones, as well as an increase in environmental requirements and health concerns, has raised the need for natural additives in food (Nitschke and Silva 2018). Therefore, the antioxidant, antimicrobial and anti-adhesive properties of the microbial biosurfactants make them possible to be used as additives in the food industry to prevent contamination during food processing.

2.3.1 Biofilm Control

Biofilms are highly organized multicellular communities composed of microorganisms enclosed within an extracellular polysaccharide matrix. The dynamic process of biofilms is an important strategy for the microbes as a mechanism of resistance and survival against antibiotics and host defence mechanisms (Gebreyohannes et al. 2019). During the formation of biofilms, microorganisms can adhere to the surfaces that come into contact with the food, thus leading to undesirable alterations in the sensory properties of the final product (Zeraik and Nitschke 2010; Sharma et al. 2018).

Biofilm formation is composed of three phases: adherence, maturation and dispersion. For example, in *P. aeruginosa*, they are (1) formation of a layer and irreversible adhesion of microorganisms to its surface, (2) microcolony formation with the appearance of multilayers and (3) dispersal of bacteria cells which may then colonize other areas (Gebreyohannes et al. 2019).

The presence of biofilms in food contact or food processing surfaces can cause transmission of food-borne diseases, contamination by non-starter cultures, advanced food deterioration and metal loss with the deterioration of pipelines used for food transport and tanks for food storage (Sharma et al. 2018). Its formation can be avoided by using biochemical and physical cleaning strategies. However, once it is formed, biofilm is resistant to antimicrobial agents and mechanical removal (Gomes and Nitschke 2012). Therefore, efficient measures are urgently needed. One example that has shown to be effective in preventing and removing biofilms is the use of biosurfactants (Sharma 2016).

Several studies have demonstrated that the prior adhesion of biosurfactants to solid surfaces decreases the amount of bacterial cells attached on a surface of stainless steel and reduces the number of bacterial cells attaching to polystyrene surfaces, as a result of their anti-adhesive properties against food-borne pathogens, such as *Listeria innocua*, *Salmonella enteritidis*, *Listeria monocytogenes* and *Enterobacter sakazakii* (Sharma 2016; Nitschke and Silva 2018).

The biosurfactant can also change the physico-chemical properties of the surfaces directly involved in the adhesion, at the beginning of the biofilm formation (Nitschke and Silva 2018). It specifically modifies the hydrophobicity of the surface, affecting the adhesion of the bacteria cells on the surface (Gomes and Nitschke 2012; Harshada 2014). De Araujo et al. (2016) has demonstrated that surfactin changes the stainless steel surfaces to become more hydrophilic by increasing the electron acceptor components. Rhamnolipids can promote the same modification in polystyrene surfaces (de Araujo et al. 2016).

Biosurfactants can also be used on biofilm removal. The mechanisms involved are poorly understood, but some can be pointed out such as membrane degradation, inhibition of the electron transport chain, the ability to make cavities within the centre of film, improvement of cell permeability, reduction in cell surface hydrophobicity, reduction of interfacial tension and attractive interaction and interference in quorum sensing leading to a decrease in the biofilm formation (Gomes and Nitschke 2012; Nitschke and Silva 2018; Paraszkiewicz et al. 2019).

The mechanisms responsible for removing biofilms by biosurfactants are correlated with their antimicrobial activity, such as anti-adhesive properties, their excellent detergency power and membrane disruption, leading to an increase in membrane permeability, leakage of metabolites and cell lysis (Campos et al. 2013; Arab and Mulligan 2014; Sharma 2016). Lipopeptides, such as surfactin, fengycin and iturin, can act similar to antibiotic, antitumour and antiviral agents and enzyme inhibitors (Harshada 2014). They can reduce the surface tension and therefore interfere with the adhesion and biofilm formation. Biosurfactants can modify the permeability of the membranes of the biofilm cells by inserting into them or chelating cations and then lead to cell disruption, swelling and death (Galié et al. 2018).

In this way, biosurfactants can be used to prevent food contamination by avoiding the biofilm formation and removing it in industrial surfaces (de Araujo et al. 2016), leading to an extension of the shelf life of unprocessed fruit and vegetables (Dilarri et al. 2016).

2.3.2 Food Preservatives

Food additives are added to the food to improve and preserve chemical, biological, physical and organoleptic properties, during production, processing, transportation, storage and packaging of the food (Campos et al. 2013). They can be classified as nutritional additives, flavouring agents, preservatives, colouring agents, miscellaneous additives and texturizing agents (Carocho et al. 2015).

Heath problems related to the consumption of synthetic additives have been increasing the demand of the food industry for natural additives. Thus, microbial biosurfactants are good alternatives as natural additives to be used in market foods, as a result of their favourable properties compared to synthetic additives used in food processing.

Biosurfactants show some potential as preservatives in the food industry. Glycolipids display certain antiviral, antibacterial and antifungal activities (Jezierska et al. 2018). *P. aeruginosa* rhamnolipid, sophorose lipid and mannosylerythritol lipids are well known by their antifungal activities. These biosurfactants are involved in the lysis of zoospore membranes of *Pythium* and *Phytophthora* fungi (Inès and Dhouha 2015a). An inhibition of mycelial growth and loss of mobility on *Phytophthora* sp. and *Pythium* sp. were observed by Yoo et al. (2005) using rhamnolipid and sophorolipid.

An antimicrobial activity of rhamnolipid biosurfactant was reported by de Freitas Ferreira et al. (2019) against the gram-positive food pathogens *S. aureus*, *B. cereus* and *L. monocytogenes*. The biosurfactant reduced the hydrophobicity of the cell surface and damaged the cytoplasmic membrane of the bacteria. The biosurfactant sophorolipids can also inhibit/disrupt the biofilms formed by *E. coli*, *P. aeruginosa* and *Bacillus subtilis* (Rienzo et al. 2016).

Sophorolipid biosurfactants are shown to be effective against various fungi such as *Aspergillus*, *Saccharomyces*, *Penicillium*, *Cladosporium*, *Schizophyllum*, *Gloeophyllum* and *Fusarium*. Kim et al. (2002) noted the sophorolipid produced by *Candida bombicola* was effective in inhibiting the cell growth of the plant pathogenic fungus *Botrytis cinerea*.

The lipopeptide biosurfactant has also presented antibacterial and antifungal activities. Yuliani et al. (2018) reported antimicrobial activity against *Escherichia coli*, *Candida albicans*, *Staphylococcus aureus*, *Salmonella enterica* Typhi, *Listeria monocytogenes* and *Pseudomonas aeruginosa* of surfactin obtained by *B. subtilis*. The marine *Bacillus* strain produced fengycin isoforms that showed to be effective against *Citrobacter freundii*, *Micrococcus flavus*, *E. coli*, *K. aerogenes*, *A. faecalis*, *P. vulgaris and Serratia marcescens* (Sivapathasekaran et al. 2009). The antimicrobial activity of the biosurfactants might act by disrupting the cell wall, affecting the permeability of the membrane and then inhibiting the synthesis and metabolism of proteins and DNA/RNA (Yuliani et al. 2018).

Antioxidative properties of biosurfactants are also of great interest in food processing. The antioxidants show great technological importance during food processing, because they act as inhibitors of non-enzymatic browning and microbial growth as well as preventing lipid oxidation, leading to an extension of the food shelf life (da Silva et al. 2016). Lipopeptide biosurfactants produced by *Bacillus* species show great anti-adhesive, antimicrobial and antioxidant activities and also present antibiofilm activities causing the disruption of pre-formed bacterial biofilms (Giri et al. 2019).

Biosurfactants have also been used for metal chelation, to prevent food contamination by heavy metals. The bioaccumulation of heavy metals in food crops and their effects on human health present a big concern worldwide. Heavy metals can cause serious human health problems, including malnutrition, gastrointestinal cancer and mental growth retardation (Wu et al. 2019). A concentration of 80 ppm of rhamnolipid produced by *P. aeruginosa* was effective in removing 53%, 62%, 56%, 28%, 20% and 7% of Cd^{2+}, Pb, Ni^{2+}, Ba, Zn and Sr, respectively, from contaminated water (Elouzi et al. 2012). Glycolipid produced by *Bacillus* sp. was also exploited for metal chelation and showed positive results in cadmium removal from different vegetables (radish, garlic, potato and onion) (Elouzi et al. 2012).

Table 2.3 shows the application of microbial biosurfactants on food processing.

Table 2.3 Food-related application of microbial biosurfactants

Biosurfactant	Producer microorganism	Target	Function	References
Rhamnolipids	*Pseudomonas aeruginosa*	*Listeria monocytogenes*	Antibiofilm and antimicrobial	Davey et al. (2003)
Surfactin	*Bacillus subtilis*	*L. monocytogenes*	Antimicrobial	de Araujo et al. (2011)
Rhamnolipids	*P. aeruginosa*	*L. monocytogenes*	Antimicrobial	de Araujo et al. (2011)
Rhamnolipids	*P. aeruginosa*	*Listeria monocytogenes*, *Staphylococcus aureus* and *Salmonella enterica* Enteritidis	Antibiofilm and anti-adhesive	Zezzi do Valle Gomes and Nitschke (2012)
Surfactin	*B. subtilis*	*Listeria monocytogenes*, *Staphylococcus aureus* and *Salmonella enterica* Enteritidis	Antibiofilm and anti-adhesive	Zezzi do Valle Gomes and Nitschke (2012)
Sophorolipids	*C. bombicola*	*Bacillus subtilis* and *Staphylococcus aureus*	Antibacterial and antibiofilm	Díaz De Rienzo et al. (2015)
Sophorolipids	*Candida bombicola*	*Escherichia coli*	Antimicrobial	Zhang et al. (2016)
Sophorolipids	*C. bombicola*	*Clostridium perfringens* and *Campylobacter jejuni*	Antimicrobial	Silveira et al. (2019)
Sophorolipids	*C. bombicola*	*Bacillus subtilis* and *Escherichia coli*	Antibacterial	Gaur et al. (2019)
Rhamnolipids	*Pseudomonas aeruginosa*	*Bacillus subtilis*	Antibiofilm and antimicrobial	Sood et al. (2020)
Surfactin	*Bacillus tequilensis*	*Escherichia coli*, *Pseudomonas aeruginosa*, *Staphylococcus aureus*, *Staphylococcus epidermidis*, *Salmonella typhi* and *Salmonella typhimurium*	Disinfectant-like activity	Singh and Sharma (2020)
Rhamnolipids	*Pseudomonas* spp.	*Staphylococcus aureus*	Antibiofilm	Silva et al. (2017)

2.4 Concluding Remarks

Biosurfactants are a growing area of research interest within the food sciences. They show properties such as anti-adhesive, antioxidant and antimicrobial activities that help on decreasing contamination during food processing. In addition, the biosurfactants present unique advantages when compared to synthetic ones, including low toxicity, high biodegradability, easy preparation, natural origins and effectiveness at different ranges of physical conditions. These particular characteristics suggest great potential application as multipurpose additives.

The production costs of the microbial biosurfactants can be significantly reduced by using inexpensive raw material with great availability such as by-products from industrial and agricultural wastes. However, the large-scale production of biosurfactant is still a challenge, since their production process is not economically viable yet when compared to the synthetic surfactants derived from petrochemical sources. Considering the vast application in food processing, along with increased global concerns in the use of products obtained from nature, the market has a great demand for biosurfactants. Novel discoveries, improvement of production conditions, development of more cost-effective recovery and downstream processes and development of new microbial strains from screening programmes may allow the use of biosurfactants in large scale to prevent contamination in food processing, among others.

References

Abdel-Mawgoud AM, Stephanopoulos G (2018) Simple glycolipids of microbes: chemistry, biological activity and metabolic engineering. Synth Syst Biotechnol 3:3–19. https://doi.org/10.1016/j.synbio.2017.12.001

Ackerley N, Sertkaya A, Lange R (2010) Food transportation safety: characterizing risks and controls by use of expert opinion. Food Prot Trends 30:212–222

Aguiar RS, Esmerino EA, Rocha RS et al (2018) Physical hazards in dairy products: incidence in a consumer complaint website in Brazil. Food Control 86:66–70. https://doi.org/10.1016/j.foodcont.2017.11.020

Alvarez VM et al (2020) Microbial enhanced oil recovery potential of surfactin-producing bacillus. Fuel 272:117730. https://doi.org/10.1016/j.fuel.2020.117730

Arab F, Mulligan CN (2014) Rhamnolipids: characteristics, production, applications, and analysis. In: Mulligan CN, Sharma SK, Mudhoo A (eds) Biosurfactants: research trends and applications. CRC, Boca Raton, FL, pp 49–104

Bages-estopa S, White DA, Winterburn JB et al (2018) Production and separation of a trehalolipid biosurfactant. Biochem Eng J 139:85–94. https://doi.org/10.1016/j.bej.2018.07.006

Banat IM, Franzetti A, Gandolfi I et al (2010) Microbial biosurfactants production, applications and future potential. Appl Microbiol Biotechnol 87:427–444. https://doi.org/10.1007/s00253-010-2589-0

Bauer A, Jesús F, Gómez Ramos MJ et al (2019) Identification of unexpected chemical contaminants in baby food coming from plastic packaging migration by high resolution accurate mass spectrometry. Food Chem 295:274–288. https://doi.org/10.1016/j.foodchem.2019.05.105

Boles BR, Thoendel M, Singh PK (2005) Rhamnolipids mediate detachment of Pseudomonas aeruginosa from biofilms. Molecular Microbiology 57: 1210-1223

Borah SN, Sen S, Goswami L et al (2019) Rice based distillers dried grains with solubles as a low cost substrate for the production of a novel rhamnolipid biosurfactant having anti-biofilm activity against Candida tropicalis. Colloids Surf B Biointerfaces 182:110358. https://doi.org/10.1016/j.colsurfb.2019.110358

Bordin K, Kunitake MT, Aracava KK, Trindade CSF (2013) Changes in food caused by deep fat frying—a review. Arch Latinoam Nutr 63:5–13

Botterweck AAM, Verhagen H, Goldbohm RA et al (2000) Intake of butylated hydroxyanisole and butylated hydroxytoluene and stomach cancer risk: results from analyses in the Netherlands cohort study. Food Chem Toxicol 38:599–605. https://doi.org/10.1016/S0278-6915(00)00042-9

Campos JM, Montenegro Stamford TL, Sarubbo LA et al (2013) Microbial biosurfactants as additives for food industries. Biotechnol Prog 29:1097–1108. https://doi.org/10.1002/btpr.1796

Carocho M, Morales P, Ferreira ICFR (2015) Natural food additives: quo vadis? Trends Food Sci Technol 45:284–295. https://doi.org/10.1016/j.tifs.2015.06.007

Carrasco E, Morales-Rueda A, García-Gimeno RM (2012) Cross-contamination and recontamination by salmonella in foods: a review. Food Res Int 45:545–556. https://doi.org/10.1016/j.foodres.2011.11.004

Chen WC, Juang RS, Wei YH (2015) Applications of a lipopeptide biosurfactant, surfactin, produced by microorganisms. Biochem Eng J 103:158–169. https://doi.org/10.1016/j.bej.2015.07.009

Choi H, Hwang BK, Kim B, Choi SH (2020) Influence of pathogen contamination on beef microbiota under different storage temperatures. Food Res Int 132:109118. https://doi.org/10.1016/j.foodres.2020.109118

Conte A, Angiolillo L, Mastromatteo M, Del Nobile MA (2013) Technological options of packaging to control food quality. In: Food industry. IntechOpen, London. https://doi.org/10.5772/53151

da Silva NKV, de Sabino LB, de Oliveira LS et al (2016) Effect of food additives on the antioxidant properties and microbiological quality of red guava juice. Rev Cienc Agron 47:77–85. https://doi.org/10.5935/1806-6690.20160009

Davey ME, Caiazza NC, O'Toole GA (2003) Rhamnolipid surfactant production affects biofilm architecture in *Pseudomonas aeruginosa* PAO1. J Bacteriol 185:1027–1036. https://doi.org/10.1128/JB.185.3.1027-1036.2003

de Araujo LV, Abreu F, Lins U et al (2011) Rhamnolipid and surfactin inhibit *Listeria monocytogenes* adhesion. Food Res Int 44:481–488. https://doi.org/10.1016/j.foodres.2010.09.002

de Araujo LV, Guimarães CR, Marquita RL da S et al (2016) Rhamnolipid and surfactin: anti-adhesion/antibiofilm and antimicrobial effects. Food Control 63:171–178. https://doi.org/10.1016/j.foodcont.2015.11.036

de Freitas Ferreira J, Vieira EA, Nitschke M (2019) The antibacterial activity of rhamnolipid biosurfactant is pH dependent. Food Res Int 116:737–744. https://doi.org/10.1016/j.foodres.2018.09.005

Díaz De Rienzo MA, Banat IM, Dolman B et al (2015) Sophorolipid biosurfactants: possible uses as antibacterial and antibiofilm agent. N Biotechnol 32:720–726. https://doi.org/10.1016/j.nbt.2015.02.009

Dilarri G, da Silva VL, Pecora HB et al (2016) Electrolytic treatment and biosurfactants applied to the conservation of *Eugenia uniflora* fruit. Food Sci Technol 36:456–460. https://doi.org/10.1590/1678-457X.00516

Dobler L, Ferraz HC, Araujo de Castilho LV et al (2020) Environmentally friendly rhamnolipid production for petroleum remediation. Chemosphere 252:126349. https://doi.org/10.1016/j.chemosphere.2020.126349

Dolman BM, Kaisermann C, Martin PJ, Winterburn JB (2017) Integrated sophorolipid production and gravity separation. Process Biochem 54:162–171. https://doi.org/10.1016/j.procbio.2016.12.021

Echavarri-Bravo V, Thygesen HH, Aspray TJ (2017) Variability in physical contamination assessment of source segregated biodegradable municipal waste derived composts. Waste Manag 59:30–36. https://doi.org/10.1016/j.wasman.2016.10.049

Elouzi AA, Akasha AA, Elgerbi AM, El-baseir M (2012) Removal of heavy metals contamination by bio-surfactants (Rhamnolipids). J Chem Pharm Res 4(9):4337–4341

Faille C, Cunault C, Dubois T, Bénézech T (2018) Hygienic design of food processing lines to mitigate the risk of bacterial food contamination with respect to environmental concerns. Innov Food Sci Emerg Technol 46:65–73. https://doi.org/10.1016/j.ifset.2017.10.002

Fasano E, Cirillo T, Esposito F, Lacorte S (2015) Migration of monomers and plasticizers from packed foods and heated microwave foods using QuEChERS sample preparation and gas chromatography/mass spectrometry. LWT- Food Sci Technol 64:1015–1021. https://doi.org/10.1016/j.lwt.2015.06.066

Gaikwad KK, Singh S, Ajji A (2019) Moisture absorbers for food packaging applications. Environ Chem Lett 17:609–628. https://doi.org/10.1007/s10311-018-0810-z

Galié S, García-Gutiérrez C, Miguélez EM et al (2018) Biofilms in the food industry: health aspects and control methods. Front Microbiol 9:1–18. https://doi.org/10.3389/fmicb.2018.00898

Gaur VK, Regar RK, Dhiman N et al (2019) Biosynthesis and characterization of sophorolipid biosurfactant by *Candida* spp.: application as food emulsifier and antibacterial agent. Bioresour Technol 285:121314. https://doi.org/10.1016/j.biortech.2019.121314

Gebreyohannes G, Nyerere A, Bii C, Sbhatu DB (2019) Challenges of intervention, treatment, and antibiotic resistance of biofilm-forming microorganisms. Heliyon 5:e02192. https://doi.org/10.1016/j.heliyon.2019.e02192

Giri SS, Ryu EC, Sukumaran V, Park SC (2019) Microbial pathogenesis antioxidant, antibacterial, and anti-adhesive activities of biosurfactants isolated from bacillus strains. Microb Pathog 132:66–72. https://doi.org/10.1016/j.micpath.2019.04.035

Gomes MZ, Nitschke M (2012) Evaluation of rhamnolipid and surfactin to reduce the adhesion and remove biofilms of individual and mixed cultures of food pathogenic bacteria. Food Control 25:441–447. https://doi.org/10.1016/j.foodcont.2011.11.025

Harshada K (2014) Biosurfactant: a potent antimicrobial agent. J Microbiol Exp 1:173–177. https://doi.org/10.15406/jmen.2014.01.00031

Henkel M, Müller MM, Kügler JH et al (2012) Rhamnolipids as biosurfactants from renewable resources: concepts for next-generation rhamnolipid production. Process Biochem 47:1207–1219. https://doi.org/10.1016/j.procbio.2012.04.018

Hennessy TW, Hedberg CW, Slutsker L et al (1996) A national outbreak of salmonella Enteritidis infections from ice cream. N Engl J Med 334:1281–1286

Hentati D, Chebbi A, Hadrich F et al (2019) Production, characterization and biotechnological potential of lipopeptide biosurfactants from a novel marine bacillus stratosphericus strain FLU5. Ecotoxicol Environ Saf 167:441–449. https://doi.org/10.1016/j.ecoenv.2018.10.036

Hussain MA (2016) Food contamination: major challenges of the future. Foods 5:21. https://doi.org/10.3390/foods5020021

Iammarino M, Di Taranto A, Cristino M (2013) Endogenous levels of nitrites and nitrates in wide consumption foodstuffs: results of five years of official controls and monitoring. Food Chem 140:763–771. https://doi.org/10.1016/j.foodchem.2012.10.094

Inès M, Dhouha G (2015a) Glycolipid biosurfactants: potential related biomedical and biotechnological applications. Carbohydr Res 416:59–69. https://doi.org/10.1016/j.carres.2015.07.016

Inès M, Dhouha G (2015b) Lipopeptide surfactants: production, recovery and pore forming capacity. Peptides 71:100–112. https://doi.org/10.1016/j.peptides.2015.07.006

Janek T, Krasowska A, Czyżnikowska Ż, Łukaszewicz M (2018) Trehalose lipid biosurfactant reduces adhesion of microbial pathogens to polystyrene and silicone surfaces: an experimental and computational approach. Front Microbiol 9:2441. https://doi.org/10.3389/fmicb.2018.02441

Jezierska S, Claus S, Van Bogaert I (2018) Yeast glycolipid biosurfactants. FEBS Lett 592:1312–1329. https://doi.org/10.1002/1873-3468.12888

Jimoh AA, Lin J (2019) Biosurfactant: a new frontier for greener technology and environmental sustainability. Ecotoxicol Environ Saf 184:109607. https://doi.org/10.1016/j.ecoenv.2019.109607

Kan SC, Lee CC, Hsu YC et al (2017) Enhanced surfactin production via the addition of layered double hydroxides. J Taiwan Inst Chem Eng 80:10–15. https://doi.org/10.1016/j.jtice.2017.06.017

Kaur G, Wang H, To MH et al (2019) Efficient sophorolipids production using food waste. J Clean Prod 232:1–11. https://doi.org/10.1016/j.jclepro.2019.05.326

Kim K, Yoo D, Kim Y et al (2002) Characteristics of Sophorolipid as an antimicrobial agent. J Microbiol Biotechnol 12:235–241

Kitamoto D, Morita T, Fukuoka T et al (2009) Self-assembling properties of glycolipid biosurfactants and their potential applications. Curr Opin Colloid Interface Sci 14:315–328. https://doi.org/10.1016/j.cocis.2009.05.009

Kumar PN, Swapna TH, Khan MY et al (2017) Statistical optimization of antifungal iturin a production from Bacillus amyloliquefaciens RHNK22 using agro-industrial wastes. Saudi J Biol Sci 24:1722–1740. https://doi.org/10.1016/j.sjbs.2015.09.014

Kuyukina MS, Ivshina IB, Baeva TA et al (2015) Trehalolipid biosurfactants from nonpathogenic Rhodococcus actinobacteria with diverse immunomodulatory activities. N Biotechnol 32:559–568. https://doi.org/10.1016/j.nbt.2015.03.006

Lau OW, Wong SK (2000) Contamination in food from packaging material. J Chromatogr A 882:255–270. https://doi.org/10.1016/S0021-9673(00)00356-3

Liu Z, Tian X, Chen Y et al (2019) Efficient sophorolipids production via a novel in situ separation technology by Starmerella bombicola. Process Biochem 81:1–10. https://doi.org/10.1016/j.procbio.2018.12.005

Liu K, Sun Y, Cao M et al (2020) Rational design, properties, and applications of biosurfactants: a short review of recent advances. Curr Opin Colloid Interface Sci 45:57–67. https://doi.org/10.1016/j.cocis.2019.12.005

Masotti F, Cattaneo S, Stuknytė M, De Noni I (2019) Airborne contamination in the food industry: an update on monitoring and disinfection techniques of air. Trends Food Sci Technol 90:147–156. https://doi.org/10.1016/j.tifs.2019.06.006

McCurdy S, Peutz J, Wittman G (2009) Storing food for safety and quality. Cooperative Extension Service Food Preservation, Moscow

McKay C, Scharman EJ (2015) Intentional and inadvertent chemical contamination of food, water, and medication. Emerg Med Clin North Am 33:153–177. https://doi.org/10.1016/j.emc.2014.09.011

Meena KR, Sharma A, Kumar R, Kanwar SS (2020) Two factor at a time approach by response surface methodology to aggrandize the *Bacillus subtilis* KLP2015 surfactin lipopeptide to use as antifungal agent. J King Saud Univ Sci 32:337–348. https://doi.org/10.1016/j.jksus.2018.05.025

Mnif I, Ghribi D (2016) Glycolipid biosurfactants: main properties and potential applications in agriculture and food industry. J Sci Food Agric 96:4310–4320. https://doi.org/10.1002/jsfa.7759

Naughton PJ, Marchant R, Naughton V, Banat IM (2019) Microbial biosurfactants: current trends and applications in agricultural and biomedical industries. J Appl Microbiol 127:12–28. https://doi.org/10.1111/jam.14243

Nerín C, Aznar M, Carrizo D (2016) Food contamination during food process. Trends Food Sci Technol 48:63–68. https://doi.org/10.1016/j.tifs.2015.12.004

Niescher S, Wray V, Lang S et al (2006) Identification and structural characterisation of novel trehalose dinocardiomycolates from n-alkane-grown Rhodococcus opacus 1CP. Appl Microbiol Biotechnol 70:605–611. https://doi.org/10.1007/s00253-005-0113-8

Nitschke M, Silva SS (2018) Recent food applications of microbial surfactants. Crit Rev Food Sci Nutr 58:631–638. https://doi.org/10.1080/10408398.2016.1208635

Niu Y, Wu J, Wang W, Chen Q (2019) Production and characterization of a new glycolipid, mannosylerythritol lipid, from waste cooking oil biotransformation by Pseudozyma aphidis ZJUDM34. Food Sci Nutr 7:937–948. https://doi.org/10.1002/fsn3.880

Oliveira MR, Magri A, Baldo C et al (2015) Review: sophorolipids a promising biosurfactant and it's applications. Int J Adv Biotechnol Res 6:161–174

Paraszkiewicz K, Moryl M, Płaza G et al (2019) Surfactants of microbial origin as antibiofilm agents. Int J Environ Health Res 3123:1–20. https://doi.org/10.1080/09603123.2019.1664729

Parthasarathi R, Subha KS (2018) Prospective biosurfactant applications in food industry-opportunities and challenges. Int J Curr Res Life Sci 7:1560–1565

Radzuan MN, Banat IM, Winterburn J (2018) Biorefining palm oil agricultural refinery waste for added value rhamnolipid production via fermentation. Ind Crop Prod 116:64–72. https://doi.org/10.1016/j.indcrop.2018.02.045

Ramya Devi KC et al (2018) Demonstration of bioprocess factors optimization for enhanced mono-rhamnolipid production by a marine pseudomonas guguanensis. Int J Biol Macromol 108:531–540. https://doi.org/10.1016/j.ijbiomac.2017.10.186

Rather IA, Koh WY, Paek WK, Lim J (2017) The sources of chemical contaminants in food and their health implications. Front Pharmacol 8:830. https://doi.org/10.3389/fphar.2017.00830

Reddy SK, Yahya Khan M, Archana K et al (2016) Utilization of mango kernel oil for the rhamnolipid production by Pseudomonas aeruginosa DR1 towards its application as biocontrol agent. Bioresour Technol 221:291–299. https://doi.org/10.1016/j.biortech.2016.09.041

Rienzo D, Stevenson P, Marchant R, Banat IM (2016) Antibacterial properties of biosurfactants against selected gram-positive and -negative bacteria. FEMS Microbiol Lett 363:1–8. https://doi.org/10.1093/femsle/fnv224

Roy A (2017) A review on the biosurfactants: properties, types and its applications. J Fundam Renew Energy Appl 08:1–5. https://doi.org/10.4172/2090-4541.1000248

Sen S, Borah SN, Bora A, Deka S (2017) Production, characterization, and antifungal activity of a biosurfactant produced by Rhodotorula babjevae YS3. Microb Cell Fact 16:1–14. https://doi.org/10.1186/s12934-017-0711-z

Sethu V, Ananth Viramuthu V (2008) Water recycling in the food industry. Woodhead Publishing, Cambridge

Sharma D (2016) Applications of biosurfactants in food. In: Biosurfactants in food. Springer International, Cham, pp 43–80

Sharma D, Gupta E, Singh J et al (2018) Microbial biosurfactants in food sanitation. In: Sustainable food systems from agriculture to industry. Elsevier, Amsterdam, pp 341–368

Sheikh Z, Pawar S, Rathod VK (2019) Enhancement of rhamnolipid production through ultrasound application and response surface methodology. Process Biochem 85:29–34. https://doi.org/10.1016/j.procbio.2019.06.023

Silva SSE, Carvalho JWP, Aires CP, Nitschke M (2017) Disruption of *Staphylococcus aureus* biofilms using rhamnolipid biosurfactants. J Dairy Sci 100:7864–7873. https://doi.org/10.3168/jds.2017-13012

Silveira VAI, Nishio EK, Freitas CAUQ et al (2019) Production and antimicrobial activity of sophorolipid against *Clostridium perfringens* and *Campylobacter jejuni* and their additive interaction with lactic acid. Biocatal Agric Biotechnol 21:101287. https://doi.org/10.1016/j.bcab.2019.101287

Singh AK, Sharma P (2020) Disinfectant-like activity of lipopeptide biosurfactant produced by *Bacillus tequilensis* strain SDS21. Colloids Surf B Biointerfaces 185:110514. https://doi.org/10.1016/j.colsurfb.2019.110514

Sivapathasekaran C, Mukherjee S, Samanta R, Sen R (2009) High-performance liquid chromatography purification of biosurfactant isoforms produced by a marine bacterium. Anal Bioanal Chem 395:845–854. https://doi.org/10.1007/s00216-009-3023-2

Sood U, Singh DN, Hira P et al (2020) Rapid and solitary production of mono-rhamnolipid biosurfactant and biofilm inhibiting pyocyanin by a taxonomic outlier *Pseudomonas aeruginosa* strain CR1. J Biotechnol 307:98–106. https://doi.org/10.1016/j.jbiotec.2019.11.004

Vandghanooni S, Forouharmehr A, Eskandani M et al (2013) Cytotoxicity and DNA fragmentation properties of butylated hydroxyanisole. DNA Cell Biol 32:98–103. https://doi.org/10.1089/dna.2012.1946

Varjani SJ, Upasani VN (2017) Critical review on biosurfactant analysis, purification and characterization using rhamnolipid as a model biosurfactant. Bioresour Technol 232:389–397. https://doi.org/10.1016/j.biortech.2017.02.047

Wei Y-H, Wang L-C, Chen W-C, Chen S-Y (2010) Production and characterization of Fengycin by indigenous *Bacillus subtilis* F29-3 originating from a potato farm. Int J Mol Sci 11:4526–4538. https://doi.org/10.3390/ijms11114526

World Health Organization (2019) Food safety. https://www.who.int/news-room/fact-sheets/detail/food-safety

Wu J, Niu Y, Jiao Y, Chen Q (2019) Fungal chitosan from Agaricus bisporus (Lange) sing . Chaidam increased the stability and antioxidant activity of liposomes modi fi ed with biosurfactants and loading betulinic acid. Int J Biol Macromol 123:291–299. https://doi.org/10.1016/j.ijbiomac.2018.11.062

Xu Y, Cai D, Zhang H et al (2020) Enhanced production of iturin a in *Bacillus amyloliquefaciens* by genetic engineering and medium optimization. Process Biochem 90:50–57. https://doi.org/10.1016/j.procbio.2019.11.017

Yoo DS, Lee BS, Kim EK (2005) Characteristics of microbial biosurfactant as an antifungal agent against plant pathogenic fungus. J Microbiol Biotechnol 15:1164–1169

Yuliani H, Perdani MS, Savitri I et al (2018) Antimicrobial activity of biosurfactant derived from Bacillus subtilis C19. Energy Procedia 153:274–278. https://doi.org/10.1016/j.egypro.2018.10.043

Zezzi do Valle Gomes M, Nitschke M (2012) Evaluation of rhamnolipid and surfactin to reduce the adhesion and remove biofilms of individual and mixed cultures of food pathogenic bacteria. Food Control 25:441–447. https://doi.org/10.1016/j.foodcont.2011.11.025

Zhang X, Fan X, Solaiman DKY et al (2016) Inactivation of *Escherichia coli* O157:H7 in vitro and on the surface of spinach leaves by biobased antimicrobial surfactants. Food Control 60:158–165. https://doi.org/10.1016/j.foodcont.2015.07.026

Zeraik AE, Nitschke M (2010) Biosurfactants as agents to reduce adhesion of pathogenic bacteria to polystyrene surfaces: effect of temperature and hydrophobicity. Current Microbiology 61: 554-559

Zheng H, Fan S, Liu W, Zhang M (2020) Production and separation of pseudomonas aeruginosa rhamnolipids using coupling technology of cyclic fermentation with foam fractionation. Chemical Engineering and Processing - Process Intensification 148: 107776

Zhi Y, Wu Q, Xu Y (2017) Production of surfactin from waste distillers' grains by co-culture fermentation of two Bacillus amyloliquefaciens strains. Bioresour Technol 235:96–103. https://doi.org/10.1016/j.biortech.2017.03.090

Antioxidant Biosurfactants

3

Sameer J. Nadaf, Vijay M. Kumbar, Anand I. Torvi, Joy H. Hoskeri, and Arun K. Shettar

Abstract

Recent decades have shown that there is increase in the considerable interest in research of biogenic surfactants (microbial surfactants, biosurfactants), products of biosynthesis of various microorganisms, as well as the possibilities for their practical use. In the present chapter, the detailed discussion was included regarding sources, isolation, and potential role and applications of biosurfactants as antioxidant agents. This chapter covers the various aspects of biosurfactants such as sources of biosurfactants with examples. It also included the different types of biosurfactants isolated from the microbial sources. It clearly showed the optimum conditions for production of microbial biosurfactants in culture conditions. Microbial-derived surfactants can replace synthetic surfactants in a great variety of industrial applications as detergents, foaming, emulsifiers, solubilizers, and wetting agents. Microbial surfactants or biosurfactants can be defined as a natural class of surface-active compounds produced by microorganisms. Several biosurfactants have strong antibacterial, antifungal, and antiviral activity. In spite of several activities, even microbial biosurfactants have the property of antioxidant activity which is discussed in detail in the present chapter. The

S. J. Nadaf
Sant Gajanan Maharaj College of Pharmacy, Chinchewadi, Maharashtra, India

V. M. Kumbar
Central Research Laboratory, Maratha Mandal's Nathajirao G. Halgekar Institute of Dental Sciences and Research Centre, Belagavi, Karnataka, India

A. I. Torvi
Department of Chemistry, Karnataka University, Dharwad, Karnataka, India

J. H. Hoskeri
Department of Bioinformatics and Biotechnology, Karnataka State Akkamahadevi Women's University, Vijayapura, Karnataka, India

A. K. Shettar (✉)
Department of Biotechnology and Microbiology, Karnataka University, Dharwad, Karnataka, India

Inamuddin et al. (eds.), *Microbial Biosurfactants*, Environmental and Microbial Biotechnology, https://doi.org/10.1007/978-981-15-6607-3_3

chemical structure and properties of various biosurfactants have been extensively discussed and presented in this chapter.

Keywords

Microbes · Biosurfactants · Plant · Screening · Optimization · CTAB assay · Emulsification · Antioxidant

3.1 Introduction

Surfactants are well-recognized, most lucrative, amphiphilic molecules and known to reduce the interfacial tension between the different phases such as liquid and gases or between solid and liquid mostly by congregating at the interface (Belhaj et al. 2020; Nakama 2017). Inclusive of its traditional role of cleansing agent, surfactants are now widely accepted for numerous routine household and industrial applications such as emulsifiers, defoaming agents, deinking agents, and so on (Nakama 2017; Makkar et al. 2011). For that reason, the surfactants are no longer limited to a particular sector or industry, rather utilized extensively in ample industries ranging from detergents to water treatment, paint to petroleum, and food items to pharmaceuticals as well as cosmeceuticals (Jimoh and Lin 2019). Due to these myriad applications, the global surfactant market is anticipated to arrive at USD 66,408 million by 2025. Sorbitan esters (spans) and ethoxylated sorbitan esters (tweens) are widely used surfactants.

Most of these frequently implemented surfactants are synthesized from either nonrenewable petroleum products or renewable chemicals (Makkar et al. 2011; Rebello et al. 2013). Continuous use of petroleum substrates for the production of surfactants has increased the environmental pollution drastically, and it also diminishes the vital nonrenewable energy sources (Jimoh and Lin 2019; Henkel et al. 2012; Freitas et al. 2016). This may create a problem in the nearby future. Wastewater discharge from surface-active agent-based industries causes a higher accumulation of surfactants or their degradation products in rivers or marine water than eco-toxicological no-effect concentration, which may impact biotic elements of ecosystems harmfully (Rebello et al. 2013; Olkowska et al. 2014). The concentration of surfactants higher than 0.01 g/L causes chronic toxicity to aquatic animals (Rodríguez-López et al. 2019).

Chemical-based surfactants are comprised of diverse category compounds. viz., inorganic phosphates, artificial colorants, and optical brighteners (Yu et al. 2008). It is well documented that, due to excessive accumulation of inorganic phosphates, water becomes exceedingly enriched with minerals and nutrients that promote rapid algae growth. This causes a depletion of oxygen in water, leaving the water incapable of supporting other aquatic life (Yang et al. 2008). Optical brighteners may cause reproductive and respiration problems in aquatic animals (Salas et al. 2019). Further, the synthetic detergents routinely employed to clean up oil spillages have destructed the environment enormously. Cationic surfactants are known to impart greater toxic effects followed by anionic, zwitterionic, and nonionic (Rebello et al. 2013; Rodríguez-López et al. 2019). The situation gets worse due to the slow biodegrade nature of chemical surfactants and hence causes hazardous environmental problems. Due to these facts, surfactants are considered as one of the most important environmental pollutants (Cierniak et al. 2020). These hazardous outcomes due to chemical-based surfactant usage have aroused a universal alert concerned to its usage, safe disposal, and urge for alternative surfactants with least toxic impact on the environment.

The idea of going green developed a special class of surfactants known as biosurfactants. Biosurfactants are tensioactive amphiphilic biomolecules obtained from plants and microorganisms (Jimoh and Lin 2019; Costa et al. 2018). Chemicals produced at the microbial cell surface or excreted extracellular with distinct hydrophobic as well as hydrophilic portion and able to reduce surface and interfacial tension by adsorbing at the interface are termed as biosurfactants (Costa et al. 2018; Fakruddin 2012). Biosurfactants are membrane attached metabolites that facilitate the growth of microorganisms on water-immiscible substrates by diminishing tension at the boundary layer and improving the availability of substrate for uptake (Huang and Tang 2007; Desai and Banat 1997a). Biosurfactant market was at USD 3.99 billion in 2016 and is expected to reach USD 5.52 billion by 2022, at a compound annual growth rate of 5.6% during the forecast period. The Asia Pacific is the fastest-growing market for biosurfactants due to the technologically advancing and emerging countries in the region demanding innovative, biodegradable, renewable, and less toxic biosurfactant products (Markets and Markets 2016).

3.2 Sources of Biosurfactants

Biosurfactants are mainly obtained from plants and microbes.

3.2.1 Plant-Based Biosurfactants

3.2.1.1 Saponins

Plant-based secondary metabolites mainly saponins are well-recognized and best-known biosurfactants of plant origin (Pradhan and Bhattacharyya 2017). They develop foam or lather when shaken with water. Saponins are abundantly distributed in different parts including roots, shoots, flowers, and seeds of an ample plant species belonging to nearly 100 families. Saponin-enriched dietary foods are leguminous plants such as soya beans (*Glycine max L. Merrill* 6.5 g/kg), haricot bean (*Phaseolus vulgaris* 4.1 g/kg), kidney beans (*Phaseolus vulgaris* 3.5 g/kg), and runner bean (*Phaseolus coccineus L.* 3.4 g/kg) (Savage 2003). *Yucca schidigera* and *Quillaja saponaria* are considered as plants with high content of saponins up to 10% of dry matter. The US Food and Drug Administration (USFDA) has approved these plants for their use in food and cosmetic products (Oleszek and Hamed 2010).

Structure, Properties, and Types of Saponins

Saponins are amphipathic glycosides in which glycosidic bond connects aglycone and one or more carbohydrate moieties (Chen et al. 2010). The aglycone portions are termed genin or sapogenins (Wisetkomolmat et al. 2019). A number of sugar fractions attached to the sapogenin core categorize saponins as monodesmosidic, bidesmosidic, and tridesmodic (Greek desmos = chain) (Benaiges and Guillén 2007).

Saponins as a Biosurfactants

An aqueous solution of saponins produces soap-like foams and micelles. Surface tension reducing the ability of chemical surfactants is assigned to amphiphilic nature, i.e., hydrophilic and lipophilic groups of them. Further, they are categorized as anionic, cationic, nonionic, and zwitterionic based on the different hydrophilic group attached to lipophiles such as long saturated or unsaturated hydrocarbon chain (Lombardo et al. 2015). Likewise, surface-active or detergent properties of saponins are attributed to water-soluble carbohydrate moieties and fat-soluble aglycone core of the molecule (Savage 2003). Hydrophilic portions of saponins are carbohydrate or sugar chains differing mainly in the length of branches, substitution, and composition. Mainly attached sugars are glucose, fructose, xylose, galactose, rhamnose, and arabinose, whereas, lipophiles are steroidal or triterpene unit (Savage 2003; Oleszek and Hamed 2010). Monodesmosidic saponins possess the best formability followed by bidesmosidic and tridesmodic. The more sugar chains attached to the sapogenin, the lesser is the foam-producing ability. Biosurfactants reduce the free energy of the system by replacing the bulkier molecules of higher energy at an interface (Mulligan 2007).

Different saponin-based plants and their role as biosurfactants are summarized in Table 3.1.

3.2.2 Microbe-Based Biosurfactants

In the late 1960s during hydrocarbon fermentation, microbial-derived surfactants or biosurfactants were first employed as extracellular compounds. Microorganisms, because of their large surface-to-volume ratio and diverse synthetic capabilities, are promising candidates for widening the present range of surfactants (Desai and Banat 1997b). Akin to synthetic surfactants, biosurfactants or microbial surfactants produced by microorganisms are also consisted of distinct hydrophiles and lipophiles causing them to reduce surface/interfacial tension by aggregating at interfaces between two immiscible fluids of different polarities or air and water interphase and subsequently solubilize polar compounds in nonpolar solvents (Sena et al. 2018). Different hydrophiles comprised of acid, alcohol, peptide cations, or anions are joined by ester linkage, amide linkage, or glycosidic linkage to lipophiles made up of unsaturated or saturated hydrocarbon chains of fatty acids. Hydrophiles cause biosurfactants to solubilize in water and lipophiles provide capillary activity (Mondal et al. 2015).

Table 3.1 Plant-based saponins as biosurfactants

Plant (botanical/ common name)	Family	Part/extract used	Saponin content/% used	Formulation	Role	Parameters/model/ assay	Key points	References
Quillaja saponaria (soap bark tree)	Quillajaceae	Bark/*Quillaja* water extract Sapnov™	65%	Emulsion/ vaccine	Emulsifier/ vaccine adjuvant	PS: 203.40 ± 1.83 nm; ZP: −51.7 ± 0.2; pH: 7.1	Higher IgG and virus-neutralizing antibodies in pigs were observed	Burakova et al. (2018)
Camellia oleifera	Theaceae	Defatted seed meal	39.5%	Defatted meal	Foam stabilizer and emulsifier	Foam height: 4.57 ± 0.10 cm	The ratio of height foam at 0 and 5 min foam was 86.0%	Chen et al. (2010)
Gleditsia australis	Caesalpinioideae	Peel/ ethanolic extract	9%	Dishwashing liquid detergent	Detergent	pH; 6.5–7.5	Saponin strongly influenced product quality	Do et al. (2019)
Orthosiphon aristatus (Blume) Miq.	Lamiaceae	Leaf/ maceration with ethanol	–	Microemulsion	Surfactants	PS: 91.21 nm	Microemulsion improved gliclazide solubility up to at 1.6 mg/mL	Abbas et al. (2018)
Morinda citrifolia L.	Rubiaceae	Leaf/ maceration with ethanol	–	Microemulsion	Surfactants	PS: 9.65 nm	⍰-Carotene optimum solubility obtained at 2.00 mg/mL	Tania et al. (2018)
Solanum dulcamara	Solanaceae	Herb/ethanol and diethyl ether	0.48 ± 0.13%	Bodywash gel	Foaming and emulsifying agents	AA: 63% ZV: 150 mgN/ 100 mL	Foaming ability was 488 mL and influenced metabolism of human skin fibroblasts	Nizioł-Łukaszewska and Bujak (2018)
Glycyrrhiza glabra	Fabaceae (Leguminosae or pea)	Root/ethanol and diethyl ether	5.49 ± 0.53%	Bodywash gel	Foaming and emulsifying agents	AA: 71% ZV: 170 mgN/ 100 mL	Foaming ability was 513 mL and influenced fibroblast proliferation	

(continued)

Table 3.1 (continued)

Plant (botanical/ common name)	Family	Part/extract used	Saponin content/% used	Formulation	Role	Parameters/model/ assay	Key points	References
Viola tricolor	Violaceae	Herb/ethanol and diethyl ether	7.26 ± 0.82%	Bodywash gel	Foaming and emulsifying agents	AA: 90% ZV: 170 mgN/ 100 mL	Foaming ability was 510 mL and influenced fibroblast proliferation	
Panax ginseng	Araliaceae	Root powder/ ethanol	66.2 wt%	–	Surface-active agent	CMC: 0.009% w/v	Produced foam lasts for 20 min	Mesgarzadeh et al. (2017)
Ilex paraguariensis	Aquifoliaceae	Unripe mate fruits/ethanol	–	–	Foaming agent	CMC: 150 mg/L and γ_{cmc}: 52.8 mN/ m	Electrolyte addition did not affect foamability	do Canto et al. (2010)
Quillaja saponaria	Quillajaceae	Q-Naturale® 200	3, 6, 9, and 12 wt%	Nanostructured lipid carriers	Emulsifier	PS: 0.82 ± 0.17 to 0.53 ± 0.06 μm	Nanoparticulate system formulated using natural surfactants	Kharat and McClements (2019)
Quillaja saponaria	Quillajaceae	Wood/spray granulated extract	68.6%	Emulsion	Emulsifiers	ZP: −55.2 ± 3.66 mV PS: 180–220 nm	Emulsion stability improved based on the concentration of extract and sodium caseinate	Salminen et al. (2019)
Yucca schidigera	Asparagaceae	Trunk part/ food grade *Yucca* saponin extract	9.5%	Emulsion	Emulsifiers	PS: 0.178 ± 0.002 μm ZP: −26.5 ± 2.0 mV	Saponin stabilized emulsion over wide pH range (5–9)	Ralla et al. (2017)
Tea saponins (TS) and Quillaja saponaria (QS)	Theaceae and Quillajaceae	Q-Naturale® 200	51.8% and 20–22%	Nanoemulsion	Surfactants	*Y*: 4.8 mN/m for TS and 2.8 mN/m for QS; PS: 186 nm for TS and 200 nm for QS	TS obtained from waste mass can stabilize O/W emulsions	Zhu et al. (2019)

Saponin	Supplied by Sigma-Aldrich		>10%	Quercetin (QT)-trapped nanoemulsion	Emulsifier	PS: 152.7 ± 12 nm ZP: −41 ± 8 mV	Saponin stabilization significantly affected QT solubilization, loading, and release properties	Kaur et al. (2016)
Quillaja saponaria (QS)	Quillajaceae	Extract supplied by Ingredion Incorporated, USA	1.5%	Oleogels/ emulsion	Emulsifier	PS: 150–870 nm ZP: −22 and −37 mV	Translucent orange oil oleogel with high oil loading of 98.2 wt% was obtained	Chen and Yang (2019)
Yucca schidigera Roezl	Asparagaceae	Liquid and dry powder extract	9%	Sunflower oil-in-water emulsions	Emulsifier	Emulsion elastic modulus: ≈ 2000 Pa	>50 times higher elastic emulsions compared to other emulsions	Tsibranska et al. (2020a)
Quillaja saponaria (QS)	Quillajaceae	Q-Naturale 200	1.5%	Vitamin E-loaded O/W emulsions	Emulsifier	ZP: > −60 mV	Saponin-based emulsions showed improved stability within pH 2 to 8 due to strong electrostatic repulsion	Lv et al. (2019)
Camellia oleifera Abel (tea saponins)	Theaceae	–	>96%	Emulsion	Emulsifier	IDE: 480 ± 110 mN/m	Interfacial elasticity impacted on the emulsion shear elasticity and dynamic yield stress. However,	Tsibranska et al. (2020b)
Sapindus mukorossi (berry saponin concentrate)	Sapindaceae	Berry	≈ 53%	Emulsion	Emulsifier	IDE: 230 ± 7 mN/m		

(continued)

Table 3.1 (continued)

Plant (botanical/common name)	Family	Part/extract used	Saponin content/% used	Formulation	Role	Parameters/model/assay	Key points	References
Quillaja saponaria Molina tree	Quillajaceae	Bark/supersap (sups) and *Quillaja* dry (QD)	≈ 91% (sups) and ≈ 38% (QD)	Emulsion	Emulsifier	IDE (sups): 300 ± 30 mN/m IDE (QD): 260 ± 1 mN/m	viscous stress remains unchanged	
Quillaja saponaria (QS)	Quillajaceae	Q-Naturale 200	2%	Curcumin-loaded nanocrystals	Emulsifier	EE: 93.2 ± 3.1%	Saponin-coated lipid droplets in the nanoemulsions were relatively stable to aggregation in the gastric fluids	Zheng et al. (2019)
				Curcumin-loaded nanoemulsion		EE: 93.2 ± 4.8% PS: 0.18 ± 0.02 μm ZP: −55.8 ± 1.5 mV		
				Curcumin-loaded oil bodies		EE: 94 ± 3.8% PS: 0.41 ± 0.01 μm ZP: −30.3 ± 1.9		

PS particle size, *ZP* zeta potential, AA antioxidant activity, *ZV* zein value (irritant potential), *CMC* critical micelles concentration, γ_{CMC} minimum attainable surface tension, *Υ* interfacial tension, *IDE* interfacial dilatational elasticity, *EE* entrapment efficiency

Biosurfactants are mainly produced by bacteria, yeast, and fungi (Sena et al. 2018). For living or feeding purpose, microbes present in the aqueous phase produce the surface-active agent to efficiently adsorb, emulsify, and disperse or solubilize the water-immiscible substrates (Desai and Banat 1997b). *Pseudomonas aeruginosa*, *Bacillus subtilis*, *Rhodococcus erythropolis*, *Candida albicans*, *Acinetobacter calcoaceticus*, *Halomonas* sp., and *Myroides* sp. strain are biosurfactants producing dominant species (Fenibo et al. 2019; Shekhar et al. 2014). Biosurfactants possess distinct properties than chemical surfactants such as lower toxicity, higher biodegradability, environment-friendly nature, possible production using fermentation, higher foaming, tolerance to extreme conditions of temperature, pH, ionic strength, and salinity (Vijayakumar and Saravanan 2015; Shekhar et al. 2014; Mukherjee et al. 2006). Owing to these outstanding characteristics, they have been widely applied in pharmaceutical, cosmetics, petroleum, food processing, agriculture, textile, paper, leather, oil spill cleanup, wastewater treatment, and biodegradation industries over the years (Fenibo et al. 2019; Shekhar et al. 2014; Mukherjee et al. 2006; Akbari et al. 2018).

3.2.2.1 Types of Microbial Surfactants

Chemical surfactants are generally classified based on the nature of the hydrophilic group. In contrast to this, biosurfactants are classified based on chemical composition and microbial origin. Mainly they are categorized into two major classes based on molecular weight as follows (Fig. 3.1):

1. Low molecular weight surfactants:
   (a) Glycolipids
   (b) Lipopeptides or lipoprotein
   (c) Phospholipids, fatty acids (mycolic acids), and neutral lipids
2. High molecular weight surfactants:
   (a) Polymeric surfactants
   (b) Particulate surfactants (Mondal et al. 2015; Santos et al. 2016; Sharma 2016)

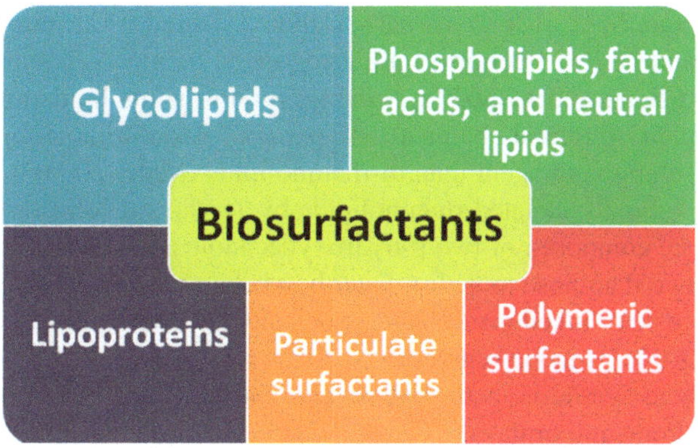

Fig. 3.1 Major types of biosurfactants

Low molecular weight surfactants reduce the surface/interfacial tension effectively, whereas, high molecular weight surfactants are effective as emulsion stabilizers (Mondal et al. 2015). Based on ionic charges, biosurfactants are further classified: (1) anionic, (2) cationic, (3) nonionic, and (4) neutral biosurfactants, while considering secretion type they are classified as (1) intracellular, (2) extracellular, and (3) adhered to microbial cells (Inès and Dhouha 2015).

Glycolipids

Glycolipids are the most studied and well-liked biosurfactant. Glycolipids are comprised of carbohydrates (hydrophiles) linked to hydroxy aliphatic acids or long-chain fatty acids (lipophiles) through either ester or ether linkage (Sharma 2016; Inès and Dhouha 2015). Glycolipids are characterized by high structural diversity and effectively diminish the surface and interfacial tension. The nature of the hydrophiles (carbohydrate moiety) and lipophiles categorizes glycolipid into different subclasses such as (Inès and Dhouha 2015; Mnif et al. 2017; Kitamoto et al. 2002):

1. Rhamnolipids
2. Sophorolipids
3. Trehalolipids
4. Succinoyl trehalolipids
5. Cellobiose lipids
6. Mannosylerythritol lipids
7. Xylolipids
8. Mannosylribitol lipids
9. Mannosylarabitol lipids
10. Lipids of oligosaccharides

Rhamnolipids

Rhamnolipids, primarily crystalline acids, are a class of biosurfactants which are produced by *Pseudomonas aeruginosa* and contain one or two rhamnose as the sugar moiety (hydrophiles) linked to one or two β-hydroxylated fatty acid chains (lipophiles) (Fig. 3.2) (Chong and Li 2017; Maier 2003). Fatty acid chains can be composed of 8 to 16 carbon atoms. Based on a number of rhamnose molecules attached, rhamnolipids are categorized as mono-rhamnolipids and di-rhamnolipids (Chong and Li 2017). Notably, mono-rhamnolipids and di-rhamnolipids differ in their physical-chemical properties. Mono-rhamnolipid is a more powerful solubilizing agent than di-rhamnolipids (Maier 2003).

Firstly isolated rhamnolipids by Jarvis and Johnson (1949) from *P. aeruginosa* are composed of two β-hydroxydecanoic acids linked through a glycosidic bond to two rhamnose moieties. Ester bond connected two β-hydroxy fatty acid and 1,3-glycosidic-linked disaccharide portion (Inès and Dhouha 2015; Abdel-Mawgoud et al. 2010). Genetic regulatory systems and central metabolic pathways involving fatty acid synthesis, activated sugars, and enzymes govern the rhamnolipid production (Pornsunthorntawee et al. 2010). These surface-active compounds can be produced

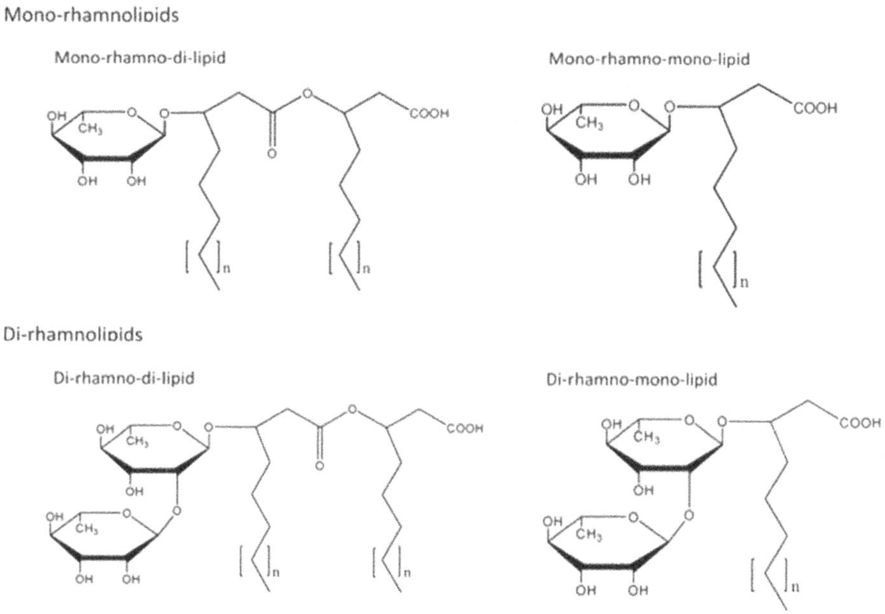

Fig. 3.2 Chemical structure of rhamnolipids (adapted with permission from (Wittgens et al. 2018))

from various types of low-cost water-miscible and water-immiscible substrates, such as carbohydrates, vegetable oils, and even industrial byproducts (Mnif et al. 2017; Pornsunthorntawee et al. 2010; Wadekar et al. 2012). However, the complexity during the synthesis and pathogenicity limits the application of *P. aeruginosa* for the industrial-scale production of rhamnolipids particularly to cosmetic and food industries (Wittgens et al. 2018; Müller and Hausmann 2011). Hence other species are also investigated.

Akin to *P. aeruginosa*, some other *Pseudomonas* species such as *P. chlororaphis*, *P. plantarii*, *P. putida*, and *P. fluorescens* are also capable of rhamnolipid production (Sekhon Randhawa and Rahman 2014). Several species of *Burkholderia* such as *Burkholderia thailandensis*, *B. glumae*, *B. kururiensis*, and *B. plantarii* (Funston et al. 2016; Nickzad et al. 2018). Organisms from *Burkholderia* species exclusively synthesize di-rhamnolipids with long-chain fatty acids usually composed of 8 to 16 carbon atoms unlike *P. aeruginosa* synthesized rhamnolipids (chains of 8 to 14 carbon atoms) (Wittgens et al. 2018). These longer fatty acid chains provide some additional properties to rhamnolipids from genus *Burkholderia*.

Sophorolipids

Sophorolipids are surface-active agents and one of the most imperative classes of glycolipid biosurfactants. They are comprised of sophorose (2-O-β-D-glucopyranosyl-D-glucopyranose), hydrophilic carbohydrate head linked to C_{16} to

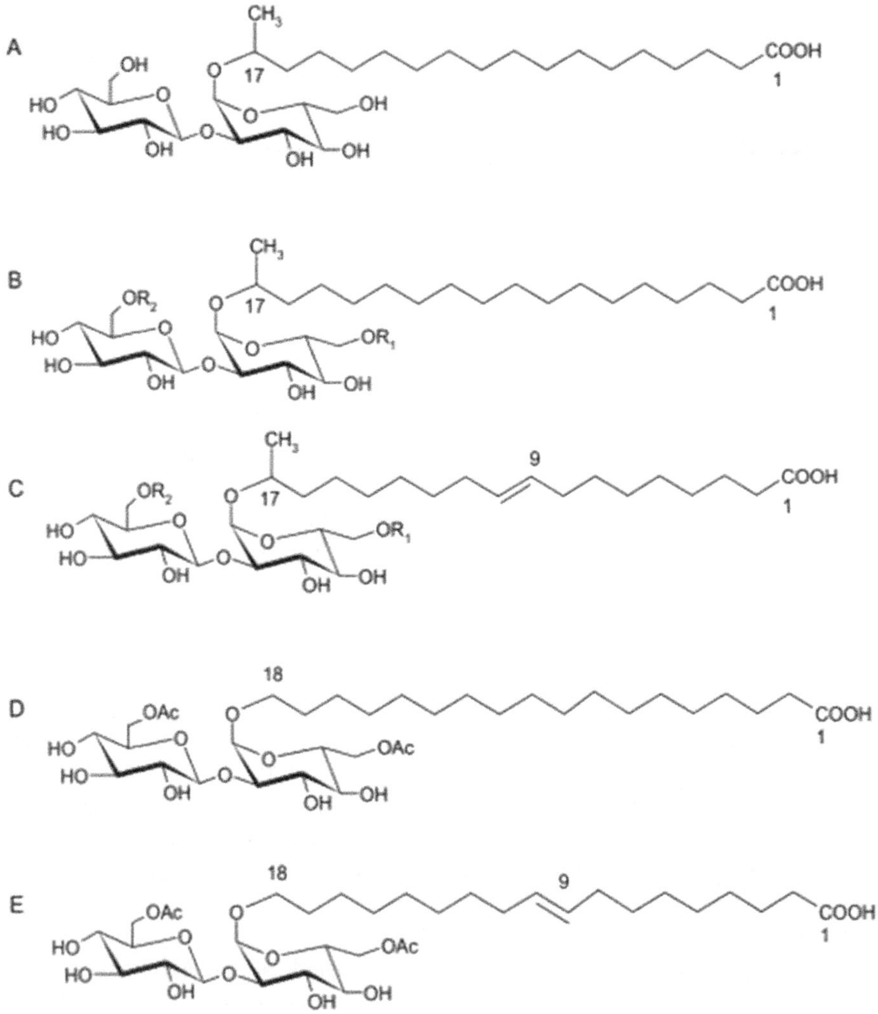

Fig. 3.3 Structures of sophorolipids in the acid form: (**a**) deacetylated sophorolipid, (**b**, **c**) major sophorolipids of *Starmerella bombicola*, and (**d**, **e**) major sophorolipids of *Candida batistae* (adapted with permission from (Kulakovskaya and Kulakovskaya 2014))

C_{18} long lipophilic hydroxy fatty acid chain (Fig. 3.3). Sophorose is the disaccharide consisting of two glucose units connected by the β-1,2′ bond (Mnif et al. 2018). During the fermentation process, sophorolipids exist in two conformations: (1) lactone form and (2) acidic form. Lactone form exists due to esterification and formation of the macrocyclic ring between the fatty acid carboxylic group and the disaccharide ring at fourth position. The acidic form is an open structure and consists of two carbohydrate head with a fatty acid carboxylic group (Nguyen and Sabatini

2011; Tulloch et al. 1968). Notably, the lactone form can exist in monomeric or dimeric forms (Kulakovskaya and Kulakovskaya 2014).

The lactonic and acidic forms of sophorolipids differ significantly in their prime physicochemical and biological properties. Lactone form of the sophorolipid exhibits superior properties, hence accepted over free acid form for ample applications in cosmetics, pharmaceuticals, and medical field (Glenns and Cooper 2006). Compared to acid sophorolipids, lactonic sophorolipids reduce the surface tension of water more efficiently, whereas acidic sophorolipids display better foaming properties (Mnif et al. 2018).

Sophorolipids are produced by yeasts, such as *Yarrowia lipolytica*, *Candida apicola*, *Rhodotorula bogoriensis*, *Starmerella bombicola*, *Rhodotorula babjevae*, *Candida gropengiesseri*, and *Candida magnolia* (Kulakovskaya and Kulakovskaya 2014; Shah et al. 2005; Cavalero and Cooper 2003; Sen et al. 2017; Delbeke et al. 2016). Up to eight to nine different structural classes of sophorolipids are observed during fermentation (Shah et al. 2005; Cavalero and Cooper 2003). Different sophorolipids differ mainly in the length of the fatty acid residues, saturations, and degree of acetylation (Kulakovskaya and Kulakovskaya 2014; Cavalero and Cooper 2003). Acetyl groups lower the hydrophilicity of sophorolipids. Sophorolipids diminish the surface tension of water from 72.8 to 30–40 mN/m and show a critical micelle concentration of 40–100 mg/L (Delbeke et al. 2016). Sophorolipids obtained by the fermentation process cannot effectively stabilize water in oil emulsions, hence not implemented as emulsifiers. However, modifications improved their wetting, cleansing, and emulsifying properties (Mnif et al. 2018; Van Bogaert and Soetaert 2011).

Trehalolipids

Trehalolipids produced by species of actinobacterial genera *Mycobacterium*, *Nocardia*, *Corynebacterium*, *Arthrobacter*, *Rhodococcus*, *Gordonia*, *Dietzia*, *Tsukamurella*, *Skermania*, and *Williamsia*. Microorganisms belonging to the mycolate group produce surfactants with notable properties (Kuyukina et al. 2015; Franzetti et al. 2010). Trehalolipids are comprised of trehalose and long-chain fatty acids joined together by an ester bond. Trehalose, a nonreducing sugar, is made up of two glycosidically linked monomeric glucose units (Sharma 2016; Kuyukina et al. 2015). Trehalolipids have gained popularity for their applicability in different fields due to their potential ability to diminish interfacial tension and increase pseudosolubility of hydrophobic compounds (Franzetti et al. 2010). Among all, biosurfactants produced by a bacterium of genus *Rhodococcus* are extensively studied due to its diverse biochemical properties and low toxic nature (Kuyukina et al. 2015; Bages et al. 2018). Contrast to rhamnolipids from *P. aeruginosa* and synthetic surfactant, *Rhodococcus erythropolis*, showed less toxicity to *Vibrio fischeri* (Munstermann et al. 1992).

These biosurfactants are considered as cell-bound biosurfactants or extracellular biosurfactants that facilitate the access of cells to water-insoluble substrates, by either direct contact or adhesion of cells to the oil drops, respectively (Franzetti et al. 2010; Bages et al. 2018). High hydrophobic cells promote direct contact, whereas lower cell hydrophobicity facilitates adhesion of microbial cells to emulsified oils (Franzetti et al. 2010; Bouchez-Naitali et al. 2001; Van Hamme et al. 2003).

Succinoyl Trehalolipids

Succinoyl trehalose lipids are mainly produced by *rhodococci* from n-alkanes and consist of one or two succinic acids and two or three fatty acids on the 1,1′-α,-α-trehalose core (Fig. 3.4) (Jana et al. 2017). *Rhodococcus erythropolis* S67 and *R. erythropolis* SD-74 have studied to produce succinoyl trehalose lipids (Zaragoza et al. 2010; Luong et al. 2018; Inaba et al. 2013). Succinoyl trehalose lipids have shown hemolytic activity, and it was mediated by colloid-osmotic mechanism (Zaragoza et al. 2010).

Cellobiose Lipids

Cellobiose lipids consist of cellobiose and fatty acid residue as an aglycone. Cellobiose is a disaccharide in which two monomeric glucose units are linked by a 1, 40-β-glycoside bond (Kulakovskaya and Kulakovskaya 2014). Different yeasts such as *Cryptococcus humicola*, *U. maydis*, *Pseudozyma fusiformata*, and *S. paphiopedili* produce cellobiose lipids and exhibit antifungal activity against different pathological strains, namely, *Cryptococcus terreus*, *Candida albicans*, *Sclerotinia sclerotiorum*, and *Phomopsis helianthi* (Kulakovskaya et al. 2004,

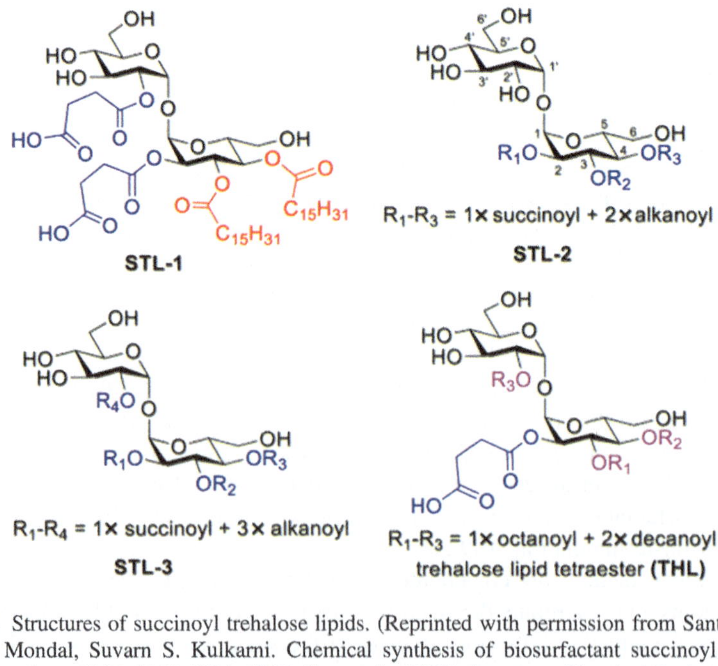

Fig. 3.4 Structures of succinoyl trehalose lipids. (Reprinted with permission from Santanu Jana, Sumana Mondal, Suvarn S. Kulkarni. Chemical synthesis of biosurfactant succinoyl trehalose lipids. Org. Lett. 2017, 19(7):1784–1787. Copyright (2017) American Chemical Society)

2009; Morita et al. 2011). *S. cerevisiae* mutants were shown to be less sensitive to cellobiose lipids from *Cryptococcus humicola* (Kulakovskaya and Mironov 2016).

Cellobiose lipid exhibits an antimycotic effect in acidic medium, where it is a weak acid owing to carboxyl group dissociation (Trilisenko et al. 2012). The antifungal activity of these compounds is because of enhanced permeability of the cytoplasmic membrane, followed by the fast expelling of ATP and potassium ions (Trilisenko et al. 2012; Kulakovskaya et al. 2008).

Mannosylerythritol Lipids

Mannosylerythritol lipids (MELs) are surface-active agents that belong to the glycolipid class of biosurfactants. Recently the MELs has drawn the attention of researchers due to its notable physicochemical and biochemical properties and application in diverse fields such as food, cosmetics, and pharmaceutical (Coelho et al. 2020; Fan et al. 2016). MELs are made up of 4-O-β-D-mannopyranosyl-meso-erythritol (hydrophiles), fatty acid, and/or an acetyl group as the lipophiles and exhibit ease in biodegradability, good stability, nontoxicity, environment compatibility, and excellent emulsifying activity (Coelho et al. 2020; Arutchelvi et al. 2008). MELs are produced by yeast strains of the genus *Pseudozyma*, viz., *Pseudozyma aphidis*, *Pseudozyma aphidis ZJUDM34*, *Pseudozyma graminicola*, *P. antarctica*, and *P. shanxiensis*, and by *Ustilago maydis* and *Schizonella melanogramma* (Fan et al. 2014, 2016; Arutchelvi et al. 2008; Morita et al. 2008; Goossens et al. 2016).

Different MELs such as MEL-A, MEL-B, MEL-C, and MEL-D produced by yeast mainly vary in degree of acylation, fatty acid chain length, and saturation (Fan et al. 2016; Goossens et al. 2016). MEL-A, MEL-B, and MEL-C are characterized by diacetylation at the C4 and C6 and monoacetylation at C6 or C4, respectively, whereas, MEL-D is devoid of acetyl group (Coelho et al. 2020). This difference in the degree of acylation also influences their antibacterial activity (Nashida et al. 2018). MEL-A is extensively studied and known to form micelles in aqueous solution and reduce surface tension up to 31.14 mN/m (Fan et al. 2014, 2016). Interestingly, the association of MEL-A with cationic liposomes has improved the effectiveness of liposomal delivery of different therapeutics followed by improved biological activities such as antioxidant and antitumor (Naughton et al. 2019; Wu et al. 2019; Sharma et al. 2015).

Xylolipids

Xylolipid is a biosurfactant and comprised of xylose as carbohydrate moiety and β-hydroxydecanoic acid as a hydrophobic part (Sharma et al. 2015). Xylolipids are mainly produced by *Lactococcus lactis* and *Pichia caribbica* (Shu et al. 2019; Saravanakumari and Mani 2010). *Enterococcus faecium* also had shown the potential to produce the xylolipid biosurfactant similar in characteristics to that of different *Lactococcus*-produced biosurfactants (Sharma et al. 2015). *Lactobacillus acidophilus*-derived biosurfactant can effectively inhibit the microorganisms causing urinary, vaginal, and gastrointestinal tract infections (Shokouhfard et al. 2015). Biosurfactants produced using *Lactobacillus jensenii* P_{6A} and *Lactobacillus gasseri* P_{65} reduced water surface tension up to 43.2 mN/m and 42.5 mN/m from

72, respectively. They also exhibited antimicrobial activity against *Escherichia coli*, *Candida albicans*, *Staphylococcus saprophyticus*, *Enterobacter aerogenes*, and *Klebsiella pneumoniae* (Morais et al. 2017). *Lactobacillus helveticus*-derived biosurfactant reduced the surface tension of phosphate buffer (pH 7.2) from 72.0 to 39.5 mN/m, and its CMC was found to be 2.5 mg/mL (Sharma et al. 2014). Biosurfactant from *L. helveticus* was found to be stable at a wide pH range (4 to 12) and high temperature (125 °C) without reducing the emulsification efficiency and showed antimicrobial and antiadhesive potential against pathogenic strains such as *Escherichia coli, Pseudomonas aeruginosa, Salmonella typhi, S. aureus*, and so on (Sharma and Singh 2014). Biosurfactants derived from *Lactobacillus casei* (Sharma and Singh 2014), *Lactobacillus delbrueckii* (Thavasi et al. 2011), *Lactobacillus fermentum* (Tahmourespour et al. 2011), and *Lactobacillus paracasei* (Gudina et al. 2011) had shown potential against different pathogenic strains.

Mannose Lipids

Monomannophosphoinositide was derived from the *Corynebacterium aquaticum*. Surface activity was attributed to a lipopeptide containing corynomycolic acids and several phospholipids (Hackett and Brennan 1975). Novel glycolipid biosurfactants, namely, mannosylribitol lipid (MRL), mannosylarabitol lipid (MAL), and mannosylmannitol lipid (MML), were derived using *Pseudozyma parantarctica*. The observed CMC value for MRL, MAL, and MML was 1.6×10^{-6} M, 1.5×10^{-6} M, and 2.6×10^{-6} M, respectively (Morita et al. 2009, 2012).

Lipopeptides or Lipoprotein

These biosurfactants are comprised of hydroxy fatty acid chains attached to a polypeptide (Vijayakumar and Saravanan 2015; Sharma 2016). Different identified lipopeptides differ mainly in peptide moiety, the length of the fatty acid chain, and the linkage between the two parts (Mnif and Ghribi 2015). Lipopeptides are known for their emulsifying, foaming, and moisturizing properties and widely applied in food processing, pharmaceutical, and cosmetic industries (Kanlayavattanakul and Lourith 2010). Lipopeptides are mainly produced by *Bacillus* sp. such as *Bacillus siamensis, B. subtilis, B. amyloliquefaciens, B. licheniformis*, etc. (Kanlayavattanakul and Lourith 2010; Ait et al. 2020; Chen et al. 2020; Varadavenkatesan and Murty 2013). Some *Pseudomonas* species are also known to produce lipopeptides (de Sousa and Bhosle 2012). They are further classified based on producing strains as follows:

1. *Bacillus*-related lipopeptides
2. *Pseudomonas*-related lipopeptides
3. *Actinomycetes*-related lipopeptides
4. Fungal-related lipopeptides (Mnif and Ghribi 2015)

Bacillus-Related Lipopeptides

Bacillus species produce different lipopeptides which differ mainly in sequence and type of amino acids, peptide cyclization, and length of fatty acid chains. They mainly belong to surfactin, iturin, and fengycin families. Fengycins and surfactins are macrolactones, whereas iturins are cyclopeptides (Mnif and Ghribi 2015). Iturin and fengycin show antifungal activities, while surfactin exhibits potent antibacterial action (Meena and Kanwar 2015).

Surfactin

Surfactin (~1.36 kDa) is one of the lipopeptides and composed of seven amino acids linked by lactone linkage to one β-hydroxy fatty acid chain (C_{13} to C_{16}) which forms a close cyclic lactone ring structure (Vijayakumar and Saravanan 2015; Mnif and Ghribi 2015). It is similar to lichenysin but only differing in the first amino acid (Chen et al. 2020). Different isoforms of surfactins mainly vary in fatty acid chain length, their branching pattern, and amino acids in the peptide ring (Nozhat et al. 2012). Surfactins in the aqueous solution or at the air/water interface form a horse-saddle conformation due to hydrophilic negatively charged glutamyl and aspartyl residues which are supposed to be responsible for its notable properties (Nozhat et al. 2012; Wu et al. 2017).

Fengycin

Fengycins are cyclic decapeptides produced by several strains of *Bacillus subtilis*. It contains the peptide chain of ten amino acids linked to a fatty acid chain of C_{14} to C_{17} carbon atoms (Meena and Kanwar 2015). Out of ten amino acids of the decapeptide chain, eight amino acids form a cyclic peptide ring via lactone linkage (Pathak et al. 2012). Different isoforms of fengycin differ mainly at the sixth position in the peptide ring and length of the β-hydroxy fatty acid chains. These variations classify fengycin family into diverse subgroups mainly fengycin A, fengycin B, and plipastatin A and B (Mnif and Ghribi 2015; Meena and Kanwar 2015). The presence of Ala or Val at sixth position of peptide chain causes the formation of fengycin A and fengycin B, respectively (Meena and Kanwar 2015; Zhang and Sun 2018). Fengycin family shows prominent antifungal activity against different filamentous fungi, such as *Aspergillus niger*, *Mucor rouxii*, *Magnaporthe grisea*, and *Fusarium graminearum* (Zhang and Sun 2018). Recently, new structural analogs of fengycin A and fengycin B have been produced using *Bacillus amyloliquefaciens* and named as fengycin X and fengycin Y, respectively. Replacement of glutamic acid residue at the eighth position of fengycin A and fengycin B by isoleucine or leucine residue causes the formation of fengycin X and fengycin Y (Ait et al. 2020).

Iturin

Iturin (~1.1 kDa), a cyclic lipopeptide comprised of a peptide chain of 7 amino acids (heptapeptides), connected to a fatty acid chain that varies from C-14 to C-17 carbon molecules. They are further classified as iturin A, C, D, and E (Mnif and Ghribi 2015; Meena and Kanwar 2015). Out of these, iturin A has been extensively synthesized and exhibits prominent antifungal activity (Dang et al. 2019).

Kurstakins

Kurstakins were isolated from the *Bacillus thuringiensis* HD-1. They are characterized by low molecular weight, and all four isolated kurstakins were differing in fatty acid chain length (Hathout et al. 2000). Recently *Bacillus* sp. P12 (Sabaté et al. 2020) and *Bacillus amyloliquefaciens* B14 (Sabaté et al. 2017) have also produced kurstakins.

Lichenysins

Bacillus licheniformis produces the lichenysins, an anionic cyclic lipoheptapeptide biosurfactant which lower the surface tension of water to 27 mN/m. Based on species-specific variations, they are named lichenysin A, B, C, D, and G and surfactant BL86. Lichenysin removes biofilms of methicillin-resistant *S. aureus*, *C. albicans*, and *Yersinia enterocolitica* (Nerurkar 2010; Coronel-León et al. 2015).

Pseudomonas-Related Lipopeptides

Cyclic lipodepsipeptides are produced by non-ribosomal peptide synthetases. In contrast to *Bacillus* species, lipopeptides isolated from *Pseudomonas* species exhibit a wide structural hetogenisty. Almost 100 cyclic lipodepsipeptides derived from *Pseudomonas* sp. have been distributed into 14 distinct groups, namely, viscosin, orfamide, amphisin, syringomycin, syringopeptin, tolaasin groups, bananamides, xantholysins, entolysins, putisolvins, pseudofactins, syringopeptins, corpeptin, and fuscopeptins (Geudens and Martins 2018). Recently milkisin, a novel lipopeptide with antimicrobial properties, has been produced by *Pseudomonas* sp. Different isomers are named as milkisin A, B, and D (Schlusselhuber et al. 2018).

Actinomycetes-Related lipopeptides

Actinomycetes particularly *Streptomyces* sp. such as *Streptomyces canus*, *Streptomyces fradiae,* and so on are known to produce diverse lipopeptide antibiotics. Some examples are amphomycin from *Actinoplanes friuliensis*, krysinomycin, glumamycin, daptomycin, etc. (Schlusselhuber et al. 2018; Genilloud 2017).

Fungal-Related Lipopeptides

Fungal strains are also known to produce lipopeptides that display notable properties (Mnif and Ghribi 2015).

Phospholipids, Fatty Acids (Mycolic Acids), and Neutral Lipids

Diverse bacteria and yeasts produce a high amount of fatty acids and phospholipid surfactants while growing on *n*-alkanes (Santos et al. 2016). Ionic surfactants like phosphatidylethanolamine, phosphatidic acids isolated from *Rhodococcus erythropolis*, showed prominent surface-active properties (Kretschmer et al. 1982).

Polymeric Surfactants

Polymeric surfactants are macromolecules that contain both hydrophilic and hydrophobic moieties in their structure (Raffa et al. 2016). Emulsan, lipomanan, alasan, liposan, and other polysaccharide-protein complexes are well-known polymeric biosurfactants (Santos et al. 2016).

Particulate Surfactants
Particulate biosurfactants are categorized into two classes, namely, (1) extracellular vesicles and (2) whole microbial cell. Extracellular membrane vesicles form microemulsions and cause the hydrocarbon uptake by microbial cells (Vijayakumar and Saravanan 2015).

3.3 Factors Affecting Biosurfactant Production

Cost-effective and higher biosurfactant production is a need of the hour. Hence, optimization of biosurfactant production is always a concern of different industries.

3.3.1 pH and Temperature

pH and temperature have shown an effect on biosurfactant production. Singh et al. studied the effect of pH and temperature on biosurfactant production. The higher yield of biosurfactant from *B. subtilis* was obtained at pH 7 and 40 °C (Nayarisseri et al. 2018). pH and temperature also influenced the biosurfactant isolation from marine *Nocardiopsis* sp. B4. Optimum growth was observed at pH 7 and 30 °C (Khopade et al. 2012).

3.3.2 Aeration and Agitation

Aeration and agitation facilitate the dissolved oxygen supply and mass transfer during the fermentation process. However, they need to be optimized as vigorous agitation and aeration lead to excessive foaming and unstabilize the fermentation process. Joshi et al. reported the higher yield of biosurfactant production by *Bacillus licheniformis* R2 when the agitation rate and aeration rate were 300 rpm and 1.0 vvm, respectively (Joshi et al. 2013). Bie et al. reported the higher surfactin production (4.7 g/L of foam) with an agitation rate of 150 rpm and an aeration rate of 1 vvm (Yao et al. 2015).

3.3.3 Effect of Salt Salinity

Maximum biosurfactant production was obtained in the presence of 3% (w/v) of NaCl, and it retained almost 80% of its activity in the presence of 12% (w/v) of NaCl (Khopade et al. 2012).

3.3.4 Optimization of Cultivation Medium

3.3.4.1 Effect of Carbon Source

The nature of the carbon substrate influences the quality and quantity of biosurfactant production (Fakruddin 2012). Olive oil, crude oil, coconut oil, *n*-hexadecane, trehalose, sucrose, fructose, maltose, glucose, and so on are used as carbon sources. Nayarisseri et al. (2018) reported that the use of sucrose, followed by crude oil, exhibited improved biosurfactant production, whereas biosurfactant isolation from marine *Nocardiopsis* sp. B4 (Khopade et al. 2012) and *Pseudomonas aeruginosa* CFTR-6 (Rashedi et al. 2005) was found to be improved using olive oil and glycerol as the carbon source, respectively. The vegetable oils have emerged as an interesting and cost-effective source to other carbon sources (Khopade et al. 2012).

3.3.4.2 Effect of Nitrogen Source

During biosurfactant production, nitrogen plays a vital role for microbial growth as protein and enzyme syntheses rely on it (Fakruddin 2012). Different inorganic and organic nitrogen sources are used for biosurfactant production, namely, urea peptone, tryptone, yeast extract, ammonium sulfate, ammonium nitrate, sodium nitrate, meat extract, and malt extract. Phenylalanine was the preferred nitrogen source for growth and biosurfactant production by *Nocardiopsis* sp. B4 (Khopade et al. 2012), whereas yeast extract and sodium nitrate improved biosurfactant yield in *Bacillus subtilis* (Nayarisseri et al. 2018) and *P. aeruginosa* CFTR-6 (Rashedi et al. 2005).

3.3.4.3 Effect of Carbon to Nitrogen (C/N) Ratio

Different studies demonstrated the effect of C/N ratio on biosurfactant production (Khopade et al. 2012; Rashedi et al. 2005). Rashedi et al. (2005) reported the optimum biosurfactant production at C/N ratio of 55/1, whereas Kokare et al. achieved the highest biosurfactant production at C/N ratio of 20 (Khopade et al. 2012).

3.4 Screening of Microorganisms for Biosurfactant Production

3.4.1 Oil Spreading Assay

To perform this test, take an appropriate quantity of water (20–30 mL) in the Petri dish, and add crude oil (10–15 μl) to form a thin layer on the surface of the water. Further, gently place 10 μl of cell-free culture broth on to the oil surface. The presence of a clear zone on the oil surface confirms the presence of biosurfactant (Nayarisseri et al. 2018; Aziz et al. 2014).

3.4.2 Drop Collapse Assay

The basic principle of this assay is to monitor the destabilization or collapse of liquid droplets by surfactants. Firstly add cell suspension drops on to the oil-coated solid or glass surface, and monitor the behavior of the droplets. Stable drops correspond to the absence of surfactants, whereas spreading or collapse of droplets on to the oil-coated surface indicates the presence of surfactants. Droplets collapse due to a reduction in interfacial tension between water and oil (Walter et al. 2010).

3.4.3 Blood Agar Method/Hemolysis Assay

Contact of erythrocytes and biosurfactants causes erythrocyte lysis. Considering this fact, inoculate the bacterial isolates on sheep blood agar plates, and incubate for 48 h at 25°C. The existence of the hemolysis zone (colorless ring around colonies) corresponds to biosurfactant presence and vice versa (Aziz et al. 2014; Walter et al. 2010).

3.4.4 Hydrocarbon Overlay Agar

Coat Zobell marine agar plates with 40 μl of kerosene, hexadecane, benzene, and toluene separately. Inoculate bacterial culture on plates and incubate for 7–10 days at 28 °C. The existence of emulsified halo around the colonies considered indicates the biosurfactant production (Nayarisseri et al. 2018).

3.4.5 Bacterial Adhesion to Hydrocarbon (BATH) Assay

BATH assay is based on the measurement of cell surface hydrophobicity, i.e., adherence ability of cells to different liquid hydrocarbons such as hexadecane or octane. Add aqueous phase containing bacterial cells to hydrocarbon and mix well. Separate the hydrocarbon phase, once the two phases separate. Measure the optical density of the aqueous phase using a spectrophotometer, and determine the percentage of cells adhered to the hydrophobic phase using the following formula (Nayarisseri et al. 2018; Walter et al. 2010):

$$H = \left(1 - \frac{A}{A_0}\right) \times 100 \tag{3.1}$$

where A_0 and A are the optical density of aqueous suspension before and after mixing with the hydrocarbon phase.

3.4.6 CTAB Agar Plate Method/Blue Agar Assay

The cetyltrimethylammonium bromide (CTAB) agar plate method is useful to detect extracellular glycolipids or anionic surfactants. Inoculate the culture on a light blue mineral salt agar plate enriched with CTAB and methylene blue, and incubate at 7 °C for 24–48 h. The formation of the dark blue zone around colonies indicates the secretion of anionic surfactants. The development of the colored zone is attributed to complex formation between secreted surfactant, CTAB, and dye used (Walter et al. 2010; Ibrahim 2018).

3.4.7 Phenol: Sulfuric Acid Method

Inoculate culture on minimal salt medium and incubate for 4–5 days at 37 °C. After incubation, centrifuge the broth and collect the supernatant. Add 1 mL of 5% phenol as well as 5 mL of conc. Add H2SO4 is added to get supernatant and mix well. The existence of orange color from yellow color corresponds to the presence of biosurfactant (Aziz et al. 2014).

3.4.8 Microplate Assay

This qualitative assay is based on the principle of measurement of optical distortion of the aqueous phase. Flat and concave fluid surfaces correspond to the absence and presence of surfactants (Walter et al. 2010).

3.4.9 Penetration Assay

Fill the cavities of a 96-well microplate with a hydrophobic paste comprised of oil and silica gel, and cover it with oil (10 μl). Add red-colored supernatant on the paste surface. Mixing of hydrophobic and hydrophilic phases and conversion of clear red hydrophilic phase to cloudy white indicate the presence of biosurfactant (Walter et al. 2010).

3.4.10 Surface/Interface Activity

The surface tension of culture supernatant can be determined using digital surface tensiometer equipped with a Du Nöuy platinum ring (Ibrahim 2018). The stalagmometric method or pendant drop shape technique has also been used (Walter et al. 2010).

Table 3.2 Research summary of antioxidant activity of biosurfactants

Strain	Substrate/ medium	Antioxidant assay	Parameters	Key findings	Reference
Lactobacillus casei	Modified MRS broth at pH of 6.4	1,1-Diphenyl-2-picrylhydrazyl (DPPH) assay	DPPH-scavenging activity: 74.6 and 77.3%	Biosurfactants from *L. casei* strains exhibited considerable antioxidant activity IC_{50}: 2.09 and 2.16 mg/mL	Merghni et al. (2017)
Yeast	Glucose, yeast extract	Phosphor molybdenum method	Carbohydrate content: PS1 (0.33 mg/mL) and PS2 (0.28 mg/mL)	Increased antioxidant activity and free radical-scavenging activity (95%)	Kharat et al. (2018)
Bifidobacterium bifidum (B-EPS) and *Lactobacillus plantarum (L-EPS)*	MRS broth supplemented with L-cysteine	Hydroxyl radical scavenging and superoxide radical scavenging	Radical-scavenging L-EPS (39.15 ± 0.58%) B-EPS (17.89 ± 3.30%)	The hydroxyl-scavenging ability of L-EPS was 2.19-fold higher than that and B-EPS	Li et al. (2014)
Bacillus methylotrophicus (lipopeptides)	Luria-Bertani agar and malt extract agar media	DPPH radical-scavenging assay Ferric-reducing activity Ferrous ion-chelating activity β-Carotene bleaching assay	DPPH radical scavenging (80.6%) at a concentration of 1 mg/mL	Strong inhibition of β-carotene bleaching by linoleic acid assay (80.8%) and showed a good chelating ability	Jemil et al. (2017)
Bacillus subtilis RW-I	Nutrient agar, glucose	Reducing power, ferrous ion-chelating and DPPH-scavenging activity	Ferrous ion chelating, 75.3%, and DPPH-scavenging activity, 85.2%, at 1.0 mg/mL; IC_{50}, 0.25 mg/mL	Compared to butylated hydroxy toluene biosurfactant (2.4 mg/mL), exhibited up to 1.2-fold weaker reducing power	Yalçın and Çavuşoğlu (2010)
Pseudozyma hubeiensis (mannosylerythritol lipids)	Soybean oil	DPPH- and superoxide-scavenging assay; H_2O_2-induced oxidative stress in cultured human skin fibroblasts	DPPH radical scavenging, 50.3% at 10 mg/mL, and superoxide anion scavenging, >50% at 1 mg/mL	MEL-C showed 2.33-fold higher protective activity against oxidative stress compared to arbutin	Takahashi et al. (2012)

(continued)

Table 3.2 (continued)

Strain	Substrate/ medium	Antioxidant assay	Parameters	Key findings	Reference
Bacillus subtilis SPB1	–	DPPH free radical-scavenging assay Ferric-reducing antioxidant power (FRAP) assay β-Carotene bleaching assay Ferrous ion-chelating assay	DPPH radical scavenging, 70.4% (IC50 = 0.55 mg/mL); β-Carotene test IC_{50} = 2.26 mg/mL; Ferrous ion-chelating activity, 80.32%	IC_{50} values of *Bacillus subtilis* biosurfactant displayed concentration-dependent scavenging activity	Zouari et al. (2016)

3.4.11 Emulsification Activity

Mix 4 mL culture supernatant with motor oil or kerosene, vortex for 3 min, and keep the mixture aside for 24 h at 25 °C. Calculate emulsification index (E_{24}) by using following equation (Walter et al. 2010; Ibrahim 2018):

$$E_{24}(\%) = \frac{H_{\text{emulsion}}}{H_{\text{total}}} \times 100 \tag{3.2}$$

where "*H*" indicates the height.

3.5 Antioxidant Properties of Biosurfactant

In recent years, the antioxidative properties of biosurfactants have gained the attention of different industries due to superior and notable properties of natural compounds than synthetic antioxidants. Antioxidants efficiently inhibit the formation of both reactive oxygen species and reactive nitrogen species and thereby prevent cancer, cardiovascular diseases, and neurodegenerative diseases (Seifried et al. 2017). Summary of research work related to antioxidant activity of biosurfactants are reported in Table 3.2.

3.6 Conclusion

The concept of going green and environmental as well as health concerns are greater than before creating the demand to replace the synthetic surfactants. In recent times, biosurfactants have gained attention and served the diverse sectors including pharmaceutical, cosmeceuticals, biotechnology, food processing, and the like due to promising properties. Biosurfactants are getting explored as foaming, emulsifying, antibacterial, solubilizers, detergents, and wetting agents. The thorough-going optimization to obtain cost-effective methods of isolation and deliberative use of sophisticate analytical techniques for biosurfactants characterization is needed to be considered.

References

Abbas M, Sari N, Utari T, Yulizar Y, Apriandanu DOB (2018) AIP Conf Proc 2023:020100. https://doi.org/10.1063/1.5064097

Abdel-Mawgoud AM, Lépine F, Déziel E (2010) Rhamnolipids: diversity of structures, microbial origins and roles. Appl Microbiol Biotechnol 86(5):1323–1336. https://doi.org/10.1007/s00253-010-2498-2. Epub 2010 Mar 25. PMID: 20336292; PMCID: PMC2854365

Ait Kaki A, Smargiasso N, Ongena M et al (2020) Characterization of New Fengycin cyclic lipopeptide variants produced by Bacillus amyloliquefaciens (ET) originating from a Salt

Lake of Eastern Algeria. Curr Microbiol 77:443–451. https://doi.org/10.1007/s00284-019-01855-w

Akbari S, Abdurahman NH, Yunus RM, Fayaz F, Alara OR (2018) Biosurfactants—a new frontier for social and environmental safety: a mini review. Biotechnol Res Innov 2:81–90

Arutchelvi JI, Bhaduri S, Uppara PV, Doble M (2008) Mannosylerythritol lipids: a review. J Ind Microbiol Biotechnol 35(12):1559–1570. https://doi.org/10.1007/s10295-008-0460-4

Aziz RNNM, Balasaheb SM, Dixit PP, Deshmukh AM (2014) Screening of biosurfactant producing microorganisms from oil contaminated soils of Osmanabad region, Maharashtra, India. Int Sci J 1(1):35–39

Bages S, White DA, Winterburn JB, Webb C, Martin PJ (2018) Production and separation of a trehalolipid biosurfactant. Biochem Eng J 139:85–94. https://doi.org/10.1016/j.bej.2018.07.006

Belhaj AF, Elraies KA, Mahmood SM et al (2020) The effect of surfactant concentration, salinity, temperature, and pH on surfactant adsorption for chemical enhanced oil recovery: a review. J Petrol Explor Prod Technol 10:125–137. https://doi.org/10.1007/s13202-019-0685-y

Benaiges A, Guillén P (2007) Botanical Extracts. Anal Cosmet Prod 2007:345–363. https://doi.org/10.1016/b978-044452260-3/50044-9

Bouchez-Naitali M, Blanchet D, Bardin V, Vandecasteele JP (2001) Evidence for interfacial uptake in hexadecane degradation by Rhodococcus equi: the importance of cell flocculation. Microbiology 147:2537–2543

Burakova Y, Madera R, Wang L, Buist S, Lleellish K, Schlup JR, Shi J (2018) Food-grade saponin extract as an emulsifier and immunostimulant in emulsion-based subunit vaccine for pigs. J Immunol Res 2018:1–8. https://doi.org/10.1155/2018/8979838

Cavalero DA, Cooper DG (2003) The effect of medium composition on the structure and physical state of sophorolipids produced by Candida bombicola ATCC 22214. J Biotechnol 103 (1):31–41. https://doi.org/10.1016/s0168-1656(03)00067-1

Chen XW, Yang XQ (2019) Characterization of orange oil powders and oleogels fabricated from emulsion templates stabilized solely by a natural triterpene saponin. J Agric Food Chem 67 (9):2637–2646. https://doi.org/10.1021/acs.jafc.8b04588

Chen YF, Yang CH, Chang MS, Ciou YP, Huang YC (2010) Foam properties and detergent abilities of the saponins from Camellia oleifera. Int J Mol Sci 11(11):4417–4425. https://doi.org/10.3390/ijms11114417. PMID: 21151446; PMCID: PMC3000090

Chen Z, Wu Q, Wang L, Chen S, Lin L, Wang H, Xu Y (2020) Identification and quantification of surfactin, a nonvolatile lipopeptide in Moutai liquor. Int J Food Prop 23(1):189–198. https://doi.org/10.1080/10942912.2020.1716791

Chong H, Li Q (2017) Microbial production of rhamnolipids: opportunities, challenges and strategies. Microb Cell Fact 16:137. https://doi.org/10.1186/s12934-017-0753-2

Cierniak D, Woźniak-Karczewska M, Parus A et al (2020) How to accurately assess surfactant biodegradation-impact of sorption on the validity of results. Appl Microbiol Biotechnol 104:1–12. https://doi.org/10.1007/s00253-019-10202-9

Coelho ALS, Feuser PE, Carciofi BAM et al (2020) Mannosylerythritol lipids: antimicrobial and biomedical properties. Appl Microbiol Biotechnol 104:2297–2318. https://doi.org/10.1007/s00253-020-10354-z

Coronel-León J, Marqués AM, Bastida J, Manresa A (2015) Optimizing the production of the biosurfactant lichenysin and its application in biofilm control. J Appl Microbiol 120(1):99–111. https://doi.org/10.1111/jam.12992

Costa JAV, Treichel H, Santos LO, Martins VG (2018) Solid-state fermentation for the production of biosurfactants and their applications. In: Current developments in biotechnology and bioengineering, pp 357–372. https://doi.org/10.1016/b978-0-444-63990-5.00016-5

Dang Y, Zhao F, Liu X et al (2019) Enhanced production of antifungal lipopeptide iturin A by Bacillus amyloliquefaciens LL3 through metabolic engineering and culture conditions optimization. Microb Cell Fact 18:68. https://doi.org/10.1186/s12934-019-1121-1

de Sousa T, Bhosle S (2012) Isolation and characterization of a lipopeptide bioemulsifier produced by Pseudomonas nitroreducens TSB.MJ10 isolated from a mangrove ecosystem. Bioresour Technol 123:256–262. https://doi.org/10.1016/j.biortech.2012.07.056

Delbeke EIP, Movsisyan M, Van Geem KM, Stevens CV (2016) Chemical and enzymatic modification of sophorolipids. Green Chem 18(1):76–104. https://doi.org/10.1039/c5gc02187a

Desai JD, Banat IM (1997a) Microbial production of surfactants and their commercial potential. Microbiol Mol Biol Rev 61:47–64

Desai JD, Banat IM (1997b) Microbial production of surfactants and their commercial potential. Microbiol Mol Biol Rev 61(1):47–64. PMID: 9106364; PMCID: PMC232600

do Canto GS, Treter J, Yang S, Borré GL, Peixoto MPG, Ortega GG (2010) Evaluation of foam properties of saponin from Ilex paraguariensis A. St. Hil. (Aquifoliaceae) fruits. Braz J Pharm Sci 46(2):237–243. https://doi.org/10.1590/S1984-82502010000200010

Do DN, Dang TT, Le QT, Lam TD, Bạch LG, Nguyen DC, Toan TQ (2019) Extraction of saponin from gleditsia peel and applications on natural dishwashing liquid detergent. Mater Today Proc 18(7):5219–5230. https://doi.org/10.1016/j.matpr.2019.07.522

Fakruddin M (2012) Biosurfactant: production and application. J Pet Environ Biotechnol 3:124. https://doi.org/10.4172/2157-7463.1000124

Fan LL, Dong YC, Fan YF, Zhang J, Chen QH (2014) Production and identification of mannosylerythritol lipid-A homologs from the ustilaginomycetous yeast Pseudozyma aphidis ZJUDM34. Carbohydr Res 392:1–6. https://doi.org/10.1016/j.carres.2014.04.013

Fan L, Li H, Niu Y, Chen Q (2016) Characterization and inducing melanoma cell apoptosis activity of mannosylerythritol lipids-A produced from Pseudozyma aphidis. PLoS One 11(2):e0148198. https://doi.org/10.1371/journal.pone.0148198. PMID: 26828792; PMCID: PMC4734834

Fenibo EO, Ijoma GN, Selvarajan R, Chikere CB (2019) Microbial surfactants: the next generation multifunctional biomolecules for applications in the petroleum industry and its associated environmental remediation. Microorganisms 7(11):581. https://doi.org/10.3390/microorganisms7110581. PMID: 31752381; PMCID: PMC6920868

Franzetti A, Gandolfi I, Bestetti G, Smyth T, Banat IM (2010) Production and applications of trehalose lipid biosurfactants. Eur J Lipid Sci Technol 112:617–627

Freitas BG, Brito JGM, Brasileiro PPF, Rufino RD, Luna JM, Santos VA, Sarubbo LA (2016) Formulation of a commercial biosurfactant for application as a dispersant of petroleum and by-products spilled in oceans. Front Microbiol 7:1–9

Funston SJ, Tsaousi K, Rudden M, Smyth TJ, Stevenson PS, Marchant R, Banat IM (2016) Characterising rhamnolipid production in Burkholderia thailandensis E264, a non-pathogenic producer. Appl Microbiol Biotechnol 100(18):7945–7956. https://doi.org/10.1007/s00253-016-7564-y. Epub 2016 May 5. PMID: 27147528; PMCID: PMC4989024

Genilloud O (2017) Actinomycetes: still a source of novel antibiotics. Nat Prod Rep 34(10):1203–1232. https://doi.org/10.1039/c7np00026j

Geudens N, Martins JC (2018) Cyclic lipodepsipeptides from Pseudomonas spp.—biological Swiss-Army Knives. Front Microbiol 9:1867. https://doi.org/10.3389/fmicb.2018.01867. PMID: 30158910; PMCID: PMC6104475

Glenns RN, Cooper DG (2006) Effect of substrate on sophorolipid properties. J Am Oil Chem Soc 83:137–145. https://doi.org/10.1007/s11746-006-1186-y

Goossens E, Wijnants M, Packet D, Lemière F (2016) Enhanced separation and analysis procedure reveals production of tri-acylated mannosylerythritol lipids by Pseudozyma aphidis. J Ind Microbiol Biotechnol 43(11):1537–1550. https://doi.org/10.1007/s10295-016-1838-3

Gudina EJ, Teixeira A, Rodrigues LR (2011) Biosurfactant-producing lactobacilli: screening, production profiles, and effect of medium composition. Appl Environ Soil Sci 2011:1–9

Hackett JA, Brennan PJ (1975) The mannophosphoinositides of Corynebacterium aquaticum. Biochem J 148(2):253–258

Hathout Y, Ho Y-P, Ryzhov V, Demirev P, Fenselau C (2000) Kurstakins: a new class of lipopeptides isolated from Bacillus thuringiensis. J Nat Prod 63(11):1492–1496. https://doi.org/10.1021/np000169q

Henkel M, Müller MM, Kügler JH, Lovaglio RB, Contiero J, Syldatk C, Hausmann R (2012) Rhamnolipids as biosurfactants from renewable resources: concepts for next-generation rhamnolipid production. Process Biochem 47:1207–1219

Huang W-C, Tang I-C (2007) Bacterial and yeast cultures—process characteristics, products, and applications. In: Bioprocessing for value-added products from renewable resources, pp 185–223. https://doi.org/10.1016/b978-044452114-9/50009-8

Ibrahim HMM (2018) Characterization of biosurfactants produced by novel strains of Ochrobactrum anthropi HM-1 and Citrobacter freundii HM-2 from used engine oil-contaminated soil. Egypt J Petroleum 27(1):21–29. https://doi.org/10.1016/j.ejpe.2016.12.005

Inaba T, Tokumoto Y, Miyazaki Y, Inoue N, Maseda H, Nakajima Kambe T et al (2013) Analysis of genes for succinoyl trehalose lipid production and increasing production in Rhodococcus sp. strain SD-74. Appl Environ Microbiol 79:7082–7090

Inès M, Dhouha G (2015) Glycolipid biosurfactants: potential related biomedical and biotechnological applications. Carbohydr Res 416:59–69. https://doi.org/10.1016/j.carres.2015.07.016

Jana S, Mondal S, Kulkarni SS (2017) Chemical synthesis of biosurfactant succinoyl trehalose lipids. Org Lett 19(7):1784–1787. https://doi.org/10.1021/acs.orglett.7b00550

Jemil N, Ben Ayed H, Manresa A et al (2017) Antioxidant properties, antimicrobial and anti-adhesive activities of DCS1 lipopeptides from Bacillus methylotrophicus DCS1. BMC Microbiol 17:144. https://doi.org/10.1186/s12866-017-1050-2

Jimoh AA, Lin J (2019) Biosurfactant: a new frontier for greener technology and environmental sustainability. Ecotoxicol Environ Saf 184:109607. https://doi.org/10.1016/j.ecoenv.2019.109607

Joshi SJ, Geetha SJ, Yadav S, Desai AJ (2013) Optimization of bench-scale production of biosurfactant by Bacillus Licheniformis R2. APCBEE Proc 5:232–236. https://doi.org/10.1016/j.apcbee.2013.05.040

Kanlayavattanakul M, Lourith N (2010) Lipopeptides in cosmetics. Int J Cosmet Sci 32(1):1–8. https://doi.org/10.1111/j.1468-2494.2009.00543.x

Kaur K, Kumar R, Mehta SK (2016) Formulation of saponin stabilized nanoemulsion by ultrasonic method and its role to protect the degradation of quercitin from UV light. Ultrason Sonochem 31:29–38. https://doi.org/10.1016/j.ultsonch.2015.11.017

Kharat M, McClements DJ (2019) Fabrication and characterization of nanostructured lipid carriers (NLC) using a plant-based emulsifier: Quillaja saponin. Food Res Int 126:108601. https://doi.org/10.1016/j.foodres.2019.108601

Kharat PP, RamsaranYadav S, Ragavan ML, Das N (2018) Isolation and characterization of exopolysaccharides fromyeast isolates. Res J Pharm Technol 11(2):537–542

Khopade A, Biao R, Liu X, Mahadik K, Zhang L, Kokare C (2012) Production and stability studies of the biosurfactant isolated from marine Nocardiopsis sp. B4. Desalination 285:198–204. https://doi.org/10.1016/j.desal.2011.10.002

Kitamoto D, Isoda H, Nakahara T (2002) Functions and potential applications of glycolipid biosurfactants. J Biosci Bioeng 94:187–201

Kretschmer A, Bock H, Wagner F (1982) Chemical and physical characterization of interfacial-active lipids from Rhodococcus erythropolis grown on n-alkanes. Appl Environ Microbiol 44 (4):864–870. PMID: 16346110; PMCID: PMC242110

Kulakovskaya E, Kulakovskaya T (2014) Structure and occurrence of yeast extracellular glycolipids. In: Extracellular glycolipids of yeasts, pp 1–13. https://doi.org/10.1016/b978-0-12-420069-2.00001-7

Kulakovskaya EV, Mironov AA (2016) Resistance to cellobiose lipids and specific features of lipid composition in yeast. Appl Biochem Microbiol 52:615–620. https://doi.org/10.1134/S0003683816060107

Kulakovskaya TV, Shashkov AS, Kulakovskaya EV, Golubev WI (2004) Characterization of an antifungal glycolipid secreted by the yeast Sympodiomycopsis paphiopedili. FEMS Yeast Res 5 (3):247–252. https://doi.org/10.1016/j.femsyr.2004.07.008

Kulakovskaya EV, Ivanov AY, Kulakovskaya TV, Vagabov VM, Kulaev IS (2008) Effects of cellobiose lipid B on Saccharomyces cerevisiae cells: K + leakage and inhibition of polyphosphate accumulation. Microbiology 77:288–292

Kulakovskaya T, Shashkov A, Kulakovskaya E et al (2009) Extracellular cellobiose lipid from yeast and their analogues: structures and fungicidal activities. J Oleo Sci 58(3):133–140. https://doi.org/10.5650/jos.58.133

Kuyukina MS, Ivshina IB, Baeva TA, Kochina OA, Gein SV, Chereshnev VA (2015) Trehalolipid biosurfactants from nonpathogenic Rhodococcus actinobacteria with diverse immunomodulatory activities. N Biotechnol 32(6):559–568. https://doi.org/10.1016/j.nbt.2015.03.006

Li S, Huang R, Shah NP, Tao X, Xiong Y, Wei H (2014) Antioxidant and antibacterial activities of exopolysaccharides from Bifidobacterium bifidum WBIN03 and Lactobacillus plantarum R315. J Dairy Sci 97(12):7334–7343. https://doi.org/10.3168/jds.2014-7912

Lombardo D, Kiselev MA, Magazù S, Calandra P (2015) Amphiphiles self-assembly: basic concepts and future perspectives of supramolecular approaches. Adv Condens Matter Phys 2015:151683. https://doi.org/10.1155/2015/151683

Luong TM, Ponamoreva ON, Nechaeva IA et al (2018) Characterization of biosurfactants produced by the oil-degrading bacterium Rhodococcus erythropolis S67 at low temperature. World J Microbiol Biotechnol 34:20. https://doi.org/10.1007/s11274-017-2401-8

Lv S, Zhang Y, Tan H, Zhang R, McClements DJ (2019) Vitamin E encapsulation within oil-in-water emulsions: impact of emulsifier type on physicochemical stability and bioaccessibility. J Agric Food Chem 67(5):1521–1529. https://doi.org/10.1021/acs.jafc.8b06347

Maier RM (2003) Biosurfactants: evolution and diversity in bacteria. Adv Appl Microbiol 52:101–121. https://doi.org/10.1016/s0065-2164(03)01004-9

Makkar RS, Cameotra SS, Banat IM (2011) Advances in utilization of renewable substrates for biosurfactant production. AMB Express 1(1):5. https://doi.org/10.1186/2191-0855-1-5

Markets and Markets (2016). https://www.marketsandmarkets.com/Market-Reports/biosurfactant-market-163644922.html. Accessed 21 Mar 2020

Meena KR, Kanwar SS (2015) Lipopeptides as the antifungal and antibacterial agents: applications in food safety and therapeutics. Biomed Res Int 2015:473050. https://doi.org/10.1155/2015/473050. Epub 2015 Jan 6. PMID: 25632392; PMCID: PMC4303012

Merghni A, Dallel I, Noumi E, Kadmi Y, Hentati H, Tobji S et al (2017) Antioxidant and antiproliferative potential of biosurfactants isolated from Lactobacillus casei and their anti-biofilm effect in oral Staphylococcus aureus strains. Microb Pathog 104:84–89. https://doi.org/10.1016/j.micpath.2017.01.017

Mesgarzadeh I, Akbarzadeh AR, Rahimi R (2017) Surface-active properties of solvent-extracted panax ginseng saponin-based surfactants. J Surfactant Deterg 20:609–614. https://doi.org/10.1007/s11743-017-1940-1

Mnif I, Ghribi D (2015) Review lipopeptides biosurfactants: mean classes and new insights for industrial, biomedical, and environmental applications. Biopolymers 104(3):129–147. https://doi.org/10.1002/bip.22630

Mnif I, Ellouz-Chaabouni S, Ghribi D (2017) Glycolipid biosurfactants, main classes, functional properties and related potential applications in environmental biotechnology. J Polym Environ 26(5):2192–2206. https://doi.org/10.1007/s10924-017-1076-4

Mnif I, Ellouz-Chaabouni S, Ghribi D (2018) Glycolipid biosurfactants, main classes, functional properties and related potential applications in environmental biotechnology. J Polym Environ 26:2192–2206. https://doi.org/10.1007/s10924-017-1076-4

Mondal MH, Malik S, Roy A, Saha R, Saha B (2015) Modernization of surfactant chemistry in the age of gemini and bio-surfactants: a review. RSC Adv 5(112):92707–92718. https://doi.org/10.1039/c5ra18462b

Morais IMC, Cordeiro AL, Teixeira GS et al (2017) Biological and physicochemical properties of biosurfactants produced by Lactobacillus jensenii P6A and Lactobacillus gasseri P65. Microb Cell Fact 16:155. https://doi.org/10.1186/s12934-017-0769-7

Morita T, Konishi M, Fukuoka T et al (2008) Identification of Pseudozyma graminicola CBS 10092 as a producer of glycolipid biosurfactants, mannosylerythritol lipids. J Oleo Sci 57(2):123–131. https://doi.org/10.5650/jos.57.123

Morita T, Fukuoka T, Konishi M et al (2009) Production of a novel glycolipid biosurfactant, mannosylmannitol lipid, by Pseudozyma parantarctica and its interfacial properties. Appl Microbiol Biotechnol 83(6):1017–1025. https://doi.org/10.1007/s00253-009-1945-4

Morita T, Ishibashi Y, Fukuoka T et al (2011) Production of glycolipid biosurfactants, cellobiose lipids, by Cryptococcus humicola JCM 1461 and their interfacial properties. Biosci Biotechnol Biochem 75(8):1597–1599. https://doi.org/10.1271/bbb.110036

Morita T, Fukuoka T, Imura T, Kitamoto D (2012) Formation of the two novel glycolipid biosurfactants, mannosylribitol lipid and mannosylarabitol lipid, by Pseudozyma parantarctica JCM 11752T. Appl Microbiol Biotechnol 96(4):931–938. https://doi.org/10.1007/s00253-012-4230-x

Mukherjee S, Das P, Sen R (2006) Towards commercial production of microbial surfactants. Trends Biotechnol 24(11):509–515. https://doi.org/10.1016/j.tibtech.2006.09.005

Müller MM, Hausmann R (2011) Regulatory and metabolic network of rhamnolipid biosynthesis: traditional and advanced engineering towards biotechnological production. Appl Microbiol Biotechnol 91(2):251–264. https://doi.org/10.1007/s00253-011-3368-2

Mulligan CN (2007) Rhamnolipid biosurfactants: solubility and environmental issues. In: Thermodynamics, solubility and environmental issues, pp 279–298. https://doi.org/10.1016/b978-044452707-3/50017-3

Munstermann B, Poremba K, Lang S, Wagner F (1992) Studies on environmental compatibility: influence of (bio)surfactants on marine microbial and enzymatic systems. In: Proceedings of international symposium soil decontamination using biological processes, pp 414–420

Nakama Y (2017) Surfactants. In: Cosmetic science and technology, pp 231–244. https://doi.org/10.1016/b978-0-12-802005-0.00015-x

Nashida J, Nishi N, Takahashi Y, Hayashi C, Igarashi M, Takahashi D, Toshima K (2018) Systematic and stereoselective total synthesis of mannosylerythritol lipids and evaluation of their antibacterial activity. J Org Chem 83:7281–7289. https://doi.org/10.1021/acs.joc.8b00032

Naughton PJ, Marchant R, Naughton V, Banat IM (2019) Microbial biosurfactants: current trends and applications in agricultural and biomedical industries. J Appl Microbiol 127:12–28. https://doi.org/10.1111/jam.14243

Nayarisseri A, Singh P, Singh SK (2018) Screening, isolation and characterization of biosurfactant producing Bacillus subtilis strain ANSKLAB03. Bioinformation 14(6):304–314. Published 2018 Jun 30. https://doi.org/10.6026/97320630014304

Nerurkar AS (2010) Structural and molecular characteristics of lichenysin and its relationship with surface activity. Adv Exp Med Biol 672:304–315. https://doi.org/10.1007/978-1-4419-5979-9_23

Nguyen TT, Sabatini DA (2011) Characterization and emulsification properties of rhamnolipid and sophorolipid biosurfactants and their applications. Int J Mol Sci 12(2):1232–1244. https://doi.org/10.3390/ijms12021232. PMID: 21541055; PMCID: PMC3083702

Nickzad A, Guertin C, Déziel E (2018) Culture medium optimization for production of rhamnolipids by *Burkholderia glumae*. Colloids Interfaces 2:49. https://doi.org/10.3390/colloids2040049

Niziol-Łukaszewska Z, Bujak T (2018) Saponins as natural raw materials for increasing the safety of bodywash cosmetic use. J Surfactant Deterg 21(6):767–776. https://doi.org/10.1002/jsde.12168

Nozhat Z, Asadi A, Zahri S (2012) Properties of Surfactin C-15 nanopeptide and its cytotoxic effect on human cervix cancer (HeLa) cell line. Biocompat Toxic Nanobiomater 2012:526580. https://doi.org/10.1155/2012/526580

Oleszek W, Hamed A (2010) Saponin-based surfactants. In: Kjellin M, Johansson I (eds) Surfactants from renewable resources. Wiley, Hoboken, NJ, pp 239–249. ISBN: 978-0-470-76041-3

Olkowska E, Ruman M, Polkowska Z (2014) Occurrence of surface active agents in the environment. J Anal Methods Chem 2014:769708. https://doi.org/10.1155/2014/769708. Epub 2014 Jan 16. PMID: 24527257; PMCID: PMC3914419

Pathak KV, Keharia H, Gupta K, Thakur SS, Balaram P (2012) Lipopeptides from the banyan endophyte, Bacillus subtilis K1: mass spectrometric characterization of a library of fengycins. J Am Soc Mass Spectrom 23(10):1716–1728

Pornsunthorntawee O, Wongpanit P, Rujiravanit R (2010) Rhamnolipid biosurfactants: production and their potential in environmental biotechnology. In: Sen R (ed) Biosurfactants. Advances in experimental medicine and biology, vol 672. Springer, New York, NY

Pradhan A, Bhattacharyya A (2017) Quest for an eco-friendly alternative surfactant: Surface and foam characteristics of natural surfactants. J Clean Prod 150:127–134

Raffa P, Broekhuis AA, Picchioni F (2016) Polymeric surfactants for enhanced oil recovery: a review. J Petrol Sci Eng 145:723–733. https://doi.org/10.1016/j.petrol.2016.07.007

Ralla T, Salminen H, Tuosto J, Weiss J (2017) Formation and stability of emulsions stabilised by Yucca saponin extract. Int J Food Sci Technol 53(6):1381–1388. https://doi.org/10.1111/ijfs.13715

Rashedi H, Jamshidi E, Mazaheri Assadi M, Bonakdarpour B (2005) Isolation and production of biosurfactant from Pseudomonas aeruginosa isolated from Iranian southern wells oil. Int J Environ Sci Tech 2(2):121–127

Rebello S, Asok AK, Jisha MS (2013) Surfactants: chemistry, toxicity and remediation. In: Lichtfouse E et al (eds) Pollutant diseases, remediation and recycling. Springer International, Cham, pp 277–320. Environmental Chemistry for a Sustainable World 4. https://doi.org/10.1007/978-3-319-02387-8_5

Rodríguez-López L, Rincón-Fontán M, Vecino X, Moldes AB, Cruz JM (2019) Biodegradability study of the biosurfactant contained in a crude extract from corn steep water. J Surfactant Deterg 23:79–90. https://doi.org/10.1002/jsde.12338

Sabaté DC, Pérez Brandan C, Petroselli G, Erra-Balsells R, Audisio MC (2017) Decrease in the incidence of charcoal root rot in common bean (Phaseolus vulgaris L.) by Bacillus amyloliquefaciens B14, a strain with PGPR properties. Biol Control 113:1–8. https://doi.org/10.1016/j.biocontrol.2017.06.008

Sabaté DC, Petroselli G, Erra-Balsells R, Carina Audisio M, Brandan CP (2020) Beneficial effect of Bacillus sp. P12 on soil biological activities and pathogen control in common bean. Biol Control 141:104131. https://doi.org/10.1016/j.biocontrol.2019.104131

Salas H, Gutiérrez-Bouzán C, López-Grimau V, Vilaseca M (2019) Respirometric study of optical brighteners in textile wastewater. Materials (Basel) 12(5):785. https://doi.org/10.3390/ma12050785. PMID: 30866454; PMCID: PMC6427247

Salminen H, Bischoff S, Weiss J (2019) Impact of concentration ratio on the formation and stability of emulsions stabilized by Quillaja Saponin—Sodium Caseinate mixtures. Food Biophys 14:109–119. https://doi.org/10.1007/s11483-018-09563-x

Santos DK, Rufino RD, Luna JM, Santos VA, Sarubbo LA (2016) Biosurfactants: multifunctional biomolecules of the 21st century. Int J Mol Sci 17(3):401. Published 2016 Mar 18. https://doi.org/10.3390/ijms17030401

Saravanakumari P, Mani K (2010) Structural characterization of a novel xylolipid biosurfactant from Lactococcus lactis and analysis of antibacterial activity against multi-drug resistant pathogens. Bioresour Technol 101(22):8851–8854. https://doi.org/10.1016/j.biortech.2010.06.104

Savage GP (2003) Saponins. In: Encyclopedia of food sciences and nutrition. Academic Press, New York, pp 5095–5098. https://doi.org/10.1016/b0-12-227055-x/01050-6

Schlusselhuber M, Godard J, Sebban M et al (2018) Characterization of Milkisin, a novel lipopeptide with antimicrobial properties produced by Pseudomonas sp. UCMA 17988 isolated from bovine raw milk. Front Microbiol 9:1030. Published 2018 May 28. https://doi.org/10.3389/fmicb.2018.01030

Seifried RM, Harrison E, Seifried HE (2017) Antioxidants in health and disease. In: Nutrition in the prevention and treatment of disease. Academic Press, Cambridge, MA, pp 321–346. https://doi.org/10.1016/b978-0-12-802928-2.00016-3

Sekhon Randhawa KK, Rahman PKSM (2014) Rhamnolipid biosurfactants—past, present, and future scenario of global market. Front Microbiol 5:454. https://doi.org/10.3389/fmicb.2014.00454

Sen S, Borah SN, Bora A et al (2017) Production, characterization, and antifungal activity of a biosurfactant produced by Rhodotorula babjevae YS3. Microb Cell Fact 16:95. https://doi.org/10.1186/s12934-017-0711-z

Sena HH, Sanches MA, Rocha DFS, Segundo WOPF, de Souza ÉS, de Souza JVB (2018) Production of Biosurfactants by Soil Fungi Isolated from the Amazon Forest [published correction appears in Int J Microbiol. 2018 Sep 6;2018:1586859]. Int J Microbiol 2018:5684261. Published 2018 Apr 24. https://doi.org/10.1155/2018/5684261

Shah V, Doncel GF, Seyoum T, Eaton KM, Zalenskaya I, Hagver R, Azim A, Gross R (2005) Sophorolipids, microbial glycolipids with anti-human immunodeficiency virus and sperm-immobilizing activities. Antimicrob Agents Chemother 49(10):4093–4100. https://doi.org/10.1128/AAC.49.10.4093-4100.2005. PMID: 16189085; PMCID: PMC1251511

Sharma D (2016) Classification and properties of biosurfactants. In: Springer briefs in food, health, and nutrition. Springer, Berlin, pp 21–42. https://doi.org/10.1007/978-3-319-39415-2_2

Sharma D, Singh SB (2014) Simultaneous production of biosurfactants and bacteriocins by probiotic Lactobacillus casei MRTL3. Int J Microbiol 2014:698713

Sharma D, Saharan BS, Chauhan N, Bansal A, Procha S (2014) Production and structural characterization of Lactobacillus helveticus derived biosurfactant. ScientificWorldJournal 2014:493548. https://doi.org/10.1155/2014/493548

Sharma D, Saharan BS, Chauhan N et al (2015) Isolation and functional characterization of novel biosurfactant produced by Enterococcus faecium. SpringerPlus 4:4. https://doi.org/10.1186/2193-1801-4-4

Shekhar S, Sundaramanickam A, Balasubramanian T (2014) Biosurfactant producing microbes and their potential applications: a review. Crit Rev Environ Sci Technol 45(14):1522–1554. https://doi.org/10.1080/10643389.2014.955631

Shokouhfard M, Kermanshahi RK, Shahandashti RV, Feizabadi MM, Teimourian S (2015) The inhibitory effect of a Lactobacillus acidophilus derived biosurfactant on biofilm producer Serratia marcescens. Iran J Basic Med Sci 18(10):1001–1007

Shu Q, Wu J, Chen Q (2019) Synthesis, characterization of liposomes modified with biosurfactant MEL-A loading betulinic acid and its anticancer effect in HepG2 cell. Molecules 24(21):3939. https://doi.org/10.3390/molecules24213939

Tahmourespour A, Salehi R, Kermanshahi RK, Eslami G (2011) The anti-biofouling effect of Lactobacillus fermentum-derived biosurfactant against Streptococcus mutans. Biofouling 27:385–392

Takahashi M, Morita T, Fukuoka T, Imura T, Kitamoto D (2012) Glycolipid biosurfactants, mannosylerythritol lipids, show antioxidant and protective effects against H(2)O(2)-induced oxidative stress in cultured human skin fibroblasts. J Oleo Sci 61(8):457–464. https://doi.org/10.5650/jos.61.457

Tania TCD, Deastri YN, Utari T, Yulizar Y, Apriandanu DOB (2018) AIP Conf Proc 2023:020103. https://doi.org/10.1063/1.5064100

Thavasi R, Jayalakshmi S, Banat IM (2011) Effect of biosurfactant and fertilizer on biodegradation of crude oil by marine isolates of Bacillus megaterium: Corynebacterium kutscheri and Pseudomonas aeruginosa. Bioresour Technol 102(2):772–778

Trilisenko LV, Kulakovskaya EV, Kulakovskaya TV, Ivanov AY, Penkov NV, Vagabov VM, Kulaev IS (2012) The antifungal effect of cellobiose lipid on the cells of Saccharomyces cerevisiae depends on carbon source. SpringerPlus 1:18. https://doi.org/10.1186/2193-1801-1-18. PMID: 23961349; PMCID: PMC3725880

Tsibranska S, Tcholakova S, Golemanov K, Denkov N, Arnaudov L, Pelan E, Stoyanov SD (2020a) Origin of the extremely high elasticity of bulk emulsions, stabilized by Yucca Schidigera saponins. Food Chem 316:126365. https://doi.org/10.1016/j.foodchem.2020.126365

Tsibranska S, Tcholakova S, Golemanov K, Denkov N, Pelan E, Stoyanov SD (2020b) Role of interfacial elasticity for the rheological properties of saponin-stabilized emulsions. J Colloid Interface Sci 564:264–275. https://doi.org/10.1016/j.jcis.2019.12.108

Tulloch AP, Hill A, Spence JFT (1968) Structure and reactions of lactonic and acidic sophorosides of 17-hydroxyoctadecanoic acid. Can J Chem 46:3337

Van Bogaert INA, Soetaert W (2011) Sophorolipids. Biosurfactants 20:179–210

Van Hamme JD, Singh A, Ward OP (2003) Recent advances in petroleum microbiology. Microbiol Mol Biol R 6:503–549

Varadavenkatesan T, Murty VR (2013) Production of a lipopeptide biosurfactant by a novel Bacillus sp. and its applicability to enhanced oil recovery. ISRN Microbiol 2013:621519. Published 2013 Sep 24. https://doi.org/10.1155/2013/621519

Vijayakumar S, Saravanan V (2015) Biosurfactants-types, sources and applications. Res J Microbiol 10(5):181–192. https://doi.org/10.3923/jm.2015.181.192

Wadekar SD, Kale SB, Lali AM, Bhowmick DN, Pratap AP (2012) Microbial synthesis of rhamnolipids by Pseudomonas aeruginosa (ATCC10145) on waste frying oil as low cost carbon source. Prep Biochem Biotechnol 142:249–266

Walter V, Syldatk C, Hausmann R (2010) Screening concepts for the isolation of biosurfactant producing microorganisms. Adv Exp Med Biol 672:1–13. https://doi.org/10.1007/978-1-4419-5979-9_1

Wisetkomolmat J, Suppakittpaisarn P, Sommano SR (2019) Detergent plants of Northern Thailand: potential sources of natural saponins. Resources 8:10. https://doi.org/10.3390/resources8010010

Wittgens A, Santiago-Schuebel B, Henkel M et al (2018) Heterologous production of long-chain rhamnolipids from Burkholderia glumae in Pseudomonas putida-a step forward to tailor-made rhamnolipids. Appl Microbiol Biotechnol 102(3):1229–1239. https://doi.org/10.1007/s00253-017-8702-x

Wu YS, Ngai SC, Goh BH, Chan KG, Lee LH, Chuah LH (2017) Anticancer activities of surfactin and potential application of nanotechnology assisted surfactin delivery. Front Pharmacol 8:761. https://doi.org/10.3389/fphar.2017.00761. PMID: 29123482; PMCID: PMC5662584

Wu J, Niu Y, Jiao Y, Chen Q (2019) Fungal chitosan from Agaricus bisporus (Lange) Sing. Chaidam increased the stability and antioxidant activity of liposomes modified with biosurfactants and loading betulinic acid. Int J Biol Macromol 123:291–299. https://doi.org/10.1016/j.ijbiomac.2018.11.062

Yalçın E, Çavuşoğlu K (2010) Structural analysis and antioxidant activity of a biosurfactant obtained from Bacillus subtilis RW-I. Turk J Biochem 35(3):243–247

Yang XE, Wu X, Hao HL, He ZL (2008) Mechanisms and assessment of water eutrophication. J Zhejiang Univ Sci B 9(3):197–209. https://doi.org/10.1631/jzus.B0710626. PMID: 18357622; PMCID: PMC2266883

Yao S, Zhao S, Lu Z, Gao Y, Lv F, Bie X (2015) Control of agitation and aeration rates in the production of surfactin in foam overflowing fed-batch culture with industrial fermentation. Rev Argent Microbiol 47(4):344–349. https://doi.org/10.1016/j.ram.2015.09.003

Yu Y, Zhao J, Bayly AE (2008) Development of surfactants and builders in detergent formulations. Chin J Chem Eng 16(4):517–527. https://doi.org/10.1016/s1004-9541(08)60115-9

Zaragoza A, Aranda FJ, Espuny MJ, Teruel JA, Marqués A, Manresa A, Ortiz A (2010) Hemolytic activity of a bacterial trehalose lipid biosurfactant produced by Rhodococcus sp.: evidence for a colloid-osmotic mechanism. Langmuir 26(11):8567–8572. https://doi.org/10.1021/la904637k

Zhang L, Sun C (2018) Fengycins, cyclic lipopeptides from marine Bacillus subtilis strains, kill the plant-pathogenic fungus Magnaporthe grisea by inducing reactive oxygen species production and chromatin condensation. Appl Environ Microbiol 84(18):e00445–e00418. https://doi.org/10.1128/AEM.00445-18

Zheng B, Zhang X, Peng S, Julian MCD (2019) Impact of curcumin delivery system format on bioaccessibility: nanocrystals, nanoemulsion droplets, and natural oil bodies. Food Funct 10 (7):4339–4349. https://doi.org/10.1039/c8fo02510j

Zhu Z, Wen Y, Yi J, Cao Y, Liu F, Julian McClements D (2019) Comparison of natural and synthetic surfactants at forming and stabilizing nanoemulsions: Tea saponin, Quillaja saponin, and Tween 80. J Colloid Interface Sci 536:80–87. https://doi.org/10.1016/j.jcis.2018.10.024

Zouari R, Moalla-Rekik D, Sahnoun Z, Rebai T, Ellouze-Chaabouni S, Ghribi-Aydi D (2016) Evaluation of dermal wound healing and in vitro antioxidant efficiency of Bacillus subtilis SPB1 biosurfactant. Biomed Pharmacother 84:878–891. https://doi.org/10.1016/j.biopha.2016.09.084

Classification and Production of Microbial Surfactants

4

Mostafa Mostafa Ahmed Abo Elsoud

Abstract

Microbial surfactants (biosurfactants) have been industrially produced and applied in many fields worldwide. There is an annual increase in biosurfactant production which reflects the world attitude toward the safe and eco-friendly products. Surfactants of biological origin have several advantages over the chemically synthesized counterparts, which include biodegradability, specificity, and less toxicity. Biosurfactants are composed of hydrophilic and hydrophobic moieties in which the hydrophobic moiety is a lipid structure, in most cases. However, sugars, amino acids, proteins, and phosphates represent the hydrophilic moiety. Many microorganisms were reported to produce surfactants. The output and type of the produced microbial surfactant are species-specific and depend on nutritional and environmental factors. Biosurfactants can be classified based on physical or chemical properties. The chemical structure of biosurfactants represents the main criteria for their classification. Through the current chapter, we aim to highlight the classes of microbial surfactants along with the factors affecting their production and methods of cultivation in lab and industrial scale.

Keywords

Biosurfactants · Classification · Market · Waste substrates · Nutritional · Environmental · Bioprocess

M. M. A. Abo Elsoud (✉)
Department of Microbial Biotechnology, National Research Centre, Cairo, Egypt

Inamuddin et al. (eds.), *Microbial Biosurfactants*, Environmental and Microbial Biotechnology, https://doi.org/10.1007/978-981-15-6607-3_4

4.1 Introduction

Biosurfactant is a portmanteau word composed of two parts: "bio" and "surfactant" which means surfactant, surface-active agent, produced from biological origin. The biological origin is mostly bacteria, fungi, and yeast (Mukherjee et al. 2006), though animal and plant cells also produce biosurfactant (Geetha et al. 2018). Structurally, biosurfactants are diverse group of amphiphiles composed of two moieties (parts): hydrophilic (polar) and hydrophobic (nonpolar) (Fig. 4.1) (Pacwa-Płociniczak et al. 2011). Hydrophilic-lipophilic balance (HLB) is one of the biosurfactant's characteristic features. This feature can specify the biosurfactant potential uses.

The microbial surfactants serve to increase the water bioavailability of hydrophobic (water-insoluble) substances, e.g., hydrocarbons, fats, and oils, to be used as nutritional substances by the increase of surface area of these substances. Due to their surface activities, biosurfactants were reported by Desai and Banat (1997) as promising and excellent dispersant, emulsifying, foaming, and wetting agents, which made them applicable in several industries (Muthusamy et al. 2008) such as food production, pharmaceutics, cosmetics, agriculture, and chemistry (Banat et al. 2010; Soberón-Chávez and Maier 2011). Compared to chemically synthesized surface-active substances, biosurfactants have many advantages (Table 4.1).

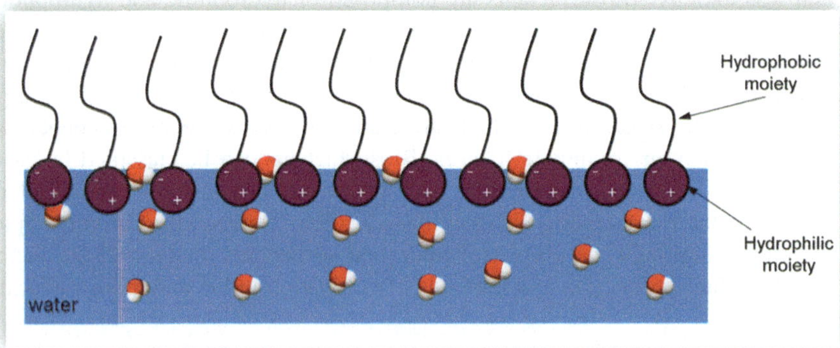

Fig. 4.1 The interface between the liquid and air and the accumulation of biosurfactants

Table 4.1 Differences between chemically synthetic surfactants and biosurfactants

Characters	Synthetic surfactants	Biosurfactants
Efficiency	Less efficient	More efficient
Tolerance to extreme conditions	Low tolerance	High tolerance
Biodegradability	Nonbiodegradable	Biodegradable
Toxicity	More toxic	Less toxic
Specificity	Nonspecific	Specific
Cost	Cheap	Expensive
Bulk production	Short process	Long process

Among bacteria, *Acinetobacter*, *Bacillus*, and *Pseudomonas* (Sobrinho et al. 2013); among yeast, *Kluyveromyces*, *Pseudozyma*, *Rhodotorula*, *Torulopsis*, and *Saccharomyces* (Zinjarde and Pant 2002; Amaral et al. 2006; Morita et al. 2015); and finally, among fungi (Niu et al. 2017; Raja et al. 2017), *Aspergillus*, *Fusarium*, *Penicillium*, *Trichoderma*, and *Ustilago* (Shah and Prabhune 2007; Bhardwaj et al. 2013; Balaji et al. 2014; Lima et al. 2016) are dominating the research publications and literature of biosurfactant production. In Table 4.2 some biosurfactant-producing microbial species are listed.

4.1.1 Global Biosurfactant Market

In 2016, the surfactant global market reached USD 30.64 billion which included USD 3.99 billion for biosurfactant market. During the forecast period (2016–2022) and according to the compound annual growth rate, which is worldwide and can be abbreviated as CAGR, of 5.6%, it was forecasted that the biosurfactant market size is going to reach USD 5.52 billion by 2022 (Markets and Markets 2016).

Middle East, Africa, and Asia-Pacific lead the worldwide surfactant production with 29% for each, while Europe leads biosurfactant production with 32% productivity. On the other hand, Europe and North Africa were the lowest surfactant-producing (14%) regions, and Middle East and Africa were the lowest biosurfactant-producing (7%) regions. Figure 4.2 shows that Europe resorts to safe and eco-friendly process for production of the surface-active agents compared with Middle East and Africa which resort to the synthetic processes.

4.2 Types of Biosurfactants

Biosurfactant molecules possess different physical and chemical properties (Amaral et al. 2010). They are composed of the hydrophilic (polar) part, referred to as "head" which is made of mono-, di-, or polysaccharides, peptide cations, or anions, and the hydrophobic (nonpolar) part, known as "tail," made of hydrocarbon chains or unsaturated or saturated fatty acids (Karanth et al. 1999). These structures are capable of forming micelles and microemulsions between different phases (Fig. 4.3) and also lowering the liquid (Smyth et al. 2010) surface tension.

Synthetic surfactants are, generally, classified according to their ionic charge into anionic, cationic, zwitterionic, and nonionic. This ionic charge appears on the polar part of the surfactant molecules (Christofi and Ivshina 2002). On the other hand, biosurfactants are quite different from that of synthetic counterparts; they can be grouped on different basis: chemical and physical properties, their mode of action, and molecular weight.

Biosurfactants (Fig. 4.4) can be represented by molecules of low or high molecular weight. The low molecular weight biosurfactants are mostly with lipid hydrophobic moiety: glycolipids and lipopeptides, and can, efficiently, lower the surface tension in liquids and form less stable emulsions (Christofi and Ivshina 2002).

Table 4.2 Some potential biosurfactant-producing microorganisms

Microorganism	Isolation sources	Carbon sources/by-products	Biosurfactant	Reference
Pseudomonas sp.	Oil spilled soil, used edible oil	Cheese whey/diesel/glucose/molasses/petrol/rice-water/used edible oil/whey	Rhamnolipid	Anandaraj and Thivakaran (2010) and Soniyamby et al. (2011)
Pseudomonas aeruginosa	Petroleum and oil contaminated soil, petrochemical factory waste water	Soap-stock, palm oil, glucose/sucrose/soybean/starch/casein/glycerol/sunflower oil/sugarcane molasses/vegetable oil/kerosene/petrol/ diesel/hexane/olive oil/oleic acid/soybean oil	Rhamnolipids	Benincasa (2002), Wei et al. (2005), Wu et al. (2008), Priya and Usharani (2009), Sarachat et al. (2010) and Abo Elsoud et al. (2017)
Bacillus subtilis	Contaminated site with crude oil	Rapeseed oil and crude oil/glucose	Iturin	Bayoumi et al. (2010)
Bordetella hinizi-DAFI	Crude oil contaminated sites	Molasses and crude oil/sucrose	Trehalose-2,3,4,2′-tetraester	Bayoumi et al. (2010)
Trichosporon asahii	Petroleum-contaminated soil sample	Diesel oil	Sophorolipids	Chandran and Das (2010)
Serratia marcescens	Petroleum contaminated soil sample	Glycerol	Lipopeptide	Anyanwu et al. (2011)
Candida sp. *SY-16*	Oil-containing soil	Soybean oil and glucose	Mannosylerythritol (Glycolipid)	Kim et al. (1999)
Rhodococcus sp.	Oil-contaminated soil sample	n-heptane/n-hexadecane/n-octane/n-paraffin/gas oil/kerosene/sucrose	Glycolipid and Extra-cellular lipids	Shavandi et al. (2011)
Bacillus subtilis	Oil contaminated soil	Diesel/kerosene/petrol/vegetable oil	Surfactin	Priya and Usharani (2009)
Bacillus brevis	Sediment of Mangrove tree	Glucose	Biosurfactant	Mouafi et al. (2016)

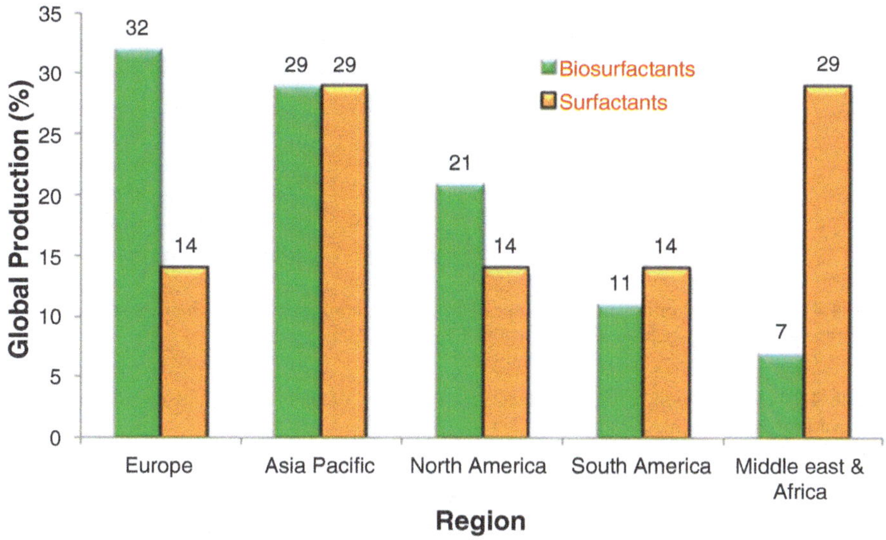

Fig. 4.2 The market size based on the region in 2016

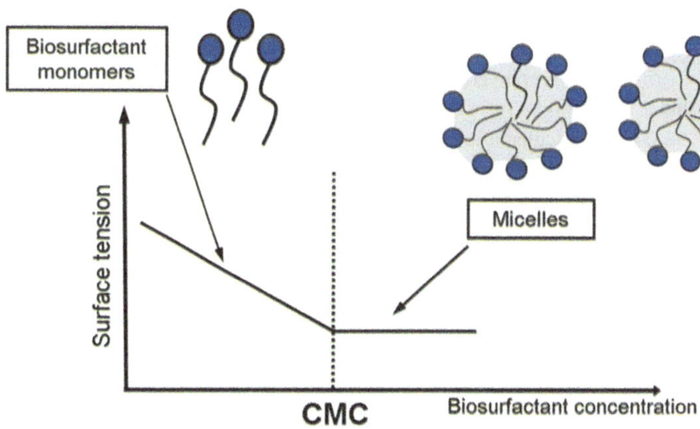

Fig. 4.3 The relationship between the surface tension, biosurfactant concentration, and micelle formation (Whang et al. 2008)

Pseudomonas and *Bacillus* species are widely known for producing low molecular weight biosurfactant (Edosa et al. 2018). On the other hand, the biosurfactants of high molecular weight are composed of biopolymers including proteins, polysaccharides, lipopolysaccharides, or complexes of mixed biopolymers and are associated with stable emulsions, although it forms inefficient liquid surface tension (Ron and Rosenberg 2001). Most high molecular weight biosurfactants can be produced by *Acinetobacter* (Edosa et al. 2018).

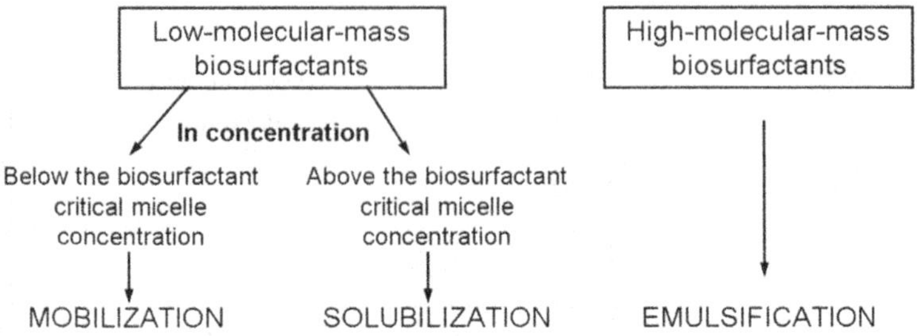

Fig. 4.4 Mechanisms of biosurfactants for hydrocarbon removal based on their molecular weight and concentration (Rosenberg and Ron 1999; Urum and Pekdemir 2004)

Table 4.3 Classification of biosurfactant with examples of potential producing microorganism (Pacwa-Płociniczak et al. 2011)

Biosurfactant		
Group	Class	Microorganism
Glycolipids	Rhamnolipids	*Pseudomonas* sp., *Pseudomonas aeruginosa*
	Trehalolipids	*Arthrobacter* sp., *Corynebacterium* sp., *Mycobacterium tuberculosis, Nocardia* sp., *Rhodococcus erythropolis*
	Sophorolipids	*Torulopsis apicola, Torulopsis petrophilum, Torulopsis bombicola*
Neutral lipids, phospholipids and fatty acids	Corynomycolic acid	*Corynebacterium lepus*
	Spiculisporic acid	*Penicillium spiculisporum*
	Phosphati-dylethanolamine	*Rhodococcus erythropolis, Acinetobacter* sp.
Polymeric biosurfactants	Emulsan	*Acinetobacter calcoaceticus* RAG-1
	Alasan	*Acinetobacter radioresistens* KA-53
	Biodisperean	*Acinetobacter calcoaceticus* A2
	Liposan	*Candida lipolytica*
	Mannoprotein	*Saccharomyces cerevisiae*
Lipopeptides	Surfactin	*Bacillus subtilis*
	Lichenysin	*Bacillus licheniformis*

Nevertheless, the main criteria for biosurfactant classification are their chemical structure (Mnif et al. 2018). Accordingly, there are many distinguished types of biosurfactants (Table 4.3): fatty acids, neutral lipids, lipopeptides, phospholipids, polymeric surfactants, and glycolipids.

4.2.1 Glycolipids

Glycolipids are composed of two combined parts: carbohydrates, viz., rhamnose, sophorose, trehalose, etc., and hydroxy form of fatty acids. Glycolipids are the most studied and known biosurfactants (Mnif et al. 2018). According to the carbohydrate (the polar moiety) in the biosurfactant, glycolipids can be subdivided into cellobiolipids, sophorolipids, trehalolipids, rhamnolipids, lipomannans, lipomannosyl-mannitols, mannosylerythritol lipids, lipoarabinomannans, diglycosyl diglycerides, galactosyl diglyceride, and monoacylglycerol. Among these, rhamnolipids, trehalolipids, and sophorolipids are the best studied and produced by *Pseudomonas* sp., *Rhodococcus*, and some yeast strains, respectively (Shoeb et al. 2013).

4.2.1.1 Rhamnolipids

The group of glycolipid surfactants that are composed of 3-hydroxy fatty acids (hydrophobic moiety) and rhamnose, a disaccharide hydrophilic moiety, is referred to as "rhamnolipids" (Lang and Wullbrandt 1999) (Fig. 4.5). Rhamnolipids have been widely studied and produced by *Pseudomonas aeruginosa* as a homologous mixture of different species (Rahman et al. 2002).

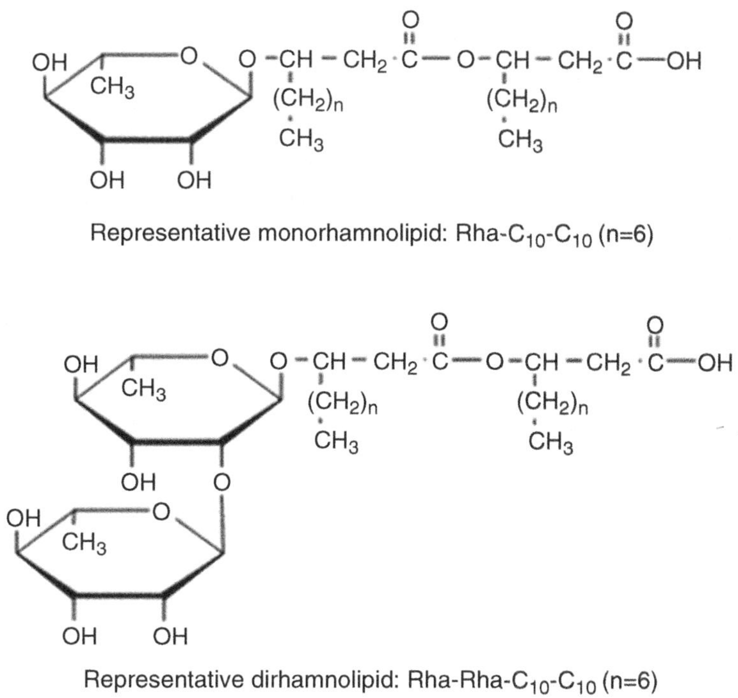

Fig. 4.5 Mono- and di-rhamnolipid structure

4.2.1.2 Sophorolipids

Sophorolipids are a group of glycolipid surfactants consisting of a hydroxyl fatty acid and sophorose, a dimeric sugar. These two moieties are linked to each other (Asmer et al. 1988) by a β-glycosidic bond. Two types of sophorolipids can be differentiated: lactonic and non-lactonic (Fig. 4.6). In lactonic sophorolipids, the hydroxyl fatty acid moiety forms a cyclic lactone ring with 4′-hydroxy group of sophorose by intramolecular esterification, while in non-lactone sophorolipids, the hydroxyl fatty acid moiety has a free carboxylic acid functional group (Hu and Ju 2001). *Torulopsis* sp. is the widely used microorganism for production of sophorolipids.

4.2.1.3 Trehalolipids

Trehalolipids are glycolipid biosurfactant composed of trehalose (hydrophilic disaccharide) linked to β-hydroxy fatty acid chain with long α-branches (Fig. 4.7), mycolic acid. The produced trehalolipids from different microorganisms can differ in many ways: structure (unsaturation degree) and size (carbon atom number) of mycolic acid (Desai and Banat 1997).

4.2.2 Lipoproteins and Lipopeptides

Generally, the lipopeptide-type biosurfactants, surfactin, are produced by the Gram-positive bacterium, *Bacillus* sp. A peptide loop represents the hydrophilic moiety of surfactin (Fig. 4.8), while a fatty acid chain of 13–15 carbons long represents the hydrophobic moiety. The peptide loop is composed of seven successive amino acids (Chen et al. 2015), namely, L-aspartic acid, L-leucine, glutamic acid, L-leucine, L-valine, and two D-leucines.

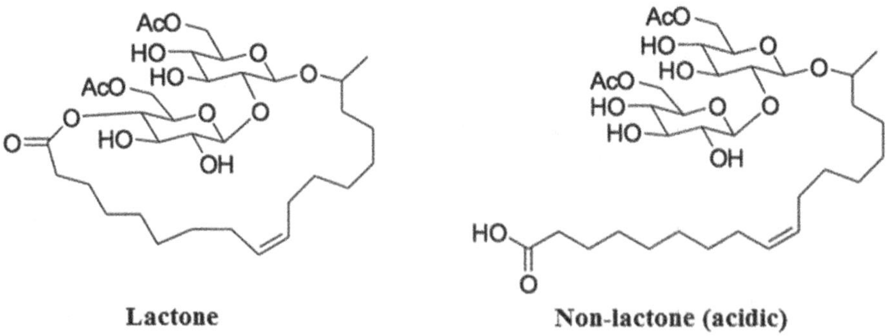

Fig. 4.6 Lactonic and non-lactonic forms of sophorolipids

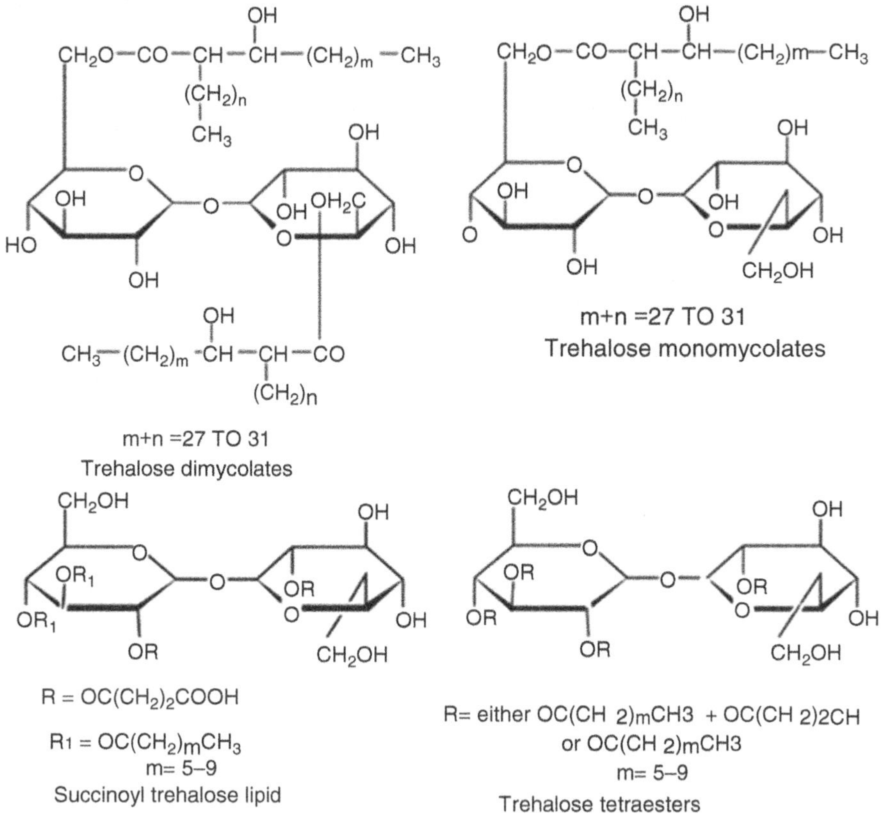

Fig. 4.7 The chemical structure of some trehalose lipids (Franzetti et al. 2010)

4.2.3 Fatty Acids

Microbial oxidation of alkanes may produce fatty acids which have been reported by Rehn and Reiff (1981) as surfactants. These fatty acids can be either straight chain or complex with alkyl branches and hydroxyl groups (Rahman and Gakpe 2008). The balance between hydrophilicity and hydrophobicity of the fatty acid is highly related to the chain length of hydrocarbon and its degree of complexity. Most surface-active fatty acids have the length between 12 and 14 carbon atoms (Rosenberg and Ron 1999).

4.2.4 Phospholipids

The amphipathic molecules of phospholipids are composed of a hydrophobic and a hydrophilic component. Phospholipids have hydrophilic phosphate group on one end and hydrophobic fatty acids on the other. Phospholipid biosurfactants are major

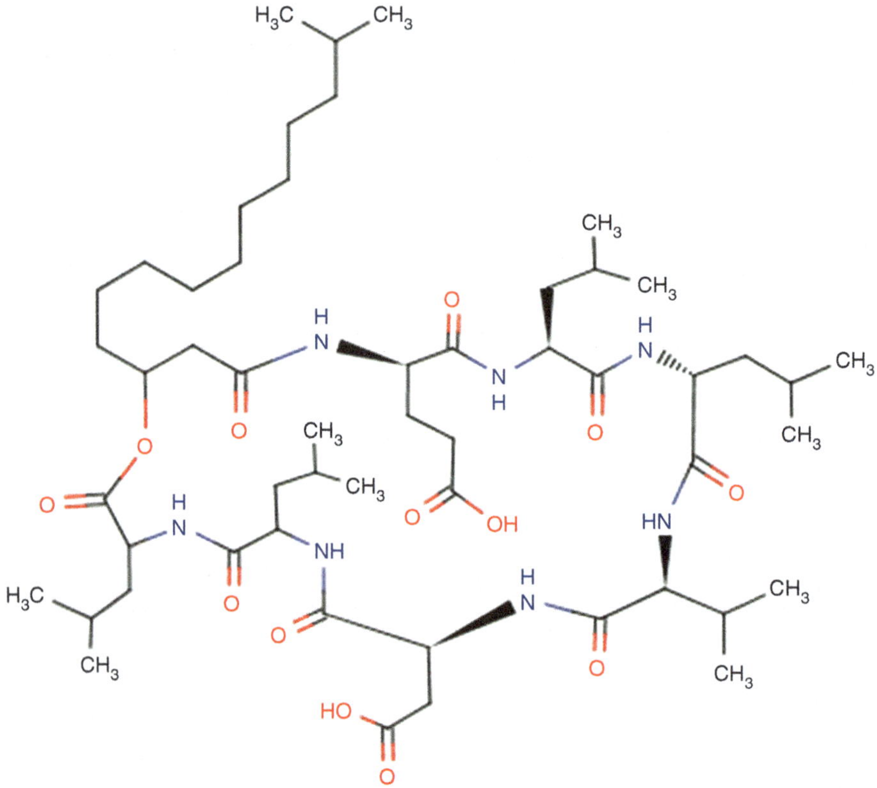

Fig. 4.8 Molecular structure of surfactin isolated from *Bacillus subtilis*

Fig. 4.9 The cell membrane consists of two adjacent layers of phospholipids

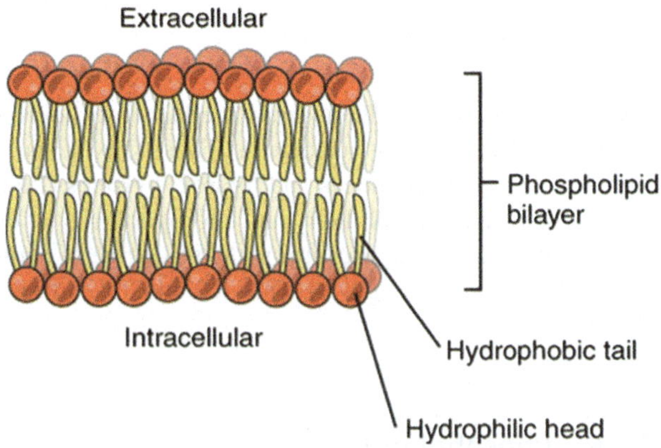

components of microbial plasma membrane (Fig. 4.9). The levels of phospholipids increase, greatly, if the microbial strain grows in the presence of hydrocarbons (Rahman and Gakpe 2008).

4.2.5 Polymeric Biosurfactants

The polymeric biosurfactants isolated from bacterial strains are exopolysaccharide (EPS) in nature and characterized in that they have a role in antibiotic resistance, as they offer many regulatory pathways to act against antibiotics. The polymeric biosurfactants that are best studied worldwide are alasan, emulsan, liposan, lipomanan, and other protein polysaccharide complexes. At very low concentrations (0.001% to 0.01%), emulsan was used as an emulsifying agent for hydrocarbons in water (Lang 2002; Hatha et al. 2007).

4.3 Factors Influencing Biosurfactant Productivity

The studies (Banat et al. 2010) on biosurfactant production confirmed that microorganisms produce various types of biosurfactants in response and adaptation to the surrounding microenvironments in addition to other physiological and natural functions. Microbial biosurfactant production is affected by several factors including genetic, physiological, nutritional, and environmental. These biosurfactants can reach maximum levels under definite nutritional and environmental conditions (Singh et al. 2018).

Nutritional factors, which can be represented by nitrogen and carbon sources, trace elements, and vitamins, and environmental factors, e.g., agitation, pH, incubation temperature, and aeration, have a crucial role in the production amount and type of biosurfactant; therefore, for a cost-effective bioprocess, primary researches should be conducted to evaluate the significance of these factor and their optimum levels.

4.3.1 Nutritional Factors

4.3.1.1 Carbon Source

The majority of biosurfactant-producing microorganisms is heterotrophs which mean that for their growth and production of different metabolites they should consume organic carbon sources. Therefore, carbon sources are critical factors that influence the quantity and quality of biosurfactants (Raza et al. 2007).

Most of the biosurfactant production cost (30–40%) comes only from the production medium preparation cost (George and Jayachandran 2013). Usually, during cultivation of the microorganisms, carbon consumption is the limiting factor for biosurfactant and cell biomass production rates (Singh et al. 2017). Generally, three categories of carbon sources, hydrocarbon groups, oils and fats, and carbohydrate, were commonly used for biosurfactant production (Nurfarahin et al. 2018). The carbon source category and concentration depend on the microbial species and type of biosurfactant. Several sources of carbon have been investigated for biosurfactant production including **water-soluble** substrates such as sucrose, glycerol, and glucose and **water-insoluble** substrates, e.g., hexane, crude oil, diesel, olive oil, oleic acid, palm oil, kerosene, and petrol (Table 4.1) (Bhardwaj et al. 2013). The

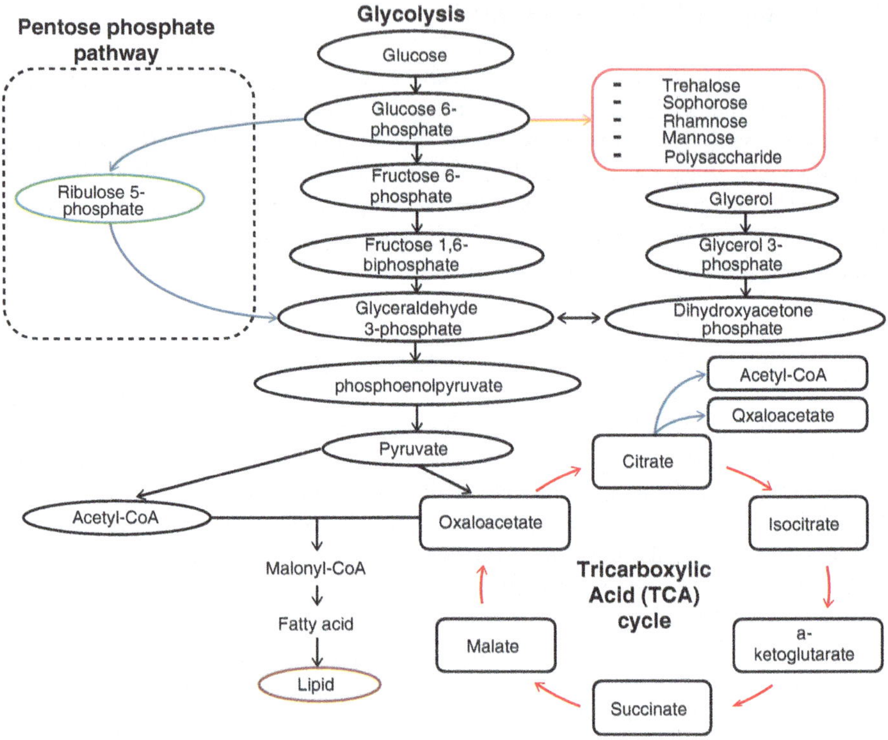

Fig. 4.10 Mechanism of surfactant biosynthesis using water-soluble substrate (Giraud and Naismith 2000; Taylor et al. 2013)

metabolic pathway associated with creation of the precursors for utilization in biosurfactant production relies on the sources of carbon that are found in the production medium (Figs. 4.10 and 4.11). Some researchers reported low biosurfactant yield in case of water-soluble (hydrophilic) substrates (Mata-Sandoval et al. 2000). However, for economic production of biosurfactants, great attention was given by researchers to the use of agro-industrial waste to cut down the processing cost of its purification.

4.3.1.2 Low-Cost and Waste Substrates

During the last few years, with the aim of reduction of the production cost coupled with waste management, low-cost and waste substrates have been utilized as sources of carbon for biosurfactant production (Banat et al. 2014) as well as other add-value products (Satpute et al. 2017). Researchers and industry men had immense scope for production of biosurfactants from a variety of food, agricultural, cheap, and renewable industrial wastes (Table 4.4). However, the raw material and waste cost is not the only limiting factor for their use, but their variability, stability, and availability are also to be considered as critical factors. Moreover, the characteristics of the material to be used, including liquid and solid forms, amount, purity, texture, size of

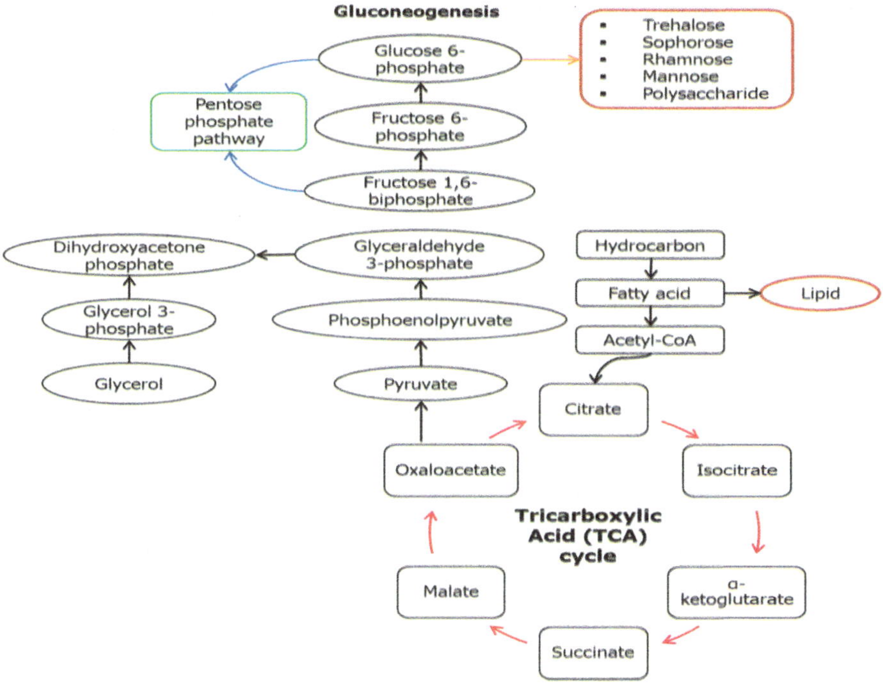

Fig. 4.11 The metabolic pathway used by hydrocarbon substrates for the synthesis of biosurfactant (Papagianni 2012; Santos et al. 2016)

the particles, storage, packaging, and transportation, are all significant factors in its final selection for the production of biosurfactants (Singh et al. 2018).

4.3.1.3 Nitrogen Source

Nitrogen is a limiting nutritional factor as it is highly critical for microbial growth and essential for production of certain primary and secondary metabolites including biosurfactants, protein, and enzymes (Saharan et al. 2012). Sources of nitrogen were investigated by many researchers for biosurfactant production, e.g., extracts of yeast, meat and malt, peptone, urea, and salts of ammonium nitrate.

Two forms of nitrogen sources, inorganic and organic nitrogen, have been categorized on the basis of their unit structures which they present. In case of organic nitrogen, the unit structure will be in organic molecules, e.g., peptone, yeast extract, tryptone, or meat extract. In the other hand, inorganic nitrogen has a unit structure that consists of positive or negative ions such as ammonium nitrate (NH_4NO_3). The organic nitrogen has an advantage over inorganic nitrogen as it also contains carbon content that significantly supports metabolite formation and cell growth (Deepika et al. 2016). However, it has been reported that ammonia, nitrates, and amino acids were used by many researchers as the nitrogen source of choice for production of biosurfactants by some *P. aeruginosa* strains (Nurfarahin et al. 2018).

Table 4.4 The used low-cost waste substrates for production of biosurfactant (Singh et al. 2018)

Waste type	Microbial species	References
Food and Agroindustrial residue (banana peel, cassava waste, corn steep liquor, date molasses, moringa residue, orange peel, peanut oil cake, potato peel, sesame peel flour, sugarcane baggasse, tuna fish residue)	*Bacillus licheniformis, Bacillus pumilis, Candida tropicalis, Cunninghamella phaeospora, Pseudomonas aeruginosa, Halobacteriaceae archaeon*	Magalhães et al. (2018), Rubio-Ribeaux et al. (2017), Lins et al. (2016), Kumar et al. (2016), Sharma et al. (2015) and Chooklin et al. (2014)
Animal Waste (fish processing waste, animal fats, slaughterhouse waste)	*Aneurinibacillus migulanus, Nocardia higoensis, Pseudomonas gessardii*	Sellami et al. (2016), Patil et al. (2016) and Ramani et al. (2012)
Mil (including refinery waste) and agroindustrial waste (olive mill waste, palm oil industry waste, soybean oil industry waste, tannery pretreated effluent)	*Bacillus pseudomycoides, Bacillus subtilis, Brachybacterium, paraconglomeratum, Pseudomonas aeruginosa,*	Radzuan et al. (2017), Moya-Ramírez et al. (2016), Li et al. (2016), Gudiña et al. (2016) and Kiran et al. (2014)
Cooking oil waste (cooking oil waste, frying coconut oil waste)	*Candida lipolytica, Pseudomonas aeruginosa*	Souza et al. (2016), Lan et al. (2015) and George and Jayachandran (2013)

4.3.1.4 Minerals

Two groups of minerals can be differentiated according to the amount needed: macronutrient and micronutrient minerals (trace elements). In medium formulation, macronutrient minerals are supplied with other nutrients. The macronutrient minerals play important roles in the active material's product, such as enzymes and biosurfactants, regulation, balancing the cell wall communication, as well as the aid in the mechanism of protein synthesize (Schobert 1992; Nurfarahin et al. 2018). It was reported that iron, manganese, and magnesium act as cofactors for the synthesis of the enzymes responsible for production of surfactin by *B. subtilis* (Gudiña et al. 2015).

Phosphates are important factors for the microbial activity and growth as it is important for energy storage and metabolic processes including nucleic acids and metabolite formation. Usually, phosphate is provided in the form triphosphates (Nurfarahin et al. 2018). Furthermore, KH_2PO_4 and K_2HPO_4 are usually used as a buffering system in the biosurfactant medium; keep the desired pH constant throughout the production process (Epstein 2003).

As macronutrient minerals, in addition to phosphorous, calcium (Ca), iron (Fe), potassium (K), and magnesium (Mg) are important for medium formulation to aid in the protein synthesizing mechanism and to balance the communication of the cell wall (Schobert 1992).

Calcium acts to mediate the processes of signal delivering between the cell surface and intracellular of the microorganism (Dominguez 2004). Although in the production medium of biosurfactant calcium is added in small amount, usually in chloride or hydrated chloride salts, it is still needed to support growth and production (Thaniyavarn et al. 2006; Thavasi et al. 2008).

Potassium is also an important macronutrient mineral for microbial growth and activity. It has been suggested by Tempest et al. (1966) that potassium ions should be added to the bacterial growth medium as it is substantial to regulate the structure of ribosomes.

Both calcium and potassium ions play vital roles in microbial cells and prevent cell lysis in the medium by controlling the membrane potential and balancing the osmotic pressure of the cells (Činátl 1969).

Magnesium ion is usually added in around 50 times higher concentrations than that of calcium ions added in the production medium of biosurfactants (Thavasi et al. 2011; Saikia et al. 2012; Rekha et al. 2014). To activate ATP, the cell's energy store, it should bind with a magnesium ion to form Mg-ATP and result in the release of ADP, magnesium, and energy. The amount of magnesium to be used in this reaction increases with detection of higher metabolic activity (Gout et al. 2014).

Iron is a well-known cofactor important for the process of metabolism in various microorganisms. Most formulations of the growth and production media utilize iron in different forms depending on the microbial mechanism of iron uptake (Nurfarahin et al. 2018).

Trace elements (micronutrients) are minerals that could have positive effect on biosurfactant production and/or microbial growth when added to the production medium in trace amounts, usually less than 0.1%. For biosurfactant production, many trace elements have been used: boron (B), cobalt (Co), copper (Cu), molybdenum (Mo), and zinc (Zn). However, the specific micronutrients required for production of biosurfactant depend on the used microorganism (Nurfarahin et al. 2018).

4.3.2 Environmental Factors

The amount and characteristics of biosurfactants are highly susceptible to the environmental factors. Therefore, it is vitally important to optimize and evaluate the different environmental factors, temperature, pH, aeration, agitation, etc., affecting the bioprocess of biosurfactant production (Fenibo et al. 2019).

Most of the studied microorganisms for biosurfactant production are mesophiles, which grow at a temperature range between 20 and 45 °C; hence the optimum temperature for biosurfactant production, usually, lies between 25 and 37 °C (Auhim and Mohamed 2013).

In parallel with incubation temperature, biosurfactant production has been reported to be influenced by the culture medium pH (Amaral et al. 2006). It was reported that pH (the concentration of the medium hydrogen ions [H+]) disrupts hydrogen bonding and modifies the ionization characteristics of the amino acid functional groups and, subsequently, promotes changes in the structure and folding of proteins and enzyme and denaturation, finally, destroying their activity. Most studies on bacterial biosurfactant were conducted at neutral or almost neutral pH, while for yeast and fungi, acidic conditions were guaranteed (Jagtap et al. 2010; Fontes et al. 2010; Morais et al. 2017).

The optimum incubation period required for biosurfactant production varies according to microbial species (Jagtap et al. 2010; Fontes et al. 2010; Bhardwaj

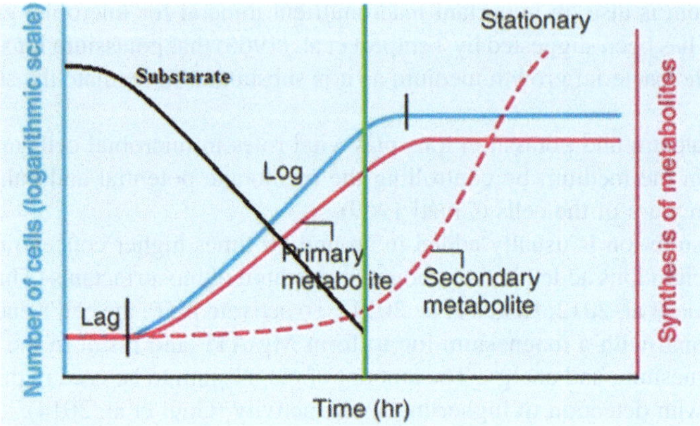

Fig. 4.12 Primary and secondary metabolite production in relation to microbial growth curve and substrate concentration

et al. 2013). During the production process, the level of biosurfactants goes through different phases including lag, exponential, stationary or apex, and finally decline phases. Hence, the incubation period should be optimized before large-scale production to obtain the maximum output at a minimum period (Fig. 4.12). As a conclusion, the incubation period of biosurfactant production process is an unpredictable factor.

Microorganisms can be aerobic, anaerobic, or facultative anaerobe meaning that the amount of oxygen influences their growth, metabolism, and activity. Aeration and agitation of the culture medium are very important factors that facilitate oxygen transfer and distribution and, subsequently, influence biosurfactant production (Roy 2017).

4.3.3 Cultivation Strategy

For the development, economic widespread, and sustained biosurfactant production, two bioprocess strategies, submerged and solid-state fermentation (SSF), are well studied and applied in both lab and pilot scales. These strategies represent the most critical part for economic production of biosurfactant and for medium optimization. Furthermore, these strategies are key determinants for the establishments of commercial and economic biosurfactants and to be able to compete with the chemically produced surfactants.

4.3.3.1 Solid-State Fermentation (SSF)

As the name may show, the solid-state fermentation (SSF) was described to be a biological process taking place in an inert solid matrix which acts as a support or support and substrate in the presence of moisture or a very few amount of free water (Fig. 4.13) (Singhania et al. 2010) to support the metabolic activity and the growth of

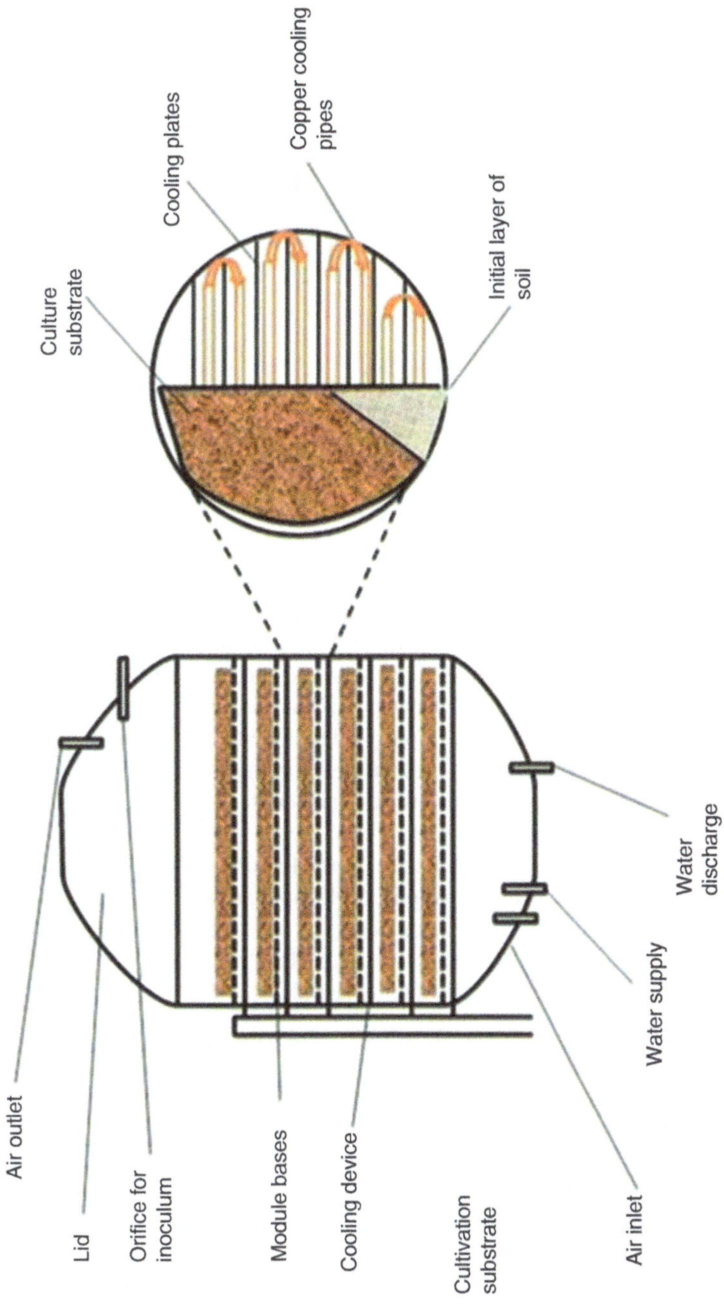

Fig. 4.13 System of solid-state fermentation (SSF)

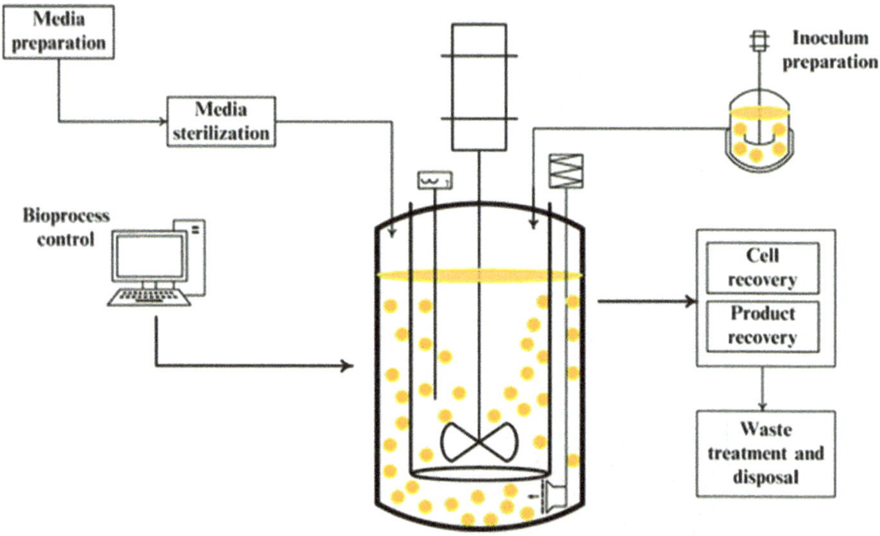

Fig. 4.14 Aerated stirred tank used for submerged fermentation

producer microorganism (Thomas et al. 2013). SSF is an emerging and promising strategy for the production of biosurfactant compared with submerged fermentation (SmF) (Das and Mukherjee 2007) as it overcomes the foam production problem (Camilios-Neto et al. 2011). In addition, it is a simple process and can produce concentrated products (Mir et al. 2017). During the past few years, the solid-state fermentation technologies have been rapidly developed by researchers and applicants. Worldwide, an attention and gradual development of the researches on SSF with the impact of cost, waste management, science, technology, and sustainable development have made a great progress.

The substrates employed during solid-state fermentation are usually cost-free renewable wastes such as wheat bran; banana peel; tapioca peel (Vijayaraghavan et al. 2011); cassava bagasse; sugarcane; rice husk; oil cakes, e.g., coffee husk and palm kernel cakes (Pandey 2008); and cassava dregs (Hong et al. 2001). All these substrates are rich in protein and carbon contents (Pandey 2008).

4.3.3.2 Submerged Fermentations (SmF)

The processes of submerged fermentation (SmF) were described as a biological process where the microbial cells are surrounded, completely, by liquid medium (Fig. 4.14) (Singhania et al. 2015). This strategy comprises large varieties of microbial processes including non-stirred or stirred. It has been reported that the production yield of different products, by solid-state fermentation, is usually more significant than that obtained by submerged fermentation. This may be due to the fact that the microbes in the SSF grow under similar conditions to their natural habitats, where they can produce certain metabolites and enzymes that are usually produced only with a low yield and not produced at all in the submerged culture

(Jecu 2000; Castilho et al. 2000). However, submerged fermentation has been used for biosurfactant production by many scientists (Sayyad et al. 2007; Iroha et al. 2015; Mouafi et al. 2016; Das and Kumar 2018). Currently, biosurfactant production is done either by batch or fed-batch processes. This can be attributed to the limitations of data provided for continuous nutritional feeding process (Winterburn and Martin 2012), including heat and mass transfer and biomass growth reactions, which limit the efficiency of the process. Therefore, more efforts and research are to be needed for conduction of integrated bioprocesses for biosurfactant continuous production and recovery using submerged fermentation strategy.

The physicochemical properties are the main differences between SSF and SmF processes (Table 4.5), e.g., (a) water content is very small in SSF, (b) substrates and products are homogenous in SmF compared with SSF, (c) oxygen in addition to other nonpolar gases is much more diffused in solid-state fermentation, and (d) heat conduction (Roussos et al. 1997) is much smaller in solid-state fermentation. Such differences comprise the main reasons for advantages and drawbacks of SSF, in relation to SmF processes.

Table 4.5 Comparison between submerged and solid-state fermentation

Submerged fermentation	Solid state fermentation
Fermentation may be carried out as batch, fed-batch or continuous processes	Fermentation may be carried out as batch
Medium is added in large vessel	Medium is added in flat vessel or trays
Surface area to height ratio is very less	Surface area to height ratio is very high
5–10% of inoculums is added	Less inoculum is added
Inoculum is usually in liquid form	Inoculum is usually sprayed on surface of medium
Product used are usually high as compared to input cost	Product yield is comparatively less
Lesser space is required	More space is required
Less contamination	More contamination
If a batch get contaminated there is a loss of entire batch	If a tray gets contaminated then there is a loss of only tray but not the batch
Entire fermentation media is utilized by microorganism for growth and product fermentation	There is wastage of fermentation media
Wastewater discharge	No wastewater discharge
Aeration and agitation of system is possible by use of sparger and impeller	Aeration is usually carried out by passing sterile air and no agitation
Power consumption is high	Power consumption is less
Controlling parameters like temperature, pH is easy	Controlling parameters like temperature, pH is difficult
Foaming occurs	Foaming doesn't occurs
Automation and use of computer is easy	Automation and use of computer is difficult
Less labor required	More labor required

References

Abo Elsoud MM, Elkhouly HI, Sidkey NM (2017) Optimization of multiple valuable biogenic products using experimental design. JIPBS 4(4):32–38

Amaral PFF, da Silva JM, Lehocky BM, Barros-Timmons AMV, Coelho MAZ, Marrucho IM, Coutinho JAP (2006) Production and characterization of a bioemulsifier from Yarrowia lipolytica. Process Biochem 41(8):1894–1898

Amaral PF, Coelho MA, Marrucho IM, Coutinho JA (2010) Biosurfactants from yeasts: characteristics, production and application. Adv Exp Med Biol 672:236–249

Anandaraj B, Thivakaran P (2010) Isolation and production of biosurfactant producing organism from oil spilled soil. J Biosci Technol 1:120–126

Anyanwu CU, Obi SKC, Okolo BN (2011) Lipopeptide biosurfactant production by Serratia marcescens NSK-1 strain isolated petroleum-contaminated soil. J Appl Sci Res 7:79–87

Asmer HJ, Slegmund L, Fritz W, Wrey V (1988) Microbial production, structure elucidation and bioconversion of sophorose lipid. JAOCS 65:1460–1466

Auhim HS, Mohamed AI (2013) Effect of different environmental and nutritional factors on biosurfactant production from Azotobacter chroococcum. Int J Adv Pharm Biol Chem 2:477–481

Balaji V, Arulazhagan P, Ebenezer P (2014) Enzymatic bioremediation of polyaromatic hydrocarbons by fungal consortia enriched from petroleum contaminated soil and oil seeds. J Environ Biol 35:3–9

Banat IM, Franzetti A, Gandolfi I, Bestetti G, Martinotti MG, Fracchia L, Smyth TJ, Marchant R (2010) Microbial biosurfactants production, applications and future potential. Appl Microbiol Biotechnol 87:427–444

Banat IM, Satpute SK, Cameotra SS, Patil R, Nyayanit NV (2014) Cost effective technologies and renewable substrates for biosurfactants' production. Front Microbiol 5:1–18

Bayoumi RA, Haroun BM, Ghazal EA, Maher YA (2010) Structural analysis and characteristics of biosurfactants produced by some crude oil utilizing bacterial strains. Aust J Basic Appl Sci 4:3484–3498

Benincasa M (2002) Rhamnolipid production by P. aeruginosa LBI growing on soap-stock as the sole carbon source. J Food Eng 54:283–288

Bhardwaj G, Cameotra SS, Chopra HK (2013) Biosurfactants from fungi: a review. J Pet Environ Biotechnol 4(06):1–6

Camilios-Neto D, Bugay C, de Santana-Filho AP, Joslin T, de Souza LM, Sassaki GL, Krieger N (2011) Production of rhamnolipids in solid-state cultivation using a mixture of sugarcane bagasse and corn bran supplemented with glycerol and soybean oil. Appl Microbiol Biotechnol 89:1395–1403

Castilho, L. R.; Alves, T. L. M. and Medronho, R. A (2000) Production and extraction of pectinases obtained by solid state fermentation of agro-industrial residues with Aspergillus niger. Bioresource Technol 71:45–50

Chandran P, Das N (2010) Biosurfactant production and diesel oil degradation by yeast species Trichosporon Asahii isolated from petroleum hydrocarbon contaminated soil. Int J Eng Sci Technol 2(12):6942–6953

Chen W, Juang R, Wei Y (2015) Applications of a lipopeptide biosurfactant, surfactin, produced by microorganisms. Biochem Eng J 103:158–169

Chooklin CS, Maneerat S, Saimmai A (2014) Utilization of banana peel as a novel substrate for biosurfactant production by Halobacteriaceae archaeon AS65. Appl Biochem Biotechnol 173:624–645

Christofi N, Ivshina IB (2002) Microbial surfactants and their use in field studies of soil remediation. J Appl Microbiol 93:915–929

Činátl J (1969) Inorganic-organic multimolecular complexes of salt solutions, culture media and biological fluids and their possible significance for the origin of life. J Theor Biol 23:1–10. https://doi.org/10.1016/0022-5193(69)90063-0

Das AJ, Kumar R (2018) Utilization of agro-industrial waste for biosurfactant production under submerged fermentation and its application in oil recovery from sand matrix. Bioresour Technol 260:233–240

Das K, Mukherjee AK (2007) Comparison of lipopeptide biosurfactants production by Bacillus subtilis strains in submerged and solid state fermentation systems using a cheap carbon source: some industrial applications of biosurfactants. Process Biochem 42:1191–1199

Deepika KV, Nagaraju GP, Bramhachari PV (2016) Optimization of cultural conditions for marine microbial biosurfactant production: future prospects from untapped marine resources. In: Naik MM, Dubey SK (eds) Marine pollution and microbial remediation. Springer Science & Business Media, Singapore, pp 105–128

Desai JD, Banat IM (1997) Microbial production of surfactant and their commercial potential. Microbiol Mol Biol R 61(1):47–64

Dominguez DC (2004) Calcium signalling in bacteria. Mol Microbiol 54:291–297. https://doi.org/10.1111/j.1365-2958.2004.04276.x

Edosa TT, Jo YH, Keshavarz M, Han YS (2018) Biosurfactants: production and potential application in insect pest management. Trends Entomol 14:79–87

Epstein W (2003) The roles and regulation of potassium in bacteria. Prog Nucleic Acid Res Mol Biol 75:293–320. https://doi.org/10.1016/S0079-6603(03)75008-9.

Fenibo EO, Douglas SI, Stanley HO (2019) A review on microbial surfactants: production, classifications, properties and characterization. J Adv Microbiol 18(3):1–22

Fontes GC, Amaral F, Filomena P, Nele M, Coelho Z, Alice M (2010) Factorial design to optimize biosurfactant production by Yarrowia lipolytica. Biomed Res Int 821306:1–8

Franzetti A, Gandolfi I, Bestetti G, Smyth TJP, Banat IM (2010) Production and applications of trehalose lipid biosurfactants. Eur J Lipid Sci Technol 112(6):617–627

Geetha SJ, Banat IM, Joshi SJ (2018) Biosurfactants: production and potential applications in microbial enhanced oil recovery (MEOR). Biocatal Agric Biotechnol 14:23–32

George S, Jayachandran K (2013) Production and characterization of rhamnolipid biosurfactant from waste frying coconut oil using a novel Pseudomonas aeruginosa D. J Appl Microbiol 114:373–383. https://doi.org/10.1111/jam.12069.

Giraud MF, Naismith JH (2000) The rhamnose pathway. Curr Opin Struct Biol 10:687–696

Gout E, Rebeille F, Douce R, Bligny R (2014) Interplay of Mg2+, ADP, and ATP in the cytosol and mitochondria: unravelling the role of Mg2+ in cell respiration. Proc Natl Acad Sci 111:4560–4567. https://doi.org/10.1073/pnas.1406251111.

Gudiña EJ, Fernandes EC, Rodrigues AI, Teixeira JA, Rodrigues LR (2015) Biosurfactant production by Bacillus subtilis using corn steep liquor as culture medium. Front Microbiol 6:1–7. https://doi.org/10.3389/fmicb.2015.00059.

Gudiña EJ, Rodrigues AI, de Freitas V, Azevedo Z, Teixeira JA, Rodrigues LR (2016) Valorization of agro-industrial wastes towards the production of rhamnolipids. Bioresour Technol 212:144–150

Hatha AAM, Edward G, Rahman KSMP (2007) Microbial biosurfactants—review. J Mar Atmos Res 3:1–17

Hong K, Ma Y, Li M (2001) Solid-state fermentation of phytase from cassava dregs. Appl Biochem Biotechnol 91-93(1-9):777–785. Janek, 018;14: 23-32

Hu Y, Ju LK (2001) Purification of lactonic sophorolipids by crystallization. J Biotechnol 87 (3):263–272

Iroha OK, Njoku OU, Ogugua VN, Okpashi VE (2015) Characterization of biosurfactant produced from submerged fermentation of fruits bagasse of yellow cashew (Anacardium occidentale) using Pseudomonas aeruginosa. Afr J Environ Sci Technol 9(5):473–481

Jagtap S, Yavankar S, Pardesi K, Chopade B (2010) Production of bioemulsifier by Acinetobacter species isolated from healthy human skin. Indian J Exp Biol 48:70–76

Jecu L (2000). Solid state fermentation of agricultural wastes for endoglucanase production. Ind Crops Prod 11:1–5

Karanth NG, Deo PG, Veenanadig NK (1999) Microbial production of biosurfactants and their importance. Curr Sci 77:116–126

Kim HS, Yoon BD, Choung DH, Oh HM, Katsuragi T, Tani Y (1999) Characterization of a biosurfactant mannosylerythritol lipid produced from Candida sp. SY 16. Appl Microbiol Biotechnol 52:713–721

Kiran GS, Sabarathnam B, Thajuddin N, Selvin J (2014) Production of glycolipid biosurfactant from sponge-associated marine actinobacterium Brachybacterium paraconglomeratum MSA21. J Surfactant Deterg 17:531–542

Kumar AP, Janardhan A, Viswanath B, Monika K, Jung JY, Narasimha G (2016) Evaluation of orange peel for biosurfactant production by Bacillus licheniformis and their ability to degrade naphthalene and crude oil. 3 Biotech 6:1–10

Lan G, Fan Q, Liu Y, Chen C, Li G, Liu Y, Yin X (2015) Rhamnolipid production from waste cooking oil using Pseudomonas SWP-4. Biochem Eng J 101:44–54

Lang S (2002) Biological amphiphiles (microbial biosurfactants). Curr Opin Colloid Interface Sci 7:12–20. https://doi.org/10.1016/S1359-0294(02)00007-9

Lang S, Wullbrandt D (1999) Rhamnose lipids-biosynthesis, microbial production and application potential. Appl Microbiol Biotechnol 51(1):22–32

Li J, Deng M, Wang Y, Chen W (2016) Production and characteristics of biosurfactant produced by Bacillus pseudomycoides BS6 utilizing soybean oil waste. Int Biodet Biodegrad 112:72–79

Lima JMS, Pereira JO, Batista IH, Neto PDQC, dos Santos JC, de Araújo SP, de Azevedo JL (2016) Potential biosurfactant producing endophytic and epiphytic fungi, isolated from macrophytes in the Negro River in Manaus, Amazonas, Brazil. Afr J Biotechnol 15(24):1217–1223

Lins AB, Bione AM, Silva TC, De Souza DG, Campos-Takaki GM (2016) Low-cost production of biosurfactant by Cunninghamella phaeospora using agro-industrial wastes. In: Vilas AM (ed) Microbes in the spotlight: recent progress in the understanding of beneficial and harmful microorganisms. Brown Walker Press, Boca Raton, FL, pp 339–343

Magalhães ERB, Silva FL, Sousa MADSB, Dos Santos ES (2018) Use of different agroindustrial waste and produced water for biosurfactant production. Biosci Biotechnol Res Asia 15:17–26

Markets and Markets (2016). Available at https://www.marketsandmarkets.com/Market-Reports/biosurfactants-market-493.html. Accessed 8 Mar 2018

Mata-Sandoval JC, Karns J, Torrents A (2000) Effect of nutritional and environmental conditions on production and composition of rhamnolipids by Pseudomonas aeruginosa UG2. Microbiol Res 155:1–8

Mir S, Jamalm P, Alama MZ, Mir AB, Ansari AH (2017) Microbial surface tensio-active compounds: production and industrial application perspectives: a review. Int J Biotech Bioeng 8:282–301. https://doi.org/10.25141/2475-3432-2017-8.0273

Mnif I, Ellouz-Chaabouni S, Ghribi D (2018) Glycolipid biosurfactants, main classes, functional properties and related potential applications in environmental biotechnology. J Polym Environ 26:2192–2206. https://doi.org/10.1007/s10924-017-1076-4

Morais IM, Cordeiro AL, Teixeira GS, Domingues VS, Nardi RM, Monteiro AS, Alves RJ, Siqueira EP, Santos VL (2017) Biological and physicochemical properties of biosurfactants produced by Lactobacillus jensenii P 6A and Lactobacillus gasseri P 65. Microb Cell Fact 16 (1):155

Morita T, Fukuoka T, Imura T, Kitamoto D (2015) Mannosylerythritol lipids: production and applications. J Oleo Sci 64(2):133–141

Mouafi FE, Abo Elsoud MM, Moharam ME (2016) Optimization of biosurfactant production by Bacillus brevis using response surface methodology. Biotechnol Rep 9:31–37

Moya-Ramírez I, Vaz DA, Banat IM, Marchant R, Alameda EJ, García-Román M (2016) Hydrolysis of olive mill waste to enhance rhamnolipids and surfactin production. Bioresour Technol 205:1–6

Mukherjee S, Das P, Sen R (2006) Towards commercial production of microbial surfactants. Trends Biotechnol 24(11):509–515

Muthusamy K, Gopalakrishnan S, Ravi TK, Sivachidambaram P (2008) Biosurfactants: properties, commercial production and application. Curr Sci 94:736–747

Niu Y, Fan L, Gu D, Wu J, Chen Q (2017) Characterization, enhancement and modelling of mannosylerythritol lipid production by fungal endophyte Ceriporia lacerate CHZJU. Food Chem 228:610–617

Nurfarahin AH, Mohamed MS, Phang LY (2018) Culture medium development for microbial-derived surfactants production—an overview. Molecules 23(5):1049

Pacwa-Płociniczak M, Płaza GA, Piotrowska-Seget Z, Cameotra SS (2011) Environmental applications of biosurfactants: recent advances. Int J Mol Sci 12(1):633–654

Pandey A (2008) Solid-state fermentation. SciTopics Archives. http://www.scitopics.com/Solid_state_fermentation.html

Papagianni M (2012) Recent advances in engineering the central carbon metabolism of industrially important bacteria. Microb Cell Fact 11:50

Patil SB, Sawant YP, Kamble LH, Raorane CJ (2016) Primary screening of actinomycetes in prospects with biosurfactant production from animal fat waste. Int J Curr Microbiol App Sci 5:92–97

Priya T, Usharani G (2009) Comparative study for biosurfactant production by using Bacillus subtilis and Pseudomonas aeruginosa. Bot Res Int 2:284–287

Radzuan MN, Banat IM, Winterburn J (2017) Production and characterization of rhamnolipid using palm oil agricultural refinery waste. Bioresour Technol 225:99–105

Rahman PKSM, Gakpe E (2008) Production, characterisation and applications of biosurfactants-review. Biotechnology 7:360–370

Rahman KSM, Rahman TJ, McClean S, Marchant R, Banat IM (2002) Rhamnolipid biosurfactant production by strains of Pseudomonas aeruginosa using lo-cost raw materials. Biotechnol Prog 18(6):1277–1281

Raja M, Praveena G, William SJ (2017) Isolation and identification of fungi from soil in Loyola college campus, Chennai, India. Int J Curr Microbiol App Sci 6(2):1789–1795

Ramani K, Jain SC, Mandal AB, Sekaran G (2012) Microbial induced lipoprotein biosurfactant from slaughterhouse lipid waste and its application to the removal of metal ions from aqueous solution. Colloids Surf B Biointerfaces 97:254–263

Raza ZA, Rehman A, Khan MS, Khalid ZM (2007) Improved production of biosurfactant by a Pseudomonas aeruginosa mutant using vegetable oil refinery wastes. Biodegradation 18 (1):115–121

Rehn HJ, Reiff I (1981) Mechanisms and occurrence of microbial oxidation of long-chain alkanes. Adv Biochem Eng 19:175–216

Rekha R, Hemen S, Debahuti D (2014) Achieving the best yield in glycolipid biosurfactant preparation by selecting the proper carbon/nitrogen ratio. J Surfactant Deterg 17:563–571. https://doi.org/10.1007/s11743-013-1520-y

Ron EZ, Rosenberg E (2001) Natural role of biosurfactants. Environ Microbiol 4:229–236

Rosenberg E, Ron EZ (1999) High- and low-molecular-mass microbial surfactants. Appl Microbiol Biotechnol 52:154–162

Roussos S, Lonsane BK, Raimbault M, Viniegra-Gonzalez G (1997) Advances in solid state fermentation. Springer-Science+Business Media, B.V. Kluwer Academic, Berlin. ISBN 978-94-017-0661-2 (eBook)

Roy AA (2017) Review on the biosurfactants: properties, types and its application. J Fundam Renew Energy Appl 8:1–14

Rubio-Ribeaux D, da Silva Andrade RF, da Silva GS, de Holanda RA, Pele MA, Nunes P, Campos-Takaki GM (2017) Promising biosurfactant produced by a new Candida tropicalis UCP 1613 strain using substrates from renewable-resources. Afr J Microbiol Res 11:981–991

Saharan BS, Sahu RK, Sharma D (2012) A review on biosurfactants: fermentation, current developments and perspectives. Genet Eng Biotechnol J 2011:1–14

Saikia RR, Deka S, Deka M, Banat IM (2012) Isolation of biosurfactant-producing Pseudomonas aeruginosa RS29 from oil-contaminated soil and evaluation of different nitrogen sources in biosurfactant production. Ann Microbiol 62:753–763. https://doi.org/10.1007/s13213-011-0315-5

Santos DKF, Rufino RD, Luna JM, Santos VA, Sarubbo LA (2016) Biosurfactants: multifunctional biomolecules of the 21st century. Int J Mol Sci 17:401

Sarachat T, Pornunthorntawee O, Chavadej S, Rujiravanit R (2010) Purification and concentration of rhamnolipid biosurfactant produced by Pseudomonas aeruginosa SP4 using foam fractionation. Bioresour Technol 101:324–330

Satpute SK, Płaza GA, Banpurkar AG (2017) Biosurfactants' production from renewable natural resources: example of innovative and smart technology in circular bioeconomy. Manag Syst Prod Eng 25:46–54

Sayyad SA, Panda BP, Javed S, Ali M (2007) Optimization of nutrient parameters for lovastatin production by Monascus purpureus MTCC 369 under submerged fermentation using response surface methodology. Appl Microbiol Biotechnol 73:1054–1058

Schobert B (1992) The binding of a second divalent metal ion is necessary for the activation of ATP hydrolysis and its inhibition by tightly bound ADP in the ATPase from Halobacterium saccharovorum. J Biol Chem 267:10252–10257

Sellami M, Khlifi A, Frikha F, Miled N, Belbahri L, Rebah FB (2016) Agro-industrial waste based growth media optimization for biosurfactant production by Aneurinibacillus migulanus. J Microbiol Biotechnol Food Sci 5:578–583

Shah S, Prabhune A (2007) Purification by silica gel chromatography using dialysis tubing and characterization of sophorolipids produced from Candida bombicola grown on glucose and arachidonic acid. Biotechnol Lett 29(2):267–272

Sharma D, Ansari MJ, Gupta S, Al Ghamdi A, Pruthi P, Pruthi V (2015) Structural characterization and antimicrobial activity of a biosurfactant obtained from Bacillus pumilus DSVP18 grown on potato peels. Jundishapur J Microbiol 8:e21257

Shavandi M, Mohebali G, Haddadi A, Shakarami H, Ashrafossadat N (2011) Emulsification potential of a newly isolated biosurfactant-producing bacterium, Rhodococcus sp. strain TA6. Colloids Surf B Biointerfaces 82:477–482

Shoeb E, Akhlaq F, Badar U, Akhter J, Imtiaz S (2013) Classification and industrial applications of biosurfactants. Part-I. Nat Appl Sci 4:243–252

Singh V, Haque S, Niwas R, Srivastava A (2017) Strategies for fermentation medium optimization: an in-depth review. Front Microbiol 7:1–16. https://doi.org/10.3389/fmicb.2016.02087.

Singh P, Patil Y, Rale V (2018) Review article: biosurfactant production: emerging trends and promising strategies. J Appl Microbiol 126:2–13

Singhania RR, Sukumaran RK, Patel AK, Larroche C, Pandey A (2010) Advancement and comparative profiles in the production technologies using solid-state and submerged fermentation for microbial cellulases. Enzyme Microb Technol 46(7):541–549. https://doi.org/10.1016/j.enzmictec.2010.03.010.

Singhania RR, Patel AK, Thomas L, Goswami M, Giri BS, Pandey A (2015) Chapter 13. Industrial enzymes. In: Industrial biorefineries and White biotechnology. Elsevier, Amsterdam, pp 473–497

Smyth TJ, Perfumo A, Marchant R, Banat IM, Chen M, Thomas RK, Penfold J, Stevenson PS, Parry NJ (2010) Directed microbial biosynthesis ofdeuterated biosurfactants and potential future application to other bioactive molecules. Applied microbiology and biotechnology 87(4), 1347–54

Soberón-Chávez G, Maier RM (2011) Biosurfactants: a general overview. In: Soberón-Chávez G (ed) Biosurfactants. Springer-Verlag, Berlin, pp 1–11

Sobrinho HB, Luna JM, Rufino RD, Porto ALF, Sarubbo LA (2013) Biosurfactants: classification, properties and environmental applications. Rec Dev Biotechnol 11:1–29

Soniyamby AR, Praveesh BV, Vimalin Hena J, Kavithakumari P, Lalitha S, Palaniswamy M (2011) Enhanced production of biosurfactant from isolated Pseudomonas sp growing on used edible oil. J Am Sci 7:50–52

Souza AF, Rodriguez DM, Ribeaux DR, Luna MA, Lima e Silva TA, Andrade RFS, Campos-Takaki GM (2016) Waste soybean oil and corn steep liquor as economic substrates for bioemulsifier and biodiesel production by Candida lipolytica UCP 0998. Int J Mol Sci 17:1608

Taylor P, Korla K, Mitra CK (2013) Modelling the Krebs cycle and oxidative phosphorylation. J Biomol Struct Dyn 32:242–256

Tempest DW, Dicks JW, Hunter JR (1966) The interrelationship between potassium, magnesium and phosphorus in potassium-limited chemostat cultures of Aerobacter aerogenes. J Gen Microbiol 45:135–146. https://doi.org/10.1099/00221287-45-1-135

Thaniyavarn J, Chongchin A, Wanitsuksombut N, Thaniyavarn S, Pinphanichakarn P, Leepipatpiboon N, Morikawa M, Kanaya S (2006) Biosurfactant production by Pseudomonas aeruginosa A41 using palm oil as carbon source. J Gen Appl Microbiol 52:215–222. https://doi.org/10.2323/jgam.52.215

Thavasi R, Jayalakshmi S, Balasubramanian T, Banat IM (2008) Production and characterization of a glycolipid biosurfactant from Bacillus megaterium using economically cheaper sources. World J Microbiol Biotechnol 24:917–925. https://doi.org/10.1007/s11274-007-9609-y

Thavasi R, Subramanyam Nambaru VRM, Jayalakshmi S, Balasubramian T, Banat IM (2011) Biosurfactant production by Pseudomonas aeruginosa from renewable resources. Indian J Microbiol 51:30–36. https://doi.org/10.1007/s12088-011-0076-7.

Thomas L, Larroche C, Pandey A (2013) Current developments in solid-state fermentation. Biochem Eng J 81:146–161. https://doi.org/10.1016/j.bej.2013.10.013

Urum K, Pekdemir T (2004) Evaluation of biosurfactants for crude oil contaminated soil washing. Chemosphere 57:1139–1150

Vijayaraghavan P, Remya CS, Vincent SG (2011) Production of α-amylase by Rhizopus microsporus using agricultural by-products in solid state fermentation. Res J Microbiol 6:366–375

Wei YH, Chou CL, Chang JS (2005) Rhamnolipid production by indigenous Pseudomonas aeruginosa J4 originating from petrochemical wastewater. Biochem Eng J 27:146–154

Whang LM, Liu PWG, Ma CC, Cheng SS (2008) Application of biosurfactant, rhamnolipid, and surfactin, for enhanced biodegradation of diesel-contaminated water and soil. J Hazard Mater 151:155–163

Winterburn JB, Martin PJ (2012) Foam mitigation and exploitation in biosurfactant production. Biotechnol Lett 34:187–195

Wu JY, Yeh KL, Lu WB, Lin CL, Chang JS (2008) Rhamnolipid production with indigenous Pseudomonas aeruginosa EM1 isolated from oil-contaminated site. Bioresour Technol 99:1157–1164

Zinjarde SS, Pant A (2002) Emulsifier from a tropical marine yeast, Yarrowia lipolytica NCIM 3589. J Basic Microbiol 42(1):67–73

Microbial Biosurfactants and Their Potential Applications: An Overview

5

Debajit Borah, Anu Chaubey, Abhijit Sonowal, Bhaskarjyoti Gogoi, and Rupesh Kumar

Abstract

Biosurfactants are amphiphilic and one of the most versatile compounds synthesized by certain plants and microorganisms. These compounds are well known for their wide range of applications from domestic, personal, food and medical up to industrial level. They are capable of emulsifying oily substances which makes them very useful as a cleaning agent for domestic or industrial purpose and at the same time makes it competent for the exploration of oil through microbial enhanced oil recovery (MEOR) process. Besides they have a major application in the form of a potential tool to fight against petroleum-based contamination of oil and water. They are also used as an anti-adhesive agent, emulsifying agent in bakery industry, drug delivery system, etc. They are synthesized by a number of microorganisms such as *Pseudomonas* sp., *Bacillus* sp., *Acinetobacter* sp., *Candida* sp., *Sphingomonas* sp., *Cryptococcus* sp., *Pseudozyma* sp., *Kurtzmanomyces* sp., *Rhodococcus* sp., *Arthrobacter* sp., *Lactococcus* sp., *Penicillium* sp., *Aspergillus* sp., etc. and are encoded by a number of genes. In this regard, this chapter focus on basic chemical properties of various types of biosurfactants, source of production, name of genes responsible for its production, and various applications.

Keywords

Biosurfactant · Industrial applications · Nanotechnology · MEOR · Bioremediation

---

Anu Chaubey and Abhijit Sonowal contributed equally to this work.

D. Borah (✉) · A. Chaubey · A. Sonowal · B. Gogoi · R. Kumar
Department of Biotechnology, Assam Royal Global University, Guwahati, Assam, India

Inamuddin et al. (eds.), *Microbial Biosurfactants*, Environmental and Microbial Biotechnology, https://doi.org/10.1007/978-981-15-6607-3_5

5.1 Introduction

Surfactants belong to one of the important classes of chemicals which are widely used in modern day industries. Roughly all the commercially available surfactants are chemically synthesized from petroleum-based substrate. However, the conventional chemical surfactants are mostly expensive as they are derived basically from non-renewable petroleum-based substrates and may also act as potential threats to the environment as petroleum-based products are mostly recalcitrants by nature. Due to such disadvantages of chemical surfactants, various microbes are nowadays widely studied and employed for the production of bio-based surfactants which are popularly referred as biosurfactants. Biosurfactants have various advantages over chemical surfactants in terms of their biodegradability in nature, low toxicity, low cost of production from renewable sources and stability at extreme conditions such as high temperature, wide range of pH, salinity, etc.

Microbial biosurfactants are known to be produced extracellularly by various bacteria, yeasts, fungi, etc. as amphiphilic compounds from their cell membrane and exhibit high surface activity just like chemical surfactants. These were first discovered in the late 1960s during hydrocarbon fermentation experiments (Khan et al. 2017). Briefly, they have the potential to reduce the surface tension as well as the interfacial tension of oily substances with the formation of foam and eventually increase the aqueous solubility and bioavailability of non-aqueous-phase liquids (NAPLS) (Fakruddin 2012). In general, biosurfactants may be chemically classified based on their ionic properties in water as anionic, non-ionic, cationic and amphoteric and are classified summarily as glycolipids, phospholipids, polymeric biosurfactants and lipopeptides (surfactin) based on their chemical properties (Fakruddin 2012). They are known to be produced by various microorganisms including various bacteria, fungi and yeast. For example, some of the strains of *Pseudomonas aeruginosa* are known for the production of rhamnolipid-based biosurfactants; *Candida bombicola*, one of the very lesser known yeasts, is known to produce sophorolipid-based biosurfactants from vegetable oils and starchy substrates; *Bacillus subtilis* is known to produce surfactin-based biosurfactants, etc. (Mulligan 2005). Biosurfactants are amphiphilic in nature which bears both hydrophobic and hydrophilic moieties which provide them the ability to accumulate themselves within the aqueous phase such as oil-water or air-water emulsion by reducing the surface tension of oil.

Microbial biosurfactants are nowadays exploited for their wide range of applications in the form of household instant detergents, personal care products such as cosmetic components and antimicrobial agents, in the field of petroleum exploration by microbial-enhanced oil recovery (MEOR) process, in the field of environmental biotechnology as a bioremediation agent, in food processing industries as bioemulsifier, in pharmaceutical and health care industries as active drug delivery molecules, etc. (Vaz et al. 2012).

5.2 Classes of Biosurfactants

Biosurfactants are classified based on their ionic properties in water and also based on their chemical nature.

Based on their ionic property, they are classified as follows:

- Anionic.
- Non-ionic.
- Cationic.
- Amphoteric.

On the other hand, based on the chemical nature of biosurfactants, they are classified as follows:

- Glycolipids.
- Lipopolysaccharides.
- Lipopeptides and lipoproteins.
- Phospholipids.
- Fatty acids.

5.2.1 Glycolipids

Glycolipids are the most widely studied class of microbial biosurfactants. Glycolipids are constituted with carbohydrate moieties associated with fatty acids connected with either ester or ether group (Fig. 5.1). The most studied types of glycolipids are rhamnolipids, trehalolipids and sophorolipids composed of mono- or oligosaccharides attached with lipid moieties to form low molecular weight surface-active molecules (Sharma 2016a). On the other hand, glucose, galactose, xylose or rhamnose constitutes the sugar part of the biosurfactant, and the lipid part may be constituted with saturated or unsaturated fatty acid (Sharma 2016b).

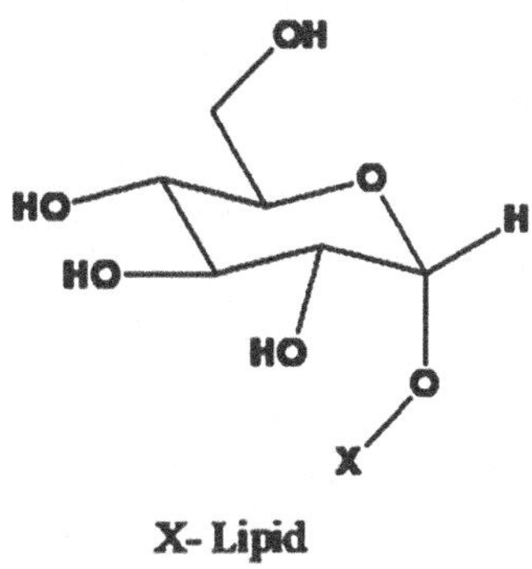

Fig. 5.1 Structure of glycolipid

5.2.2 Lipopolysaccharides

Lipopolysaccharide class of biosurfactants usually bears cyclic lipopeptides attached to fatty acids which is also known as surfactin mostly synthesized by *Bacillus* sp. (Fig. 5.2). Lipopolysaccharide is released by most of the Gram-negative microbes as one of the primary component of the outer membrane (Stromberg et al. 2017).

5.2.3 Lipopeptides and Lipoproteins

These are cyclic peptides with 7 amino acids, viz., L-aspartic acid, L-leucine, glutamic acid, L-leucine, L-valine, and two D-leucines which are finally attached to an acylated fatty acid chain which is known for its antimicrobial and more often haemolytic activities. This class of biosurfactant is also known as surfactin and are well known for their antibacterial, antifungal as well as antiviral activities (Fig. 5.3). This class of biosurfactant is known to be produced by mostly endospore-producing Gram-positive bacteria such as *Bacillus* sp., *Lactobacillus* sp., *Actinomycetes*, etc.

5.2.4 Phospholipids

Phospholipids are the derivatives of glycerol molecules with two molecules of fatty acids and a negatively charged phosphate group (Fig. 5.4). The negatively charged phosphate group in phospholipid is hydrophilic in nature, and non-polar fatty acid chains are hydrophobic in nature. During the formation of mycelium in water, they arrange themselves in a spherical manner with the hydrophilic head facing towards water and the hydrophobic tails hiding themselves within the core against water.

5.2.5 Fatty Acids

Fatty acids are the fat-soluble components of all living cells which bear long straight chain of carbon atoms along with hydrogen atoms along the length of the chain and a –COOH (carboxyl) group at other end. If the fatty acid molecule bears single bonds between carbon-to-carbon molecules, then they are considered as saturated (e.g., butyric acid, lauric acid, myristic acid, etc.), and if it bears double or triple bonds, then they are considered as unsaturated fatty acids (e.g., palmitoleic acid, oleic acid, myristoleic acid, linoleic acid, arachidonic acid, etc.). Some of the fatty acids also bear branched chains and ring structures (e.g., prostaglandins).

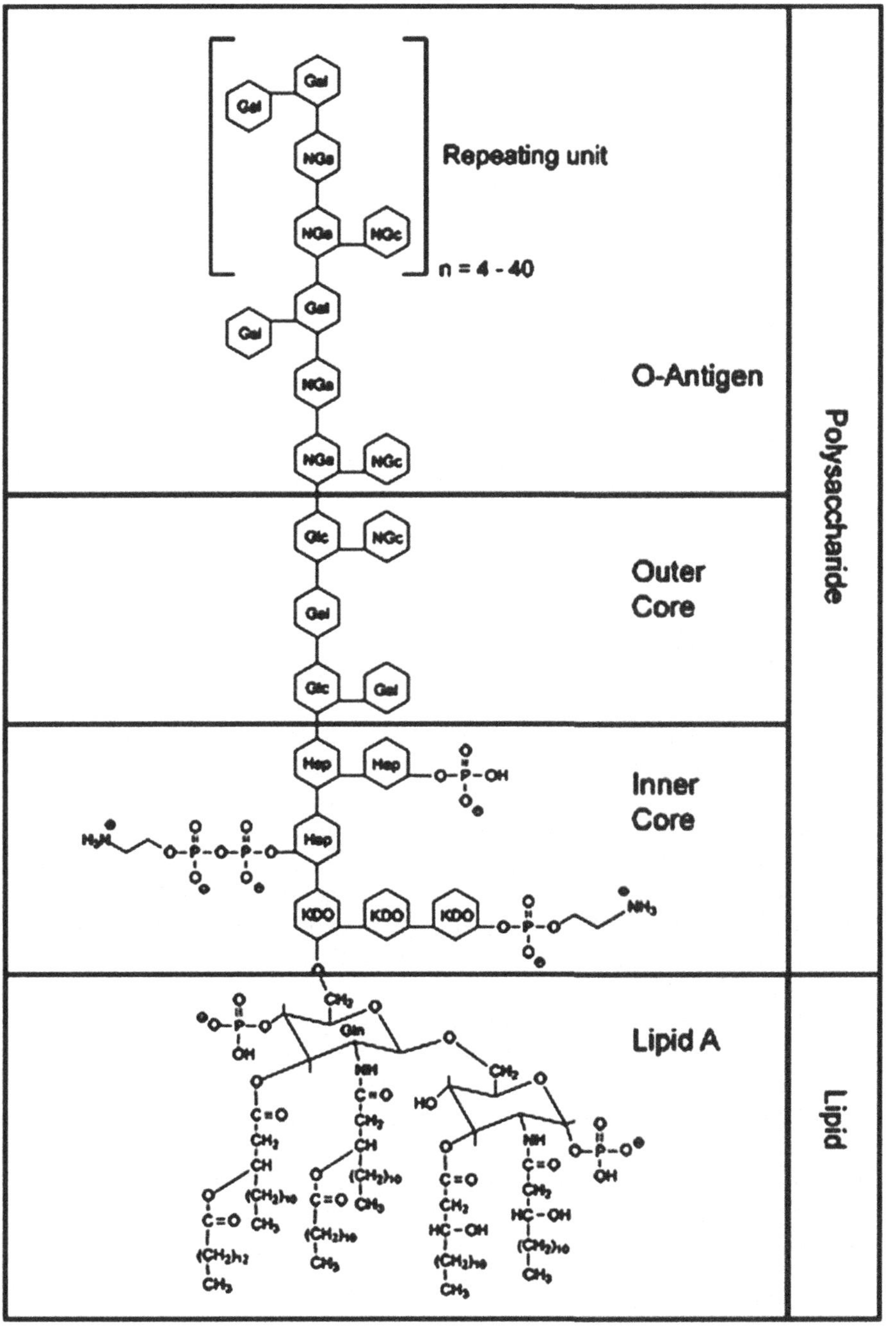

Fig. 5.2 Basic structure of bacterial polysaccharide (https://www.sigmaaldrich.com/technical-documents/articles/biology/glycobiology/lipopolysaccharides.html 2020)

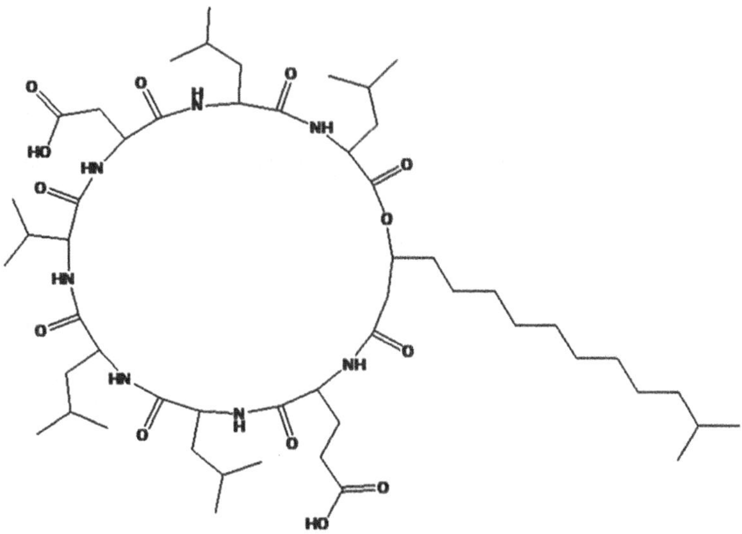

Fig. 5.3 Molecular structure of surfactin

5.3 Microbial Production of Biosurfactants

Biosurfactant is produced by a number of bacteria, most popularly, various strains of *Pseudomonas aeruginosa*, *Pseudomonas putida*, *Pseudomonas cepacia*, *Serratia marcescens*, *Bacillus subtilis*, *Bacillus licheniformis*, *Bacillus mycoides*, *Bacillus mojavensis*, *Acinetobacter bouvetii*, etc., and a number of fungi, viz., *Candida petrophilum*, *Candida lipolytica*, *Candida tropicalis*, *Penicillium* sp., *Aspergillus* sp., etc. (Bordoloi and Konwar 2008; Chaprão et al. 2015; Tugrul and Cansunar 2005; Aparna et al. 2012; Mulligan 2005; Ghosh et al. 2015; Shekhar et al. 2015; Vijayakumar and Saravanan 2015; Saravanakumari and Mani 2010; Vijayakumar and Saravanan 2015; Youssef et al. 2004; Chan et al. 2013; Shekhar et al. 2015; Wei et al. 2007; Saravanakumari and Mani 2010; Khan et al. 2017; Luna et al. 2012; Fooladi et al. 2016; Ghosh et al. 2015; Ron and Rosenberg 2002; Shekhar et al. 2015). Different types of microorganisms produce different types of biosurfactant in terms of their chemical property. Table 5.1 shows various types of biosurfactants produced by different types of microorganisms which include both bacteria and fungi. However, the efficacy of the production of biosurfactant is largely governed by growth conditions such as temperature, pH, salinity, rate of agitation and also media components such as N, Fe, Mg, etc.

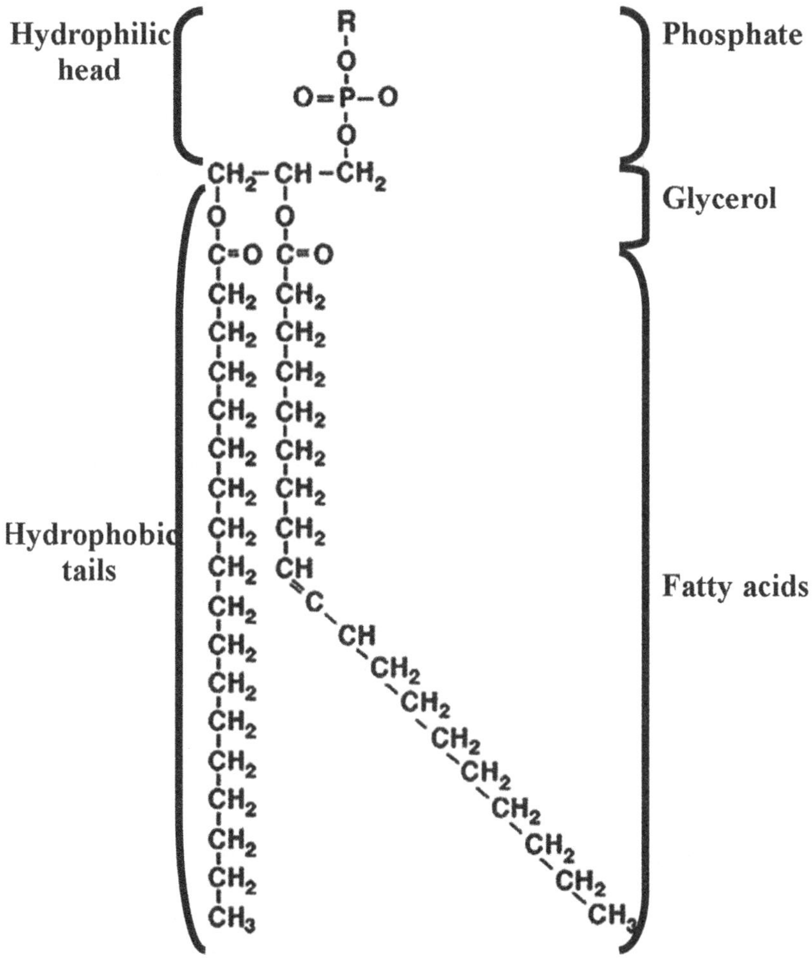

Fig. 5.4 Structure of phospholipid

5.4 Genes Involved in the Production of Microbial Biosurfactants

Among all the reported types of biosurfactants, the molecular biology for *Pseudomonas aeruginosa*-mediated biosynthesis of rhamnolipids and *Bacillus subtilis*-mediated synthesis of surfactin was first reported (Das et al. 2008). Besides, biosynthesis of arthrofactin, iturin and lichenysin, mannosylerythritol lipids and emulsan synthesized, respectively, by *Pseudomonas* sp., *Bacillus* sp., *Candida*

Table 5.1 Production of biosurfactants from various microbes

Biosurfactant types	Subdivision	Microbial species	Type	Reference
Glycolipids	Rhamnolipids	*Pseudomonas aeruginosa*	Gram-negative rod-shaped bacteria	Tugrul and Cansunar (2005)
		P. aeruginosa S7B1 (type R-1) *P. aeruginosa* KY4025 (type R-2) *P. aeruginosa* PG201	Gram-negative rod-shaped bacteria	Beal and Betts (2000)
		Pseudomonas sp. DSM 2874 (Rha–C10 and Rha–Rha–C10) *P. aeruginosa* L2–1	Gram-negative rod-shaped bacteria	Bordoloi and Konwar (2008)
		P. aeruginosa UG2 *Pseudomonas putida*	Gram-negative rod-shaped bacteria	Chaprão et al. (2015)
		Pseudomonas sp. 2B	Gram-negative rod-shaped bacteria	Mulligan (2005)
		Thermus thermophilus HB8 *P. aeruginosa* GS3	Gram-negative rod-shaped bacteria	Aparna et al. (2012)
	Sophorolipids	*Torulopsis bombicola* (or *Candida bombicola*)	Fungi	Bordoloi and Konwar (2008)
		Sphingomonas yanoikuyae	Gram-negative rod-shaped bacteria	Mulligan (2005)
		Candida batistae *Cryptococcus curvatus* ATCC 20509	Fungi	Ghosh et al. (2015)
		Torulopsis petrophilum	Fungi	Shekhar et al. (2015)
	Cellobiolipids	*Ustilago maydis*	Fungi	
	Mannosylerythritol lipids	*Pseudozyma Aantarctica*	Gram-negative rod-shaped bacteria	
		Kurtzmanomyces sp. *I-1*	Mould	
		Ustilago maydis *Schizonella melanogramma*	Fungi	
	Trehalolipids	*Rhodococcus* sp.	Gram-positive rod-shaped bacteria	Mulligan (2005)

		Mycobacterium sp.	Acid-fast-positive rod-shaped bacteria	Vijayakumar and Saravanan (2015)
		Nocardia sp. *Corynebacterium* sp.	Gram-positive rod-shaped bacteria	
		Rhodococcus erythropolis	Gram-positive cocci-shaped bacteria	
		Arthrobacter sp.	Gram–positive rod-shaped bacteria in exponential phase and cocci shaped in stationary phase	
	Xylolipids	*Lactococcus lactis*	Gram-positive cocci-shaped bacteria	Saravanakumari and Mani (2010)
	–	*Pseudomonas cepacia* (GL-K12) *Alcanivorax borkumensis*	Gram-negative rod-shaped bacteria	Mulligan (2005)
		Candida antarctica	Fungi	Shekhar et al. (2015)
		Thiobacillus thiooxidans	Gram-negative rod-shaped bacteria	
Phospholipids	–	*Acinetobacter* sp. 1-N *Thiobacillus thiooxidans* *Acinetobacter* sp. *Pseudomonas* sp.	Gram-negative rod-shaped bacteria	Vijayakumar and Saravanan (2015)
		Nocardia erythropolis	Gram-positive rod-shaped bacteria	
		Micrococcus sp. *Myxococcus* sp.	Gram-positive cocci-shaped bacteria	
		Candida lepus *Candida* sp.	Fungi	
		Penicillium sp. *Aspergillus* sp.	Fungi	Youssef et al. (2004)
Lipopeptides or lipoproteins	Serrawettin	*Serratia marcescens*	Gram-negative rod-shaped bacteria	Shekhar et al. (2015)
	Surfactin	*B. subtilis*	Gram-positive rod-shaped bacteria	Chan et al. (2013)
		B. subtilis EG1 *B. subtilis* ATCC 21332	Gram-positive rod-shaped bacteria	Tugrul and Cansunar (2005)
	Lichenysin	*B. licheniformis* (lichenysin A)	Gram-positive rod-shaped bacteria	Vaz et al. (2012)

(continued)

Table 5.1 (continued)

Biosurfactant types	Subdivision	Microbial species	Type	Reference
	Iturin	*B. subtilis*	Gram-positive rod-shaped bacteria	Wei et al. (2007)
	Subtilisin	*B. subtilis*	Gram-positive rod-shaped bacteria	Saravanakumari and Mani (2010)
	Polymyxins	*Bacillus polymyxa*	Gram-positive rod-shaped bacteria	Vijayakumar and Saravanan (2015)
		Brevibacterium polymyxa	Gram-positive coccobacilli bacteria	
	Fengycin	*Bacillus* sp.	Gram-positive rod-shaped bacteria	
	Viscosin	*Pseudomonas fluorescens*	Gram-negative rod-shaped bacteria	Shekhar et al. (2015)
	Gramicidins	*Bacillus brevis*	Gram-positive rod-shaped bacteria	
		Brevibacterium brevis	Gram-positive coccobacilli bacteria	
	Arthrofactin	*Arthrobacter* sp.	Gram-positive rod-shaped bacteria in exponential phase and cocci shaped in stationary phase	
	Ornithine	*Myroides* sp. *Pseudomonas* sp. *Thiobacillus* sp. *Agrobacterium* sp. *Gluconobacter* sp.	Gram-negative rod-shaped bacteria	
	–	*Pseudomonas poae*BA1 *Acinetobacter bouvetii* BP18 *Stenotrophomonas rhizophila* BG32	Gram-negative rod-shaped bacteria	
		B. thuringiensis BG3	Gram-positive rod-shaped bacteria	
		B. subtilis C9	Gram-positive rod-shaped bacteria	
		Klebsiella sp. Y6–1	Gram-negative rod-shaped bacteria	Khan et al. (2017)
		B. mojavensis *B. Mycoides* *B. subtilis* LB5 *B. mycoides* SH2 *B.pumilus*2IR	Gram-positive rod-shaped bacteria	Luna et al. (2012)

		B. subtilis ATCC 6633	Gram-positive rod-shaped bacteria	De et al. (2015)
		C. petrophilum	Fungi	
Polymeric compounds	Liposan	*C. lipolytica*	Fungi	Fooladi et al. (2016)
	Emulsan	*Acinetobacter* RAG-1 *A. Calcoaceticus*	Gram-negative coccobacilli-shaped bacteria	Ghosh et al. (2015)
	Biodispersan	*A. Calcoaceticus*	Gram-negative coccobacilli-shaped bacteria	Vijayakumar and Saravanan (2015)
	Mannan-lipid-protein	*Candida tropicalis*	Fungi	Ron and Rosenberg (2002)
	Carbohydrate-protein complex	*Yarrowia lipolytica*	Fungi	Velraeds et al. (1996)
	Polysaccharide lipid complex	*Acinetobacter* sp.	Gram-negative coccobacilli-shaped bacteria	Shekhar et al. (2015)
Particulate compounds	Vesicles and fimbriae	*Acinetobacter* sp. HO1-N	Gram-negative coccobacilli-shaped bacteria	

sp. and *Acinetobacter* sp. is also well documented. For instance, *srfA* operon consists of four open reading frames (ORFs) namely *srfAA*, *srfAB*, *srfAC* and *srfAD*, is responsible for the synthesis of amino acid moieties of surfactin, whereas *sfp* is the gene required for the synthesis of phosphopantetheinyl transferase enzyme that activates surfactin by post-translational modification (Das et al. 2008). Names of a wide number of genes responsible for the synthesis of biosurfactants in various microorganisms are shown in Table 5.2.

5.5 Applications

The versatility of biosurfactants makes them a potential substrate for commercial exploitation in petrochemical, pharmaceutical, food and cosmetics-based industries. Moreover, they have various domestic applications in the form of detergents, foaming agents, emulsifiers, etc. (Aparna et al. 2012). The detailed applications of biosurfactant in various fields are discussed below.

5.5.1 In Petroleum Industry

Petroleum-based oil recovery is achieved in three major steps: primary, secondary and tertiary.

- *Primary oil recovery process*. This involves mechanical drilling and pumping devices to recover petroleum oil which naturally emerges to the earth's surface due to very high pressure of gas.
- *Secondary recovery process*. This employs injection of gas and water through the drilling rigs for the displacement of petroleum oil up to the surface for easy recovery. US Department of Energy states that more than 75% of oil may not be possible to recover even after employing these two methods (Pacwa-Płociniczak et al. 2011).
- *Tertiary recovery process*. Tertiary recovery process is also popularly known as enhanced oil recovery process (EOR).

EOR widely involves the use of thermal recovery processes, gas and chemical injection processes, microbial enhanced oil recovery process (MEOR), etc. and is briefly defined below:

- *Thermal recovery*. This method involves the use of hot steam through the reservoir to heat the well which minimizes its viscosity and allows easy up-flow to the surface. Alternatively, less commonly it is practiced to burn some part of the oil to heat the rest which is also known as in situ burning or fire flooding.
- *Gas injection*. This is one of the most popular tertiary oil recovery methods which involve the injection of gases such as N_2, CO_2 or even natural gases through the drilling rigs. After mixing the gas with oil increases the viscosity, but

Table 5.2 Genes responsible for the production of biosurfactants in bacteria

Name of the organism	Name of biosurfactant produced	Genes responsible	References
Ustilago maydis	Mannosylerythritol lipid	*emt1*	Hewald et al. (2005)
	Cellobiose lipid	*cyp1*	
B. licheniformis ATCC10716	Lichenysin	*lic*A *lic*B *lic*C	Das et al. (2008)
B. subtilis RB14	Iturin A	*itu*D *itu*A *itu*B *itu*C	
B. subtilis	Surfactin	*srf*A *srf*B *srf*C *srf*D	
Pseudomonas sp. MIS38	Arthrofactin	*arf*A *arf*B *arf*C	
P. aeruginosa PG 201	Rhamnolipid	*rhl*A *rhl*B *rhl*R *rhl*I	
Pseudomonas sp. DSS73	Amphisin	*gac*S *ams*Y	
P. putida PCL1445	Pultisovin I Pultisovin II	*dna*K *dna*J *grp*E	
Acinetobacter radioresistens KA53	Alasan	*aln*A	
Acinetobacter lwoffii RAG-1	Emulsan	*wza* *wzb* *wzc* *wzx* *wzy*	
Serratia marcescens	Serrawettin W1	*psw*P	
Serratia marcescens 274	Serrawettin W1	*swr*W	
Serratia liquefaciens MG1	Serrawettin W2	*swr*A *swr*I *swr*R	
Trichoderma reesei	Hydrophobins	*hfb*1 *hfb*2	
P. aeruginosa spp.	Rhamnolipid	*las*R *las*I	Sullivan (1998)
Dietzia maris As-13-3	Di-rhamnolipid	*alk*B *cyp*153 *alg*C *rml*A *rml*B *rml*C	Wang et al. (2014)

(continued)

Table 5.2 (continued)

Name of the organism	Name of biosurfactant produced	Genes responsible	References
		*rml*D *rhl*A *rhl*B *rhl*C *acc*A *acc*B *acc*D *fab*B *fab*D *fab*G	
Starmerella bombicola	Sophorolipid	*adh* *ugt*B1 *mdr* *at* *ugt*A1 *cyp52m*1 *orf*	Van Bogaert et al. (2013)

simultaneously it increases the pressure to push the oil up to the surface for its easy recovery.

- *Chemical injection.* This is one of the least commonly used EOR methods for petroleum recovery. It employs chemical surfactant which reduces the surface tension of oil and causes free flow of oil and finally is recovered by water flooding followed by mechanical with lesser effort.
- *Microbial enhanced oil recovery (MEOR).* This is probably the most recent method of oil recovery which involves the use of biosurfactant-producing microbes to minimize the surface tension of oil which further enhances the free flow of oil to the surface with the injection of water (Fig. 5.5).

5.5.1.1 Mechanism of MEOR

A considerable amount of oil retained inside porous surface throughout the drilling rig remains unrecovered. Such porous surface which retains oil inside is called "thief zones". MEOR employs hydrocarbon degrading and biosurfactant-producing microorganism to reduce the viscosity of oil by lowering its surface tension causing the free flow of oil (Pacwa-Płociniczak et al. 2011). MEOR has attracted the attention of scientific communities in recent time mainly because of its cost-effectiveness and environment benign nature. A number of microorganisms are reported till date with a potential of producing biosurfactants. Out of all the microorganisms, various species and strains of *Pseudomonas* are most commonly reported by various researchers worldwide (Tugrul and Cansunar 2005; Beal and Betts 2000; Bordoloi and Konwar 2008; Chaprão et al. 2015). They are reported as most commonly capable of producing rhamnolipid- and phospholipid-based biosurfactants, whereas less

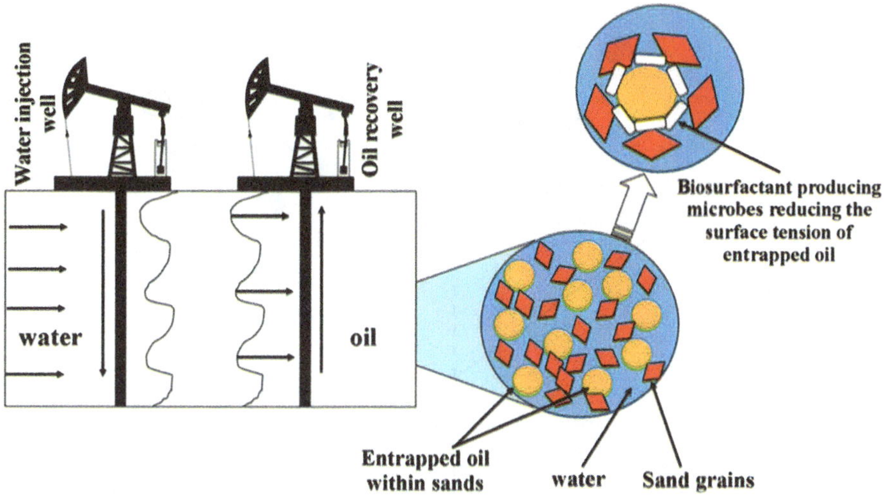

Fig. 5.5 The process of microbial enhanced oil recovery (MEOR) process

commonly reported as capable of producing viscosin- and ornithine-based biosurfactants (Tugrul and Cansunar 2005; Shekhar et al. 2015). On the other hand, various species of *Bacillus* are also widely reported for their capability in the production of surfactin-, lichenysin-, iturin-, polymyxin-, fengycin- and gramicidin-based biosurfactants (Tugrul and Cansunar 2005; Vaz et al. 2012; Wei et al. 2007; Saravanakumari and Mani 2010; Shekhar et al. 2015).

The remaining oil after primary and secondary recovery process often entrapped inside sand pores is very difficult to recover due to capillary pressure. The surface tension between oil-water and oil-rock can be reduced with the help of biosurfactants which further leads to the reduction of capillary forces preventing petroleum oil to reach the surface (Pacwa-Płociniczak et al. 2011) (Fig. 5.5). Biosurfactants also form emulsion by tightly binding with oil-water interface and thus stabilize the desorbed oil in water and enable easy exploration of oil along with the injection water (Pacwa-Płociniczak et al. 2011) (Fig. 5.5).

The critical micelle concentration (CMC) and emulsification (E) index are the important characteristics of a surfactant. CMC is defined as the concentration of surfactants in bulk phase above which micelles are formed and above which the additional surfactants added to the system go to micelles (Fig. 5.6). On the other hand, emulsification index is defined as the percentage of height of emulsified layer (mm) divided by total height of the liquid column (mm) in a given time (T) (Fig. 5.7).

E_T index can be expressed as below:

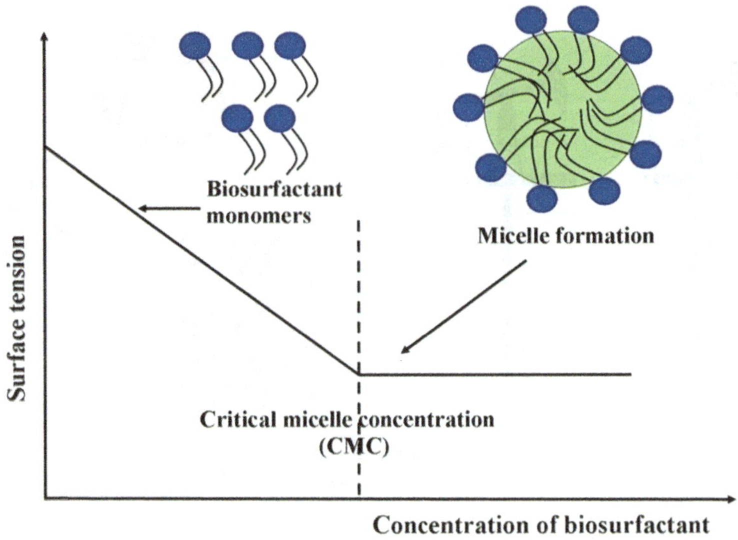

Fig. 5.6 Relationship among concentration of biosurfactant, surface tension of hydrocarbon and micelle formation

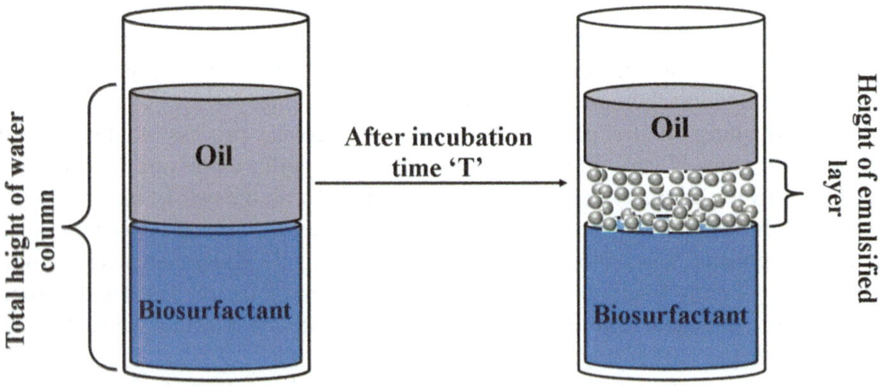

Fig. 5.7 Formation of emulsion layer in presence of biosurfactant after incubation

$$\frac{\text{Height of the emulsified layer (mm)}}{\text{Total height of the water column (mm)}} \times 100$$

Biosurfactants released by microorganisms reduce the surface tension of oil droplets, thus converting them into microdroplets (micelle) and making them easier for microorganisms for their uptake (Ibrahim et al. 2013). A number of enzymes take part in the degradation of oil droplets. Oxygenase is one of the most important microbial enzymes found to be responsible for petroleum degradation which converts complex hydrocarbon chains into much simpler forms which enter through

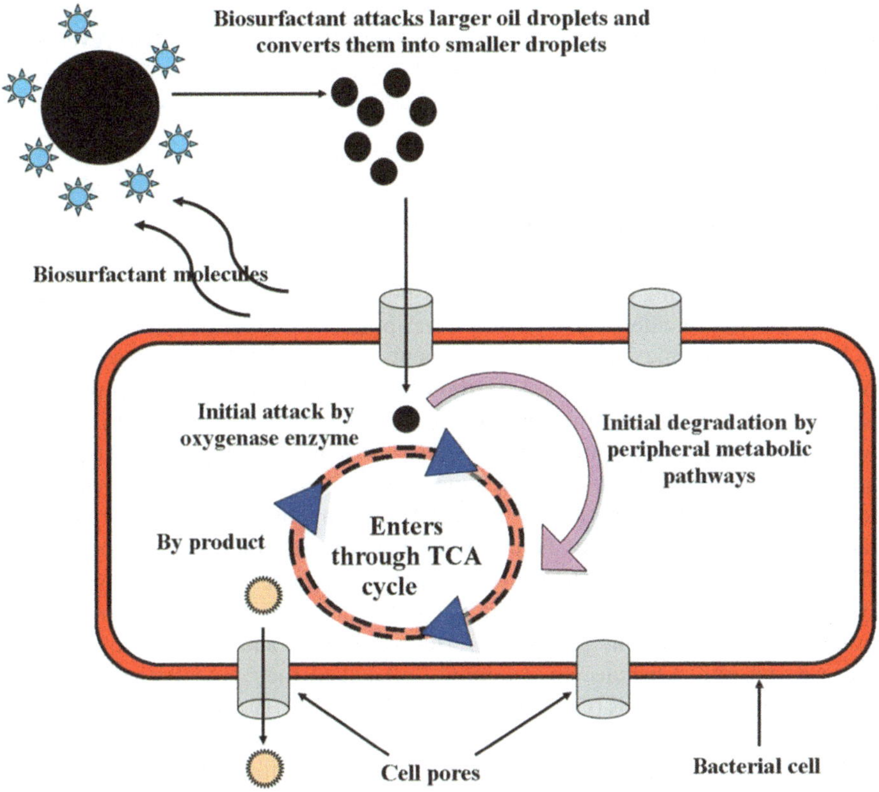

Fig. 5.8 Uptake of oil droplets by microbial cells and their degradation through metabolic processes

peripheral metabolic cycles before undergoing complex metabolic changes (Hommel 1990; Cerniglia 1992; Yakimov et al. 1995; Adebusoye et al. 2008; Ibrahim et al. 2013; Behera and Prasad 2020) (Fig. 5.8).

A large amount of carbon dioxide (CO_2) gas is produced as a by-product of metabolism over time, and this accumulated gas eventually drives up the entrapped oil through the recovery site, whereas the by-product generated by the microorganisms in the form of biomass accumulates between oil and sand surface inside the drilling rig physically helping in displacing the oil. Moreover, some microorganisms are capable of producing exopolysaccharide (EPS) which is a slimy substance along with the biosurfactants they produce. Such EPS helps in sealing the pores present in rocks and thief zones and facilitates the movement of oil throughout the recovery process (Khire 2010). Successful well-documented MEOR field trial has been reported as early as the year 2007 by Youssef et al. (2007), with the help of biosurfactant produced by a mixture of *Bacillus subtilis* subsp. *spizizenii* NRRL B-23049 and *Bacillus* strain RS-1. These strains were supplemented with nutrients such as glucose, sodium nitrate and trace metals to facilitate the growth and proliferation of the strains. In brief, MEOR is now proving itself an eco-friendly, non-toxic and commercially viable technology. However, such practices are largely governed by abiotic factors such as temperature, increasing pressure with the increase in the depth of the drilling well, pH, salinity, relative humidity (RH), etc.

5.5.2 Biosurfactant-Mediated Bioremediation

Spillage caused during oil exploration, transportation or by deliberate illegal dumping increases the incidence of soil and water pollution in recent times in the entire globe. Toxic inorganic and organic components such as volatile organic components (VOCs), polycyclic aromatic hydrocarbons (PAHs), heavy metals, etc. present in petroleum products are the most important contributors to environment toxicity and pose potential threat to aquatic animals and also to human health. Due to the growing demand in eco-friendly, biological treatments, bioremediation provides environment benign and cost-effective solution to bring back contaminated soil and water to its native condition with the help of hydrocarbon-degrading microorganisms. Microorganisms employ quite a few strategies to augment bioavailability of those hydrophobic components present in the pollutants which include formation of biofilm and biosurfactant production. Hence, growth and proliferation of such microbes on oily substrate are commonly related to their potential of producing surface-active compounds (Cameotra and Makkar 2010). The rate of bioremediation in a contaminated site can be augmented by adding additional microbes with a potential to synthesize biosurfactant. On the other hand, fortification of nutrients in the contaminated site by adding additional nutrient supplements can also be carried out to promote the growth of native indigenous microbes in the environment which is often referred to as biostimulation or bioaugmentation (Felix et al. 2019). Washing of petroleum-contaminated soil was earlier practiced with chemical surfactants but is losing popularity due the release of toxic residues. Therefore, biosurfactants which chemically include rhamnolipids, surfactin, sophorolipids, etc. are mostly reported as better alternative to such existing washing technology. However, saponin, a plant-based biosurfactant which is widely distributed in the plant kingdom, is also reported as a potential agent to treat hydrocarbon-contaminated soils and water (Zhou et al. 2013; Befkadu and Quanyuan 2018).

5.5.3 In Food Industry

Food regulation organizations in the entire world are unremittingly implementing restrictions on the use of chemical emulsifiers over suitable bio-based food emulsifiers which are extracted from plants and microorganisms. The growing concern about natural and organic products in food processing industry now has explored the use of biosurfactants not only as a bioemulsifier but also as antimicrobial agents. The basic role of emulsions in bakery industry is to attain consistency in dough making. Biosurfactants emulsifies edible oil used during food processing by lowering the surface tension of oil and thus improves texture, cheesiness and creaminess during dairy cheese preparation and in salad dressings. Moreover, it may also be used as an antimicrobial agent in food industry to prevent microbial contamination. Biosurfactants are also claimed with antioxidant potential which further extends its futuristic applications as a food ingredient (Sharma et al. 2016).

Literature survey has shown yeast-based biosurfactants, particularly from *Kluyveromyces marxianus*, with a great potential as a bioemulsifier against corn oil, whereas *Candida utilis* derived biosurfactants with a potential in salad dressing. Yeast-derived biosurfactants also show their potential application as probiotic because of their GRAS (generally regarded as safe) status in various food industries (Sharma et al. 2016).

5.5.4 In Agriculture

Biosurfactants are also used in agricultural practices as a vital component of pesticide and herbicide. They act as bioemulsifiers which makes the plant leaves hydrophobic in nature which prevents direct contact between microbe and plant leaves which minimizes the chances of infection in plants and may also prevent microbial growth by acting as antimicrobial agent.

5.5.5 In Cosmetics

Cosmetic products take a major role in our everyday life, and it possesses a big worldwide market potential. A large variety of surfactant-based cosmetic products ranging from soap, shampoo, toothpaste, skin moisturizer, etc. are available which are now gradually substituted with biosurfactant-based components to cater the demand of natural ingredients in cosmetics among customers. It is also believed that biosurfactants also act as "prebiotic" agent that facilitates the growth of healthy skin microbiota.

Some study suggests that the composition of biosurfactant such as CMC and hydrophilic-lipophilic balance (HLB) determines the use of biosurfactants in cosmetic formulations. It has been reported as glycolipids (e.g., rhamnolipid) and lipopeptides (e.g., surfactin) and exhibits lowest CMC value (Gudina et al. 2010). CMC denotes the minimum concentration of biosurfactant in the medium which undergoes micelle formation. On the other hand, HLB value of biosurfactant determines its emulsifying ability and wettability while formulating cosmetics (Vecino et al. 2017). Wettability is defined as the ability of a liquid to sustain contact with a solid surface and is governed by the intramolecular surface tension of the liquid. The wettability is more often measured in terms of contact angle which may be defined as the angle between the surface of the liquid and the outline of the contact surface (Fig. 5.9). The lesser the contact angle represents, the more the wettability due to the loss of surface tension.

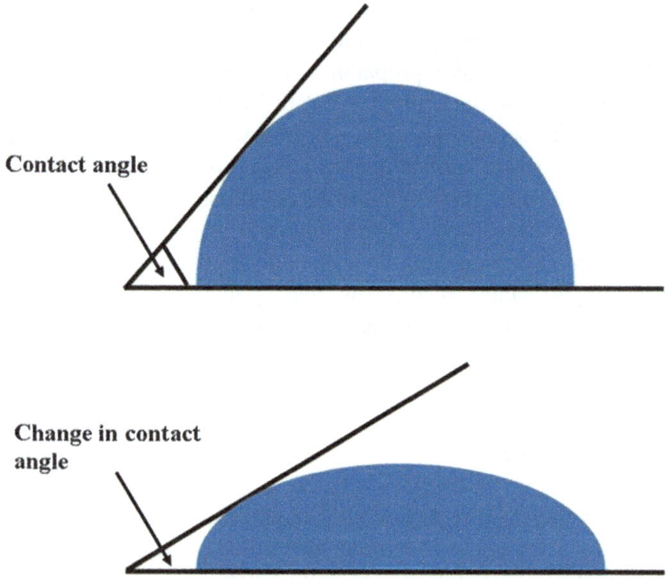

Fig. 5.9 Contact angle reduces with the decrease in surface tension

5.5.6 Biosurfactant in Nanotechnology

Green synthesis of metal nanoparticles such as gold, silver, zinc, titanium, copper, iron, palladium, platinum, etc., with the help of plant extract, is a new trend in applied chemistry (Jimoh and Lin 2019; Prasad 2014; Prasad et al. 2018; Shrivastava et al. 2021). However, very recently, the synthesis of metal nanoparticles by reducing metal salts with microbial biosurfactant emerges as a substitute for plant-based synthesis of nanoparticles due to its easy and bulk production potential and because at the same time it acts as a stabilizing agent by capping the nanoparticles. For instance, various strains and species of *Pseudomonas*, *Bacillus*, *Lactobacillus*, etc. are largely exploited for the synthesis of silver, gold, cadmium and zinc nanoparticles but are yet very limited (Table 5.3).

Not only as a reducing or stabilizing agent, biosurfactant finds its way also as drug delivery agent in nanoscale which is applicable to carry and to release the drug in a controlled manner. Such roles of biosurfactants are discussed in detail hereafter.

5.5.7 Biosurfactants as Drug Delivery Agents

With the advent of various drug delivery systems (DDS), biosurfactants also caught the eyes of the researchers for its possible role in the form of a drug-delivering agent. An ideal drug delivery system is expected to hold two important characteristics, viz., capability to carry the loaded drug to the target site followed by enhanced

Table 5.3 Synthesis of various metal nanoparticles by reducing metal salts by microbial biosurfactants

Name of organism	Name of the nanoparticles synthesized	References
Bacillus subtilis BBK006	Silver	Eswari et al. (2018)
B. subtilis ANR 88	Silver and gold	Rane et al. (2017)
Pseudomonas aeruginosa	Zinc	Kulkarni et al. (2019)
P. aeruginosa PBSC1	Silver	
P. fluorescens MFS-1	Silver	
P. aeruginosa BS-161R	Silver	
B. Ttequilensis	Silver	
Lactobacillus brevis MTCC 4463	Silver	
P. aeruginosa UCP0992	Silver	Farias et al. (2014)
Brevibacterium casei MSA19	Silver	Płaza et al. (2014)
Starmerella bombicola	Cobalt	
B. subtilis T-1	Silver	Liwarska-Bizukojc et al. (2018)
P. aeruginosa MKVIT3	Silver	Liwarska-Bizukojc et al. (2018)
B. subtilis BBK006	Gold	Reddy et al. (2009)
P. aeruginosa MTCC 424	Gold	Tomar et al. (2015)
B. amyloliquefaciens KSU-109	Cadmium	Singh et al. (2011)
B. subtilis BRS-07	Cadmium	Singh et al. (2012)
P. aeruginosa BS01	Zinc	Hazra et al. (2013)

bioavailability of the drug and releasing the drug in a controlled manner. To achieve this, polymeric, particulate or cellular carriers are most widely used. Nowadays, lipid-based microspheres, micelles, liposomes, sphingosomes, etc. are also used as effective DDS due to the non-cytotoxic nature of biosurfactants (Gudiña et al. 2013). Surfactant is the principal ingredient of microemulsion-based drug delivery system and also comprises with an aqueous and oily phase that encapsulates the drug within the core. As per literature, a microemulsion system may possess a variety of geometric shapes such as spherical, crystalline, hexagonal, bicontinuous, etc. (Gudiña et al. 2013). Even though hydrocarbon oils such as heptanes, dodecane, cyclohexanes with sodium dodecyl sulphate (SDS) or tetraethylene glycol monododecyl ether were initially practiced by researchers for the production of microemulsion system, SDS shows cytotoxicity in the long run. Hence biosurfactants have emerged as a potential substitute for such applications (Gudiña et al. 2013).

5.5.8 Antimicrobial Activity of Biosurfactants

Growing antibiotic resistance among pathogenic microbes due to frequent use of antibiotic emerges itself as a major challenge in health science to treat. Some lipopeptide-based microbial biosurfactants such as surfactin, fengycin, iturin, bacillomycins, mycosubtilins, etc. synthesized by *Bacillus subtilis* have been reported by some researchers showing antimicrobial potential against drug-resistant microbes. Literature also shows cyclic lipopeptide-based biosurfactants such as daptomycin produced by *Streptomyces roseosporus* and viscosin from *Pseudomonas* sp., rhamnolipids produced by *P. aeruginosa*, sophorolipids produced by *Candida bombicola*, mannosylerythritol lipids from *Candida antarctica*, lipopeptides from *Bacillus circulans*, flocculosin from *Pseudozyma flocculosa*, etc. with potential antimicrobial activity (Banat et al. 2010).

5.5.9 Biosurfactant as Anti-Adhesive Agent

Biosurfactants are also reported as an anti-adhesive agent who helps in removing harmful microorganisms adsorbed on solid surface by inhibiting the adhesion. One such practical example has shown surfacting coated vinyl urethral catheters inhibiting the biofilms formed by *Salmonella typhimurium*, *Salmonella enterica*, *E. coli* and *Proteus mirabilis* (Mireles et al. 2001). *Salmonella* sp. is one of the major groups of microorganisms responsible for causing urinary tract infection (UTI) in human (Mireles et al. 2001), whereas *Lactobacillus* sp. contributes to the normal microbiota of female urogenital tract and maintains the healthy microbiota by releasing lactic and biosurfactants that inhibit the growth of harmful microbes. Based on this, Heinemann et al. (2000), have reported the inhibition of *Enterococcus faecalis* and yeast-mediated biofilm on silicone rubber with the help of biosurfactants released by *Lactobacillus acidophilus* and *L. fermentum* RC-14 by lowering the adhesion between the substrate and the pathogen (Heinemann et al. 2000). Rodrigues et al. (2006), explained the ability of *L. lactis* 53- and *S. thermophilus*-mediated biosurfactants to inhibit the adhesion of pathogens from colonizing on silicone rubber voice prostheses (Salusjärvi et al. 2004). However, such reports are very preliminary and require more research and development for developing such products in commercial level.

5.5.10 In Fabric Washing

Biosurfactants may also have future applications as a substitute for chemical detergents due to their surface-active behaviour and antimicrobial potential. Only a few such literatures are available explaining such applications of biosurfactants in real time. For instance, biosurfactants from *Bacillus subtilis* SPB1 were reported for its application as laundry detergent, biosurfactants from *Pseudomonas aeruginosa* BioS as a remover for white board marker and *Bacillus subtilis* SPB1 BioS-mediated biosurfactant as a cleaning agent for oil and tea stains, etc. (Bouassida et al. 2018; Khaje Bafghi and Fazaelipoor 2012). It was also reported that sophorolipids produced by *Candida bombicola* (ATCC22214) showed its ability to remove stains equivalent to some of the commercially available detergents (Joshi-Navare et al. 2013; Jimoh and Lin 2019).

5.6 Conclusions

Biosurfactants are one of the most important microbial products with a large number of applications ranging from domestic, pharmaceutical, cosmetics and medical up to industrial level due to their easy and cost-effective production process. Besides, their environment-friendly nature gives it an additionally added value. A wide range of microorganisms are reported for their ability to produce biosurfactants with desired properties for further exploitation. Even though they have been reported for a wide range of medical applications such as anti-adhesive agents for coating prosthetic apparatus meant for internal use and drug delivery agents, they are very preliminary to be exploited commercially. Till now they may be most widely used for the purpose of MEOR and bioremediation studies as compared to their other possible applications. Hence extensive research is required in this field to explore its potential for commercial exploitation.

References

Adebusoye SA, Amund OO, Ilori MO, Domeih DO, Okpuzor J (2008) Growth and biosurfactant synthesis by Nigerian hydrocarbon-degrading estuarine bacteria. Int J Quot Trop Biol 56:1603–1611

Aparna A, Srinikethan G, Smitha H (2012) Production and characterization of biosurfactant produced by a novel *Pseudomonas sp.* 2B. Colloids Surf B Biointerfaces 95:23–29

Banat IM, Franzetti A, Gandolfi I, Bestetti G, Martinotti MG, Fracchia L, Smyth TJ, Marchant R (2010) Microorganism in environmental management: microbes and environment. Appl Microbiol Biotechnol 87:427–444

Beal R, Betts WB (2000) Role of rhamnolipid biosurfactants in the uptake and mineralization of hexadecane in *Pseudomonas aeruginosa*. J Appl Microbiol 89(1):158–168

Befkadu AA, Quanyuan CHEN (2018) Surfactant-enhanced soil washing for removal of petroleum hydrocarbons from contaminated soils: a review. Pedosphere 28(3):383–410

Behera BK, Prasad R (2020) Environmental technology and sustainability. Elsevier (ISBN: 9780128191033). https://www.elsevier.com/books/environmental-technology-and-sustainability/behera/978-0-12-819103-3

Bordoloi NK, Konwar BK (2008) Microbial surfactant-enhanced mineral oil recovery under laboratory conditions. Colloids Surf B Biointerfaces 63(1):73–82

Bouassida M, Fourati N, Ghazala I, Ellouze-Chaabouni S, Ghribi D (2018) Potential application of Bacillus subtilis SPB1 biosurfactants in laundry detergent formulations: compatibility study with detergent ingredients and washing performance. Eng Life Sci 18(1):70–77

Cameotra SS, Makkar RS (2010) Biosurfactant-enhanced bioremediation of hydrophobic pollutants. Pure Appl Chem 82(1):97–116

Cerniglia CE (1992) Biodegradation of polycyclic aromatic hydrocarbons. Int Biodegrad 3:351–368

Chan XY, Chang CY, Hong KW, Tee KK, Yin WF, Chan KG (2013) Insights of biosurfactant producing *Serratia marcescens* strain W2.3 isolated from diseased tilapia fish: a draft genome analysis. Gut Pathog 5(1):29

Chaprão MJ, Ferreira IN, Correa PF, Rufino RD, Luna JM, Silva EJ, Sarubbo LA (2015) Application of bacterial and yeast biosurfactants for enhanced removal and biodegradation of motor oil from contaminated sand. Electron J Biotechnol 18(6):471–479

Das P, Mukherjee S, Sen R (2008) Genetic regulations of the biosynthesis of microbial surfactants: an overview. Biotechnol Genet Eng 25(1):165–186

De S, Malik S, Ghosh A, Saha R, Saha B (2015) A review on natural surfactants. RSC Adv 5 (81):65757

Eswari JS, Dhagat S, Mishra P (2018) Biosurfactant assisted silver nanoparticle synthesis: a critical analysis of its drug design aspects. Adv Nat Sci Nanosci 9(4):045007

Fakruddin MD (2012) Biosurfactant: production and application. J Pet Environ Biotechnol 3(4):124

Farias CB, Ferreira Silva A, Diniz Rufino R, Moura Luna J, Gomes Souza JE, Sarubbo LA (2014) Synthesis of silver nanoparticles using a biosurfactant produced in low-cost medium as stabilizing agent. Electron J Biotechnol 17(3):122–125

Felix AKN, Martins JJ, Almeida JGL, Giro MEA, Cavalcante KF, Melo VMM, Pessoa ODL, Rocha MVP, Gonçalves LRB, de Santiago Aguiar RS (2019) Purification and characterization of a biosurfactant produced by *Bacillus subtilis* in cashew apple juice and its application in the remediation of oil-contaminated soil. Colloids Surf B Biointerfaces 175:256–263

Fooladi T, Moazami N, Abdeshahian P, Kadier A, Ghojavand H, Yusoff WMW, Hamid AA (2016) Characterization, production and optimization of lipopeptide biosurfactant by new strain *Bacillus pumilus* 2IR isolated from an Iranian oil field. J Petrol Sci Eng 145:510–519

Ghosh S, Ghosh A, Riyajuddin S, Sarkar S, Chowdhury AH, Ghosh K, Islam SM (2015) Silver nanoparticles architectured HMP as a recyclable catalyst for tetramic acid and propiolic acid synthesis through CO2 capture at atmospheric pressure. Chem Cat Chem 12(4):1055–1067

Gudina EJ, Rocha V, Teixeira JA, Rodrigues LR (2010) Antimicrobial and antiadhesive properties of a biosurfactant isolated from *Lactobacillus paracasei* sp. paracasei A20. Lett Appl Microbiol 50:419–424

Gudiña EJ, Rangarajan V, Sen R, Rodrigues LR (2013) Potential therapeutic applications of biosurfactants. Trends Pharmacol Sci 34(12):667–675

Hazra C, Kundu D, Chaudhari A, Jana T (2013) Biogenic synthesis, characterization, toxicity and photocatalysis of zinc sulfide nanoparticles using rhamnolipids from *Pseudomonas aeruginosa* BS01 as capping and stabilizing agent. J Chem Technol Biotechnol 88(6):1039–1048

Heinemann C, van Hylckama Vlieg JE, Janssen DB, Busscher HJ, van der Mei HC, Reid G (2000) Purification and characterization of a surface-binding protein from *Lactobacillus fermentum* RC-14 that inhibits adhesion of *Enterococcus faecalis* 1131. FEMS Microbiol Lett 190 (1):177–180

Hewald S, Josephs K, Bölker M (2005) Genetic analysis of biosurfactant production in *Ustilago maydis*. Appl Environ Microbiol 71(6):3033–3040

Hommel RK (1990) Formation and physiological role of biosurfactants produced by hydrocarbon utilizing microorganisms. Biodegradation 1:107–119

https://www.sigmaaldrich.com/technical-documents/articles/biology/glycobiology/lipopolysaccharides.html. Accessed 16 Feb 2020

Ibrahim ML, Ijah UJJ, Manga SB, Bilbis LS, Umar S (2013) Production and partial characterization of biosurfactant produced by crude oil degrading bacteria. Int Biodeter Biodegr 81:28–34

Jimoh AA, Lin J (2019) Biosurfactant: a new frontier for greener technology and environmental sustainability. Ecotox Environ Safe 184:109607

Joshi-Navare K, Khanvilkar P, Prabhune A (2013) Jatropha oil derived sophorolipids: production and characterization as laundry detergent additive. Biochem Res Int 2013:1–11

Khaje Bafghi M, Fazaelipoor MH (2012) Application of rhamnolipid in the formulation of a detergent. J Surfactant Deterg 15(6):679–684

Khan AHA, Tanveer S, Alia S, Anees M, Sultan A, Iqbal M, Yousaf S (2017) Role of nutrients in bacterial biosurfactant production and effect of biosurfactant production on petroleum hydrocarbon biodegradation. Ecol Eng 104:158–164

Khire JM (2010) Bacterial biosurfactants and their role in microbial enhanced oil recovery (MEOR). Adv Exp Med Biol 672:146–157

Kulkarni P, Chakraborty R, Chakraborty S (2019) Biosurfactant mediated synthesis of silver nanoparticles using *Lactobacillus brevis* (MTCC 4463) and their antimicrobial studies. Int J Pharma Sci Res 10(4):1753–1759

Liwarska-Bizukojc E, Galamon M, Bernat P (2018) Kinetics of biological removal of the selected micropollutants and their effect on activated sludge biomass. Water Air Soil Pollut 229(11):356

Luna J, Rufino R, Campos G, Sarubbo L (2012) Properties of the biosurfactant produced by *Candida sphaerica* cultivated in low-cost substrates. Chem Eng 27:67–72

Mireles JR, Toguchi A, Harshey RM (2001) *Salmonella enterica* serovar typhimurium swarming mutants with altered biofilm-forming abilities: surfactin inhibits biofilm formation. J Bacteriol 183(20):5848–5854

Mulligan CN (2005) Environmental applications for biosurfactants. Environ Pollut 133(2):183–198

Pacwa-Płociniczak M, Płaza GA, Piotrowska-Seget Z, Cameotra SS (2011) Environmental applications of biosurfactants: recent advances. Int J Mol Sci 12(1):633–654

Płaza GA, Chojniak J, Banat IM (2014) Biosurfactant mediated biosynthesis of selected metallic nanoparticles. Int J Mol Sci 15(8):13720–13737

Prasad R (2014) Synthesis of silver nanoparticles in photosynthetic plants. J Nanoparticles. Article ID 963961. https://doi.org/10.1155/2014/963961

Prasad R, Jha A, Prasad K (2018) Exploring the realms of nature for nanosynthesis. Springer International Publishing (ISBN: 978-3-319-99570-0). https://www.springer.com/978-3-319-99570-0

Rane AN, Baikar VV, Kumar VR, Deopurkar RL (2017) Corrigendum: agro-industrial wastes for production of biosurfactant by *Bacillus subtilis* ANR 88 and its application in synthesis of silver and gold nanoparticles. Front Microbiol 8:878

Reddy AS, Chen CY, Chen CC, Jean JS, Fan CW, Chen HR, Wang JC, Nimje VR (2009) Synthesis of gold nanoparticles via an environmentally benign route using a biosurfactant. J Nanosci Nanotechnol 9(11):6693–6699

Rodrigues LR, Teixeira JA, Oliveira R (2006) Low-cost fermentative medium for biosurfactant production by probiotic bacteria. Biochem Eng J 32:135–142

Ron EZ, Rosenberg E (2002) Biosurfactants and oil bioremediation. Curr Opin Biotechnol 13 (3):249–252

Salusjärvi T, Povelainen M, Hvorslev N, Eneyskaya EV, Kulminskaya AA, Shabalin KA, Neustroev KN, Kalkkinen N, Miasnikov AN (2004) Cloning of a gluconate/polyol dehydrogenase gene from Gluconobacter suboxydans IFO 12528, characterisation of the enzyme and its use for the production of 5-ketogluconate in a recombinant *Escherichia coli* strain. Appl Microbiol Biotechnol 65(3):306–314

Saravanakumari P, Mani K (2010) Structural characterization of a novel xylolipid biosurfactant from *Lactococcus lactis* and analysis of antibacterial activity against multi-drug resistant pathogens. Bioresour Technol 101(22):8851–8854

Sharma D (2016a) Applications of biosurfactants in food. In: Biosurfactants in food. Springer, Cham, pp 43–80

Sharma D (2016b) Classification and properties of biosurfactants. In: Biosurfactants in food. Springer, Cham, pp 21–42

Sharma D, Saharan BS, Kapil S (2016) Biosurfactants of lactic acid bacteria. Springer, Cham

Shekhar S, Sundaramanickam A, Balasubramanian T (2015) Biosurfactant producing microbes and their potential applications: a review. Crit Rev Environ Sci Technol 45(14):1522–1554

Singh BR, Dwivedi S, Al-Khedhairy AA, Musarrat J (2011) Synthesis of stable cadmium sulfide nanoparticles using surfactin produced by *Bacillus amyloliquefaciens* strain KSU-109. Colloids Surf B Biointerfaces 85(2):207–213

Singh BR, Singh BN, Khan W, Singh HB, Naqvi AH (2012) ROS-mediated apoptotic cell death in prostate cancer LNCaP cells induced by biosurfactant stabilized CdS quantum dots. Biomaterials 33(23):5753–5767

Srivastava S, Usmani Z, Atanasov AG, Singh VK, Singh NP, Abdel-Azeem AM, Prasad R, Gupta G, Sharma M, Bhargava A (2021) Biological nanofactories: using living forms for metal nanoparticle synthesis. Mini-Rev Med Chem 21:245–265. https://doi.org/10.2174/1389557520999201116163012

Stromberg LR, Heather MM, Harshini M (2017) Detection methods for lipopolysaccharides: past and present. In: Escherichia coli—recent advances on physiology, pathogenesis and biotechnological applications, pp 141–168

Sullivan ER (1998) Molecular genetics of biosurfactant production. Curr Opin Biotechnol 9 (3):263–269

Tomar RS, Banerjee S, Kaushik S, Singh S, Vidyarthi AS (2015) Microbial synthesis of gold nanoparticles by biosurfactant producing *Pseudomonas aeruginosa*. Int J Adv Lif Sci 8 (4):520–527

Tugrul T, Cansunar E (2005) Detecting surfactant-producing microorganisms by the drop-collapse test. World J Microbiol Biotechnol 21(6–7):851–853

Van Bogaert IN, Holvoet K, Roelants SL, Li B, Lin YC, Van de Peer Y, Soetaert W (2013) The biosynthetic gene cluster for sophorolipids: a biotechnological interesting biosurfactant produced by *Starmerella bombicola*. Mol Microbiol 88(3):501–509

Vaz DA, Gudina EJ, Alameda EJ, Teixeira JA, Rodrigues LR (2012) Performance of a biosurfactant produced by a *Bacillus subtilis* strain isolated from crude oil samples as compared to commercial chemical surfactants. Colloids Surf B Biointerfaces 89:167–174

Vecino X, Cruz JM, Moldes AB, Rodrigues LR (2017) Biosurfactants in cosmetic formulations: trends and challenges. Crit Rev Biotechnol 37(7):911–923

Velraeds MM, van der Mei HC, Reid G, Busscher HJ (1996) Physicochemical and biochemical characterization of biosurfactants released by *Lactobacillus* strains. Colloids Surf B Biointerfaces 8(1–2):51–61

Vijayakumar S, Saravanan V (2015) Biosurfactants-types, sources and applications. Res J Microbiol 10:181–192

Wang W, Cai B, Shao Z (2014) Oil degradation and biosurfactant production by the deep sea bacterium *Dietzia maris* As-13-3. Front Microbiol 5:711

Wei YH, Lai CC, Chang JS (2007) Using Taguchi experimental design methods to optimize trace element composition for enhanced surfactin production by *Bacillus subtilis* ATCC 21332. Process Biochem 42(1):40–45

Yakimov MM, Timmis KN, Wray V, Freddrickson HL (1995) Characterization of a new lipopeptide surfactant produced by thermotolerant and halotolerant subsurface *Bacillus licheniformis* BAS50. Appl Environ Microbiol 61:1706–1713

Youssef NH, Duncan KE, Nagle DP, Savage KN, Knapp RM, McInerney MJ (2004) Comparison of methods to detect biosurfactant production by diverse microorganisms. J Microbiol Methods 56: 339 – 347

Youssef N, Simpson DR, Duncan KE, McInerney MJ, Folmsbee M, Fincher T, Knapp RM (2007) *In situ* biosurfactant production by *Bacillus* strains injected into a limestone petroleum reservoir. Appl Environ Microbiol 73(4):1239–1247

Zhou W, Wang X, Chen C, Zhu L (2013) Enhanced soil washing of phenanthrene by a plant-derived natural biosurfactant, *Sapindus saponin*. Colloids Surf A Physicochem Eng Asp 425:122–128

Biodegradation of Hydrophobic Polycyclic Aromatic Hydrocarbons

6

Daniel Chikere Ali and Zhilong Wang

Abstract

Polycyclic aromatic hydrocarbons (PAHs) are hydrophobic, toxic, and carcinogenic compounds which comprise of high molecular weight (HMW) and low molecular weight (LMW) compounds and are classified based on different aromatic rings present. PAHs have been reported on its negative implications by the US Center for Children's Environmental Health (CCEH) on consistent exposure to polluted environment among the pregnant woman such as premature child delivery and delayed in child development. In this review, biodegradation pathway of PAHs (naphthalene, fluoranthene, and pyrene) and its proceeding enzymes involved for effective degradation of PAHs are briefly discussed. However, the biodegradation efficiency is limited because the compounds are highly lipophilic and therefore very insoluble in an aqueous solution. The production of biosurfactants by microorganisms and its contribution to ongoing degradation of PAHs are properly discussed.

Keywords

Biodegradation · Polycyclic aromatic hydrocarbons · Hydrophobic compound · Biosurfactants · *Pseudomonas* species

6.1 Introduction

PAHs are heteroaromatic hydrocarbons with carbon and hydrocarbon atoms. PAHs can be substituted by sulfur, oxygen, and nitrogen (Kafilzadeh 2015; Abdel-Shafy and Mansour 2016). They are among the complex mixture of potentially ranging to

D. C. Ali · Z. Wang (✉)
State Key Laboratory of Microbial Metabolism, School of Pharmacy, Shanghai Jiao Tong University, Shanghai, China
e-mail: zlwang@sjtu.edu.cn

Inamuddin et al. (eds.), *Microbial Biosurfactants*, Environmental and Microbial Biotechnology, https://doi.org/10.1007/978-981-15-6607-3_6

hundreds of different chemicals including saturated, unsaturated, and aromatic groups (Wang et al. 2007). PAHs enventually accumulate in large quantity in the soil and may be released into the atmosphere following the anthropogenic activities, for example, combustion of fossil fuels (Bamforth and Singleton 2005; Dean-Ross et al. 2002), volcano eruptions, wild fires, and erosions of ancient sediment (Boada et al. 2015; Sousa et al. 2017). The release of these organic contaminants into freshwater remains a big health concern across the globe (Reizer et al. 2019; Wu et al. 2018), because they are toxic and carcinogenic compounds that occur in the environment. These compounds are chemically stable and poorly degradable. Once ingested, the compounds and their metabolites have the capacity to form DNA adduct and induce mutations which can cause cancer. The carcinogenic characteristics of PAH is a big health-related problem (Boada et al. 2015; Kafilzadeh 2015; Abramsson-Zetterberg and Maurer 2015; Seo et al. 2009).

Presently, fluorene, anthracene, acenaphthene, naphthalene, chrysene, benzo[b] fluoranthene, pyrene, benzo[k]fluoranthene, acenaphthylene, benzo[ghi]perylene, dibenzo[a,h]anthracene, indeno[1,2,3-cd]pyrene, benzo[a]pyrene, phenanthrene, benzo[a]anthracene, and benzo[k]fluoranthene (Wang et al. 2007) had been recognized among the 16 priority pollutants which are regarded as a threat to human life by the US Environmental Protection Agency (US EPA) (Zhuo et al. 2017; Seo et al. 2009; Arun et al. 2011). Benzo[α]pyrene as a member of PAHs is believed to undergo metabolism and therefore referred as carcinogenic, teratogenic, and mutagenic (Guo et al. 2019; Jelena et al. 2015; Lily et al. 2009). In this regard, the World Health Organization (WHO) has warned that continuous exposure to air pollutants associated with PAHs has caused about seven million deaths, constituting high environmental risk to human health across the globe (Sosa et al. 2017). According to WHO in 2006, it was estimated that environmental pollution (air) associated with PAHs represents 23% to 24% of the world's morbidity/mortality rate (Montaño-Soto et al. 2014).

PAH's chemical and physical features are basically varying with the number of rings and their molecular weight which show that its chemical reactivity, volatility, and aqueous solubility decrease with an increase in molecular weight and thus contribute to their distribution, transportation, and eco-biological effects (Seo et al. 2009). PAHs are made up of high molecular weight (HMW) and low molecular weight (LMW) compounds, and they are classified based on different aromatic rings present. PAHs that contain 4 to 6 aromatic rings are HMW, and their degradability is less with native microbes, whereas PAHs that contain 2–4 aromatic rings are called LMW and are less carcinogenic when compared to HMW (Nwaichi and Ntorgbo 2016). Those having five rings "benzo(α)anthracene, anthracene, and fluoranthene" involve consortia in the soil and are effectively degraded. The ability to metabolize two or more organic substrates at the same time with respect to their concentration within suitable substrate level and bioavailable carbons contributes to biodegradable efficiency (Toräng 2004). In this review, the toxic effect of PAHs, the PAH pathways, and the use of biosurfactants produced by bacteria for application in environmental degradations are exclusively reviewed.

6.2 Health Related to PAHs

6.2.1 Consequences of Consistent of PAH Exposure by Human

PAHs can easily be absorbed from the gastrointestinal tract of a mammal because they are highly lipid soluble with a devastating effect in bone marrow cells, e.g., non-Hodgkin lymphoma, leukemia, and multiple myeloma (Kafilzadeh 2015; Montaño-Soto et al. 2014). Therefore, consistent exposure to high levels of PAH pollutant conglomeration possesses several dangers to human health condition including the eye, vomiting, irritation, diarrhea, and nausea. Nowadays, an increased health risk of skin, bladder, gastrointestinal, and lung cancer among worker prior to exposure to PAHs has become a public health concern. Naphthalene is among the PAHs that causes skin irritants, whereas anthracene and benzo[α]pyrene causes allergy to the skin in both animal and human (Rand and Petrocelli 1985).

According to epidemiological evaluation, there is a clear relationship between PAHs associated with human lung and bladder cancer on exposure to the organic compound at the workplace. Meanwhile, the relationship between PAH and cancer is very crucial for determining occupational and environmental standards. Exposure to PAH at the workplace is considered to be predominantly causing lung or bladder cancer. The workplaces (sources) are rubber industries, through steel works and diesel exhaust (Guo et al. 2019; Armstrong et al. 2004). And occupational exposure in these types of industries encounters negative impact of PAH via inhalation and has been considered a huge health threat such as smoking cigarettes from open fireplaces because tobacco contains benzo[α]pyrene suspected as human carcinogen (Kafilzadeh 2015; Guo et al. 2019).

The US Center for Children's Environmental Health (CCEH) has warned that pregnant woman consistently exposed to environment polluted with PAH may likely experience complication/advert effect in birth such as premature delivery or delayed child development and low birth weight. Benzo[a]pyrene when gets in contact with a pregnant woman could cause congenital disorder. Prenatal exposure to PAH causes low intelligence quotient (IQ) during the age of three (3) and an increased behavioral problem at six (6) to eight (8) years of childhood (Kim et al. 2013).

6.2.2 Problems Associated with PAHs Via Cytochrome P450

Human population and its activities affect freshwater ecosystem, thereby posing health threats to the society. The health threats are due to the adverse impacts on aquatic organisms carried by human population or activities with some organic and trace compounds. Organisms living in PAH-contaminated environments have enormous potentials to alter their metabolism and their cellular components (Balcıoğlu 2016). The role of microsomal cytochrome system (cytochrome P450 and flavoprotein monooxygenases) has been established in invertebrates and fish which are familiar with mammalian systems. It further shown that cytochrome P450 is induced by chemical substrates in fish and mammals, though invertebrates possess lower rate

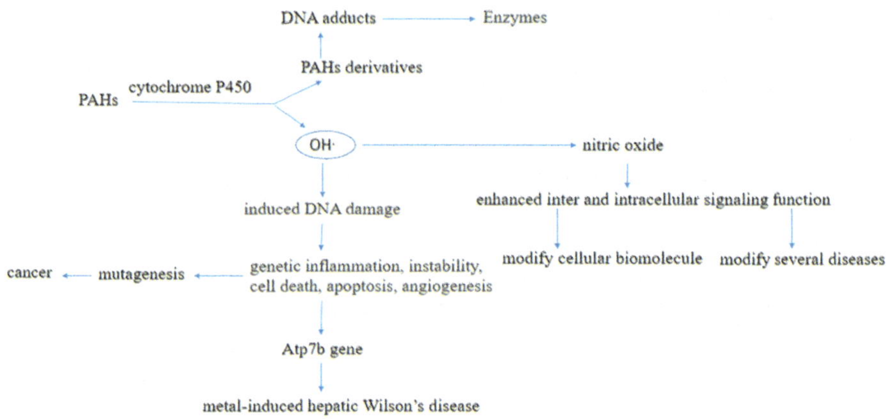

Fig. 6.1 Oxidatively induced DNA damage via cytochrome P450

of PAH metabolism than fish. Studies reveal that after treating hepatic microsomes of fish with aromatic hydrocarbon enhance the catalytic activities with selected substrates. The induction of cytochrome P450 in fish increases the disposition of hydrocarbon but also is capable of enhancing the formation of PAH's derivatives (Stegeman and Lech 1991; Gagnaire et al. 2010). The formation of stable DNA adducts by PAHs and their derivatives has been reported (Luch and Baird 2010), and this is because PAH derivatives are metabolized in the living organism leading to an oxidatively induced DNA damage (Michel and Vincent-Hubert 2015). Additionally, it is also reported that these oxygen species known as hydroxyl radical (OH·) may be activated via inflammation reactions enhancing nitric oxide production in the process (Cadet et al. 2012). The generation of nitric oxide through OH· enhances inter- and intracellular signaling function which in turn can modify cellular biomolecules and other accumulation associated with several diseases (Evans et al. 2004). The ability of invertebrates to metabolize PAHs via the activities of cytochrome P450-dependent oxidases varies. The reaction of OH· causes biological damage by inducing significant molecules such as DNA, protein, and lipids which can lead to genetic inflammation, instability, cell death, apoptosis, and angiogenesis. It can result to mutation and a development leading to cancer and even death, if not repaired (Park et al. 2005; Capó et al. 2015). Furthermore, the oxidatively induced DNA damage can undergo repair in living cell via different mechanisms involved in several DNA repair proteins (Dizdaroglu et al. 2015; Von 2006; Dizdaroglu and Jaruga 2012) (Fig. 6.1), though the DNA damage repair using deletion in the Atp7b gene present in some animals as a promising tool in studying the effect of oxidatively induced DNA damage in the pathogenesis of transition metal-induced hepatic Wilson's diseases has been established (Evans et al. 2004; Dizdaroglu et al. 2015; Wang et al. 2012; Loeb 2011; Friedberg Errol et al. 2006). DNA oxidatively generated damage enhances a process leading to cancer as a result of mutagenesis (Cadet et al. 2012; Dizdaroglu and Jaruga 2012).

6.3 Biodegradation of PAHs

6.3.1 Challenges of Limited Aqueous Solubility in Water

Most PAHs exist as hybrids enclosed with different structural components, e.g., benzo[α]pyrene (BαP). When there is an increase in the size of PAH's molecule, it will enhance and increase PAH's hydrophobicity and electrochemical stability. Thus, hydrophobicity and molecule's stability of PAHs are the two primary factors that determine whether HMW PAHs are capable of persisting in the environment (Kanaly and Harayama 2000). Some basic characters of PAHs with HMW and LMW are shown in Table 6.1. PAHs are found in a wide number of range of molecular weight especially in vegetable oil which most of them are alkylated compounds (Hossain and Salehuddin 2012). The solubility of PAHs in water decreases with an increase in molecular weight which enables PAHs to settle out of the water and accumulate in the lowest sediment. This implies that the PAH concentration is high in the sediments of the polluted environment such as aquatic organisms (Kafilzadeh 2015). The high molecular weight and low water solubility of PAH have an impact in lowering the bioavailability, and this causes resistance to microbial degradation (Bhattacharya et al. 2014). The bioavailability of a compound determines the rate of mass transfer and soil biota intrinsic activities of a particle compound such as PAHs (Snežana 2013). In addition, the high resonance energies of HMW PAHs make them recalcitrant to degradation because of the dense cloud of pi-electrons surrounding the aromatic rings (Ukiwe et al. 2013). LMW PAHs, such

Table 6.1 Characters of PAHs with HMW and LMW

No.	HMW PAH	LMW PAH	Reference
1	It contains 4 to 6 aromatic rings	It contains 2–4 aromatic ring	Abdel-Shafy and Mansour (2016), Soberón-Chávez and Maier (2011), Uzoigwe et al. (1999) and Rosenberg and Ron (1999)
2	It originated from pyrolytic PAHs	Occurs from petrogenic PAH	Kafilzadeh (2015)
3	Degradation of benzo[b] fluoranthene and benzo [a]pyrene is resistant to bacteria	Phenanthrene, naphthalene, and fluoranthene are degraded by the individual strains	Daugulis and McCracken (2003) and Cui et al. (2011)
4	They are more carcinogenic	They are less carcinogenic	Nwaichi and Ntorgbo (2016), Soberón-Chávez and Maier (2011), Uzoigwe et al. (1999) and Rosenberg and Ron (1999)
5	Mineralization is higher in HMW PAH-degrading bacteria	Mineralization is lower in LMW PAH-degrading bacteria	Raquel et al. (2013)
6	They are less susceptible to biodegradation	They are more susceptible to biodegradation	Soberón-Chávez and Maier (2011)

as phenanthrene, naphthalene, and fluoranthene biodegradability decrease, are determined by the number of rings (Daugulis and McCracken 2003). Generally, LMW is more volatile and soluble in water, and they are widely found in all environment, and it possibly helps to detect PAH-contaminated environment (Ghosal et al. 2016a). High melting and boiling point, low vapor pressure, very low aqueous solubility, and resistance to oxidation and reduction are the crucial features of HMW PAHs (Boada et al. 2015). Benzene is an example of PAHs with high vapor pressure, but the viability of vapor pressure in different PAH compound causes the distribution of different concentration in the vapor by individual PAHs (Kafilzadeh 2015). There is no doubt that when the molecular weight of PAHs is high, it may likely absorb more to the soil organic matter (Ukiwe et al. 2013).

Interestingly, the effectiveness of PAH bioremediation is limited due to their failure to effectively remove HMW PAHs and PAH resistance to microbial degradation as a result of its hydrophobic features (Potin et al. 2004). Several evidence has emerged that the microbial consortium is more effective in the degradation of PAHs with 4 to 5 rings occurring at faster rate (Daugulis and McCracken 2003), whereas PAHs that contain 2–3 rings are degraded at slow rate under anaerobically (Vaidya et al. 2017). PAHs are stable when they are absorbed into sediments because their non-pore structures can limit its dissolution in water. LMW PAHs are not completely insoluble, although small amounts of PAHs are capable of dissolving, and they become induced in the pore water. An increase in PAH concentration above aqueous solubility is enhanced by the presence of pore water organic colloids. This is because PAHs will be water-ice onto the organic colloids and can be easily transported via pore spaces of the sediment, thereby encouraging an increase in mobility and bioavailability of PAHs in the sediments (Kafilzadeh 2015). An increase in molecular weight enhances an increase in PAH's carcinogenicity and at same time acute toxicity reduction (Kim et al. 2013). Additionally, the proportion of LMW PAHs to HMW determines PAH's origin (Nwaichi and Ntorgbo 2016).

6.3.2 Biodegradation Pathway of PAHs

Biodegradation has long been applied to address contaminated environment with PAHs because of its abilities to treat different types of pollutant, low cost of operation, and no secondary by-product (toxic product) (Vaidya et al. 2018; Ghosal et al. 2016b). Bacteria are capable of using many toxic hydrocarbons in pollutant environment as a potential substrate for their removal (Heipieper et al. 2010). Degradation can be induced when the mechanisms are able to facilitate the proper use of waste disposal strategies (Wu et al. 2018; Singh and Sharma 2008). Normally, the nature of microbe and chemical structure of the chemical compounds are the two basic factors which determine the extent PAHs can be degraded (Haristash and Kaushik 2009).

6.3.2.1 Naphthalene

Naphthalene is a member of PAH which occurs when two aromatic rings can share two carbon atoms (Tomás-Gallardo et al. 2014) and remain the simplest PAHs which are generally used as a vibrant instrument to study the enzymatic aromatic degradation pathways (Selifonov et al. 1996; Jerina et al. 1976; Garrido-Sanz et al. 2019). Albeit naphthalene possesses a relatively low aqueous solubility (32 mg/L), it is very hazardous. Water has become the main target in the environmental contamination issues because 15% of contaminants such as naphthalene is discharged into the environments (water) where most aquatic animal dwell thereby posing health challenges.

The ration of naphthalene concentration in wastewater from the radioisotope manufacturing facilities is 1.65 mg/L, 6–220 ng/L in municipal wastewater, dyeing and textile wastewater 0.1–2.1 mg/L, and naphthalene sulfonic acid 285 mg/L in effluents (Karimi et al. 2015). Naphthalene degradation via metabolic diversities by *Mycobacterium* sp., *Pseudomonas putida*, *Rhodococcus opacus*, *Bacillus pumilus*, and *Nocardia otitidiscaviarum* has been established by different researchers. *Pseudomonas aeruginosa* is the most studied isolate from soil, plants, and water (Karimi et al. 2015; Akhtar and Husain 2006). They are gram-negative bacteria which are capable of causing infections such as cystic fibrosis in human (Soberón-Chávez et al. 2005). The bacteria are known with its ability to utilize hydrocarbons as sources of carbon and energy. The ability of the *Pseudomonas* sp. to produce biosurfactant enhances the effective uptake of hydrophobic compounds (Calfee et al. 2005) to degrade naphthalene which basically depends on difference in temperature range. The study revealed that at pH 8, *Pseudomonas* sp. degraded 96% of naphthalene, whereas at pH 7, 90% of naphthalene was degraded after 3 days (Karimi et al. 2015).

The degradation of naphthalene by bacteria is effective via catabolic enzymes encoded by the plasmid *Pseudomonas* sp. (Seo et al. 2009). The production of 1,2-dihydroxynaphthalenes is produced through the dehalogenation of 1,2-dihydro-1,2-dihydroxynaphthalenes by *Escherichia coli* recombinant strain. The *Escherichia coli* recombinant strain consists of dihydrodiol naphthalene dehydrogenase gene cloned from *Pseudomonas fluorescens* N_3 (Cavallotti et al. 1999). The production of 1,2-dihydroxynaphthalene by *pseudomonads* is known as an intermediate in the metabolism of naphthalene which is oxidized by oxygen in a reaction catalyzed by 1,2-dihydroxynaphthalene oxygenase. 1,2-Dihydroxynaphthalene have been differentiated from catechol-2,3-dioxygenase (C230) by virtue of the greater stability at 50 °C and the differences in control of induction (Patel and Barnsley 1980). Naphthalene metabolic pathway via 1,2-dihydroxynaphthalene is cleaved by a dioxygenase to an unstable ring cleavage product by *Pseudomonas species* and can degrade naphthalene (Eaton and Chapman 1992).

Normally, naphthalene is degraded by the hydroxylation of phenanthrene, anthracene, and fluorene and monooxygenation of acenaphthene (Resnick 1996; Ferraro et al. 2017). The degradation pathway of naphthalene remains the base for cell-to-cell communication via specific regulatory system, enzymes and transporters (electron) (Díaz et al. 2012). Subsequently, the oxidation of naphthalene and other PAHs reactions is further classified as central aromatic degradation pathways popularly

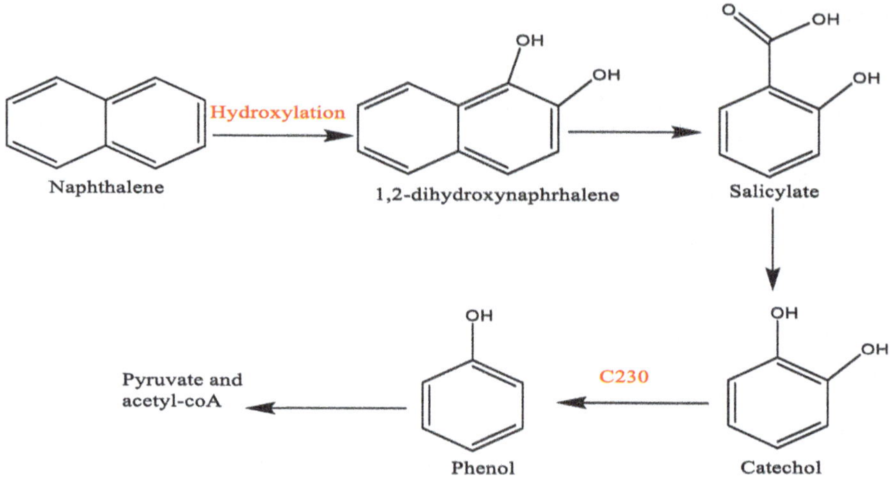

Fig. 6.2 The degradation pathway of naphthalene

called catechol and as an intermediate depending on the specific PAHs (Selifonov et al. 1996), before ring cleavage occurs (Jõesaar et al. 2017).

The microbial (*Pseudomonas*) degradation of naphthalene (Yen et al. 1988) via the metabolic steps usually proceeded via hydroxylation of one of the aromatic rings to produce 1,2-dihydroxynaphthalene. 1,2-digydroxynaphtalene later undergo further reaction to metabolized to salicylic acid, and further metabolized through catechol cleaved either meta- or ortho-rings shown in Fig. 6.2 (Phale et al. 2019). The resulting enzyme 1,2-dihydroxynaphthalene is regarded as a toxic organic solvent with aromatic and alicyclic ring that shares two carbon atoms (Tomás-Gallardo et al. 2009). Naphthalene degradation with bacteria *nah gene* is arranged in two operons in plasmid pNAH7, where one operon codes for enzymes and can covert naphthalene to salicylate (Tomás-Gallardo et al. 2014; Phale et al. 2019), based on the possibility of phenol and naphthalene to induce salicylate by *Pseudomonas* sp. which further converted to C230. This showed that C230 expression with phenol- and naphthalene-induced salicylate by *Pseudomonas*. Therefore, these enzymes involved in the process can convert salicylate over catechol to pyruvate and acetyl-coA (Jõesaar et al. 2017; Eaton and Chapman 1992).

6.3.2.2 Pyrene

Pyrene is a member of HMW PAHs and remains one of the compounds with simplest four-fused benzene ring. They are among the most abundant PAHs in the environment that occurred as a result of pyrolytic processes (Kafilzadeh 2015). Pyrene consists of high water solubility of 0.4 mg per ring structure (benzo[a] pyrene) with 1.7×10^{-3} mg/L of water solubility (Husain 2008a). Pyrene-contaminated environment remains a public health concern, and its toxicity to microinvertebrate *Gammarus pulex* capable of transmitting into quinone metabolites termed as the agent of mutagenic and toxicity to organisms in their respective habitant (Kim et al. 2007; Ma et al. 2013). The microbial degradation of pyrene with more than three rings encounter challenges (Ghosh et al. 2014), though bacteria remain one of the fastest degrading pyrene mechanisms (Shusheng et al. 2014) to effectively study the biodegradation of HMW PAHs because it has similar structure with other carcinogenic PAHs (Kim et al. 2007).

The microbial degradation of pyrene using metabolized substrate as sole source of carbon and energy for the degradation of pyrene has been established (Juhasz et al. 1997). The isolation of *Pseudomonas* sp. for the degradation of pyrene has been successfully done by several researchers, but the effectiveness of its degradation of pyrene remains a challenge because it is not a very effective degrader of pyrene. Most work currently relied on using the six consortia such as *Pseudomonas* and *Burkholderia* which can degrade various PAHs (Vaidya et al. 2017). The utilization of pyrene as sole carbon source by microbes depends on their ability to produce its metabolite (Xing-Fang et al. 1996). Pseudomonas is the promising bacteria genus for the catabolism of aromatic compounds (Nogales et al. 2017).

The enzyme degradation pathway of pyrene was determined using an environmental microbial isolate identified as *Pseudomonas* sp. grown in mineral medium containing pyrene (14.7 mg/L/day) and was able to degrade 82.38% of the pyrene in the medium used (Husain 2008b). This is because PAH degradation bacteria via the action of intracellular dioxygenase enable PAHs to be taken up by the cells and therefore enhance effective degradation (Obayori et al. 2008).

Pyrene degradation proceeds with oxidization via the monooxygenases and dioxygenases, thereby encouraging the cleavage of the oxidized ring. The pyrene can also be oxidized or mineralized by various types of microbes through oxygenases into the carbons on the PAHs, enhancing the C-C covalent bonds to cleave, and hydroxyl- and carboxyl-substituted moieties will be produced (Husain 2008a; Priyangshu et al. 2004; Mishra and Singh 2014). The metabolite produced by *Pseudomonas aeruginosa* will be converted to dihydroxypyrene, thereby causing the initial ring oxidation or cleavage at C-4, C-5 (K -region), C-1, and C-2 positions, and this can form pyrene 4,5-dihydrodiol via ortho-cleavage due to ring fission leading to the production of *cis*-3,4-phenanthrene dihydrodiol-4-carboxyclic acid, 4-phenanthroic acid, and phenanthrene (Ghosh et al. 2014; Obayori et al. 2008).

The dioxygenase present is relatively high since the dihydroxylation is a nonspecific step for metabolic pathways and the dioxygenase involved may be cleaved via ortho or meta of the aromatic nucleus to produce catechol 1,2-dioxygenase as illustrated in Fig. 6.3 (Cenci et al. 1999; Sugimoto et al. 1999),. The structural gene pcaH of protocatechuate 3,4-dioxygenase was considered having the ability to dissimilate aromatic growth substrate via the β-ketoadipate pathway (Gerischer et al. 1995; Yamanashi et al. 2015). The decarboxylation of 4-phenanthroic acid can produce monoaromatic and phthalic acid via cleavage. Then, the phthalic acid was further converted to an intermediate called pyruvate which enters the TCA cycle, and once the pyruvate enters the TCA, it will be converted to carbon dioxide and water (Obayori et al. 2008).

6.3.2.3 Fluoranthene

Fluoranthene is a very toxic PAH mainly found in many factories especially wood preservation plants (Herwijnen et al. 2003). Fluoranthene is made up of naphthalene

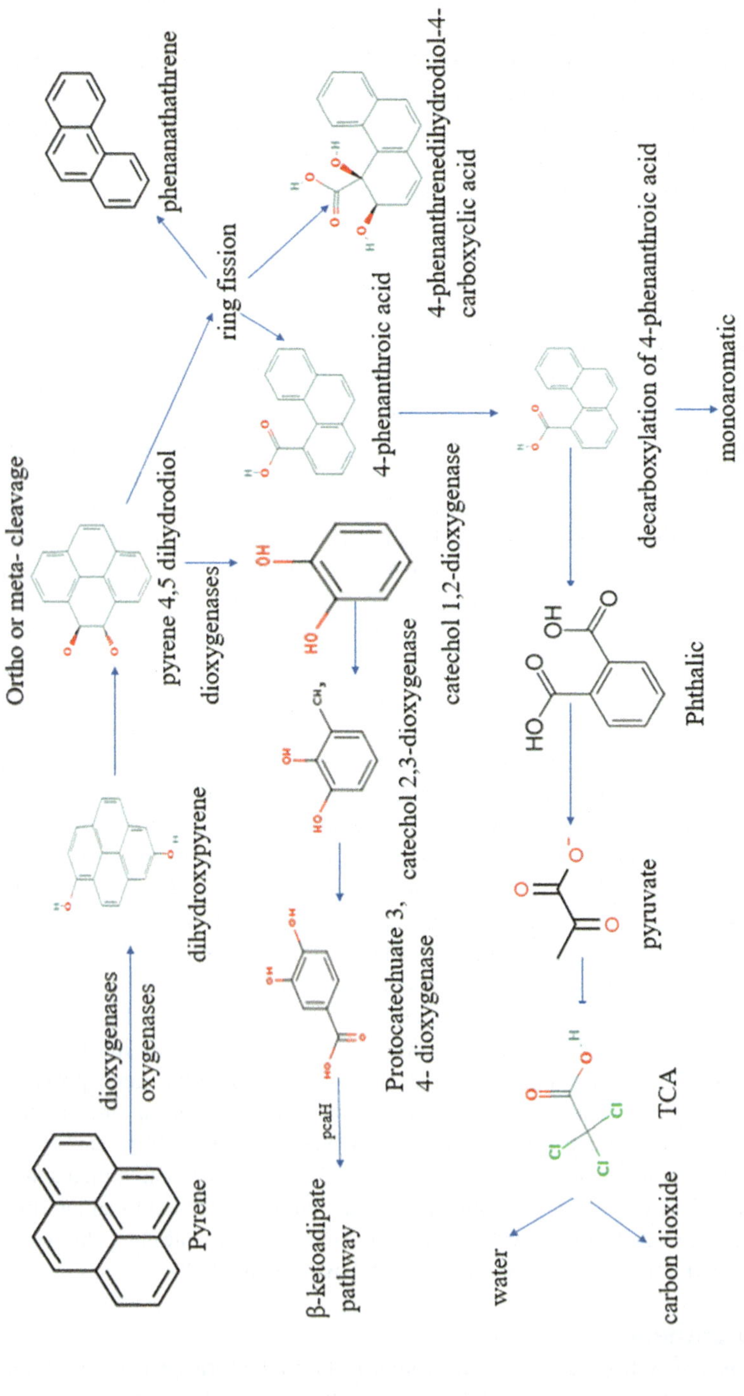

Fig. 6.3 The degradation pathway of pyrene

benzene rings usually condensed with a five-member ring, widespread in the environment, and was referred as genotoxic, mutagenic, and carcinogenic (Samanta et al. 2002) especially benzo[α]pyrene. Report showed that consistent fluoranthene exposure by laboratory animal negatively affects human such as decreased body weight, decreased blood chemistry tests, and increased liver weight. Fluoranthene can cause neurobehavioral toxicity, lung airway anion transport defects, and suppression of the immune system (Saunders 2003). They occur as a result of incomplete combustion of fossil fuels or as a result of pyrolysis of organic material with high temperature. Human exposure through inhalation of particulate due to tobacco smoking, air, or ingestion from water or food contamination remains a health threat. They are commonly identified in complex mixtures of PAHs in soil surface, water, and sediments (Yu 2002; Bisht et al. 2015).

The metabolic pathways by isolated strains of bacteria's ability to metabolize individual PAHs have been fully established by various researchers (Dean-Ross et al. 2002). This is because the isolated bacteria strain can utilize fluoranthene as carbon and energy source as was first described by Weissenfels (Weissenfels et al. 1990), whereas fluoranthene-degrading pathway was discovered in *Pseudomonas* sp. by Grifoll (Rehmann et al. 2001; Tersagh et al. 2016). *Pseudomonas aeruginosa* on the other hand can cause different types of diseases which may be harmful to human. Currently, *P. aeruginosa* has become promising bacteria used in degradation of PAH because it can easily decompose hydrocarbons and lives in oil field (Patel et al. 2014; Yan and Wu 2017).

Bacteria produces enzymes for degradation of PAH compound reported to possess a broad substrate range which is one of the desirable features of bacteria degrading PAHs (Juhasz et al. 1997). The microbial study of removal of PAH-contaminated environment through the action of enzymes remains a promising reward to the current focus on biodegradation of various PAHs. Interestingly, the enzymes produced by participating bacteria are referred as catabolic enzymes involving different mechanisms with members of PAHs. This shows that the degradation of PAH is usually enhanced and initiated via hydroxylation, especially deoxygenation in which oxygenase can be catalyzed (Simarro et al. 2013; Kweon et al. 2011; Somtrakoon et al. 2008).

Furthermore, fluoranthene metabolism also initiates deoxygenation of the fluoranthene molecule (Rehmann et al. 2001), which further produces 1,2- and 2, 3-dioxydrofluoranthene. The production of an intermediate 9-fluorenone-1-carboxylic acid by 1,2- and 2, 3-dioxydrofluoranthene cleavage through *meta*- or *ortho*-pathway was established (Somtrakoon et al. 2008), in which upon decarboxylation yields 9-hydroxyfuorene (Reddy et al. 2018). Thus, the production of 9-fluorenone-1-carboxylic acid further undergoes angular deoxygenation leading to the production of benzene-1,2,3-tricarboxylic acid in Fig. 6.4 (Dean-Ross et al. 2002; López et al. 2006). The benzene-1,2,3-tricarboxylic acid produced also showed that the degradation of 9-fluorenone1-carboxylic acid occurred via angular deoxygenation. Further decarboxylation to phthalate and degradation of phthalate by benzene-1,2,3-tricarboxylic enhance central metabolism (López et al. 2006).

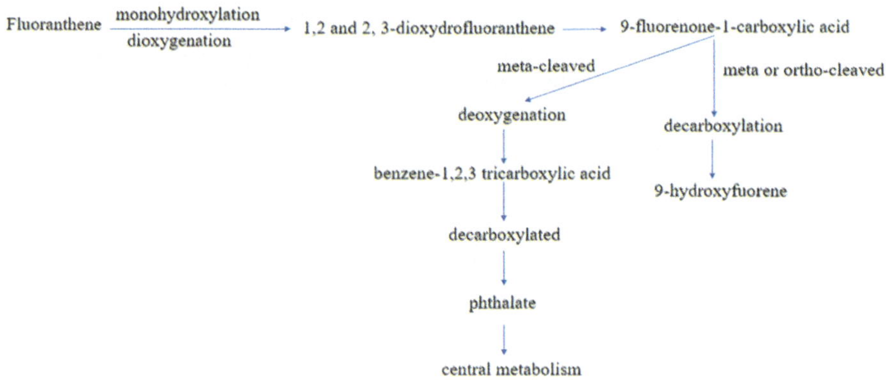

Fig. 6.4 The degradation pathway of fluoranthene

6.4 Biosurfactants

6.4.1 Biosurfactants

6.4.1.1 Glycolipid

Glycolipids are active compounds with the presence of carbohydrate moiety coupled to fatty acids. Glycolipid biosurfactant remains the most studied microbial surfactant and is the best known biosurfactant including mannosylerythritol lipids, sophorolipids, rhamnolipids, and trehalolipids. Glycolipids are also made up of mono- or disaccharides which are incorporated with long-chain aliphatic acids and sometimes hydroxy aliphatic acids as seen in Fig. 6.5 (Sen et al. 2017). Glycolipids are made up of carbohydrate moiety referred as microbial surface-active compounds (Mnif and Ghribi 2016; Irorere et al. 2017).

Glycolipid biosurfactants have a strong fungicidal activity (Morita et al. 2013), and they are used in cosmetic industries because of its amazing moisturizing and liquid-crystal-forming features (Yamamoto et al. 2012). They are well known by their ability to lower the surface and interfacial tension (Mnif and Ghribi 2016; Irorere et al. 2017). It is being classified as low molecular weight biosurfactant produced by many companies for commercial utilization (Irorere et al. 2017) and has been regarded as promising biosurfactant with respect to its adaptability such as low toxicity, biodegradability, and chemical stability (Paulino et al. 2016; Imura et al. 2014). They are very useful in their application in biodegradation, oil recovery, food, and pharmaceutical industries (Sen et al. 2017). Glycolipids of microbial origin are ranging from viruses to human cells such as antiparasitic, anticancer, and immunomodulatory activities and antimicrobial (Abdel-Mawgoud and Stephanopoulos 2017).

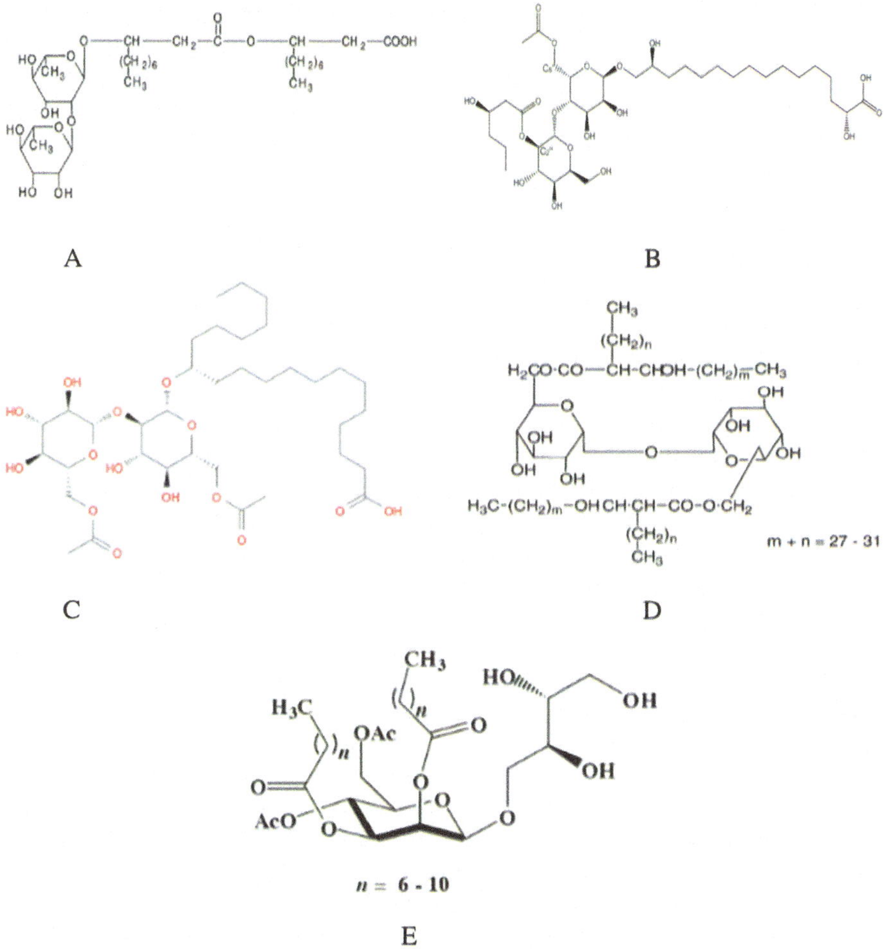

Fig. 6.5 The molecular structure of glycolipids. (**a**) rhamnolipids; (**b**) cellobiose lipids; (**c**) sophorolipids; (**d**) trehalolipids; (**e**) mannosylerythritol lipids

Rhamnolipids

Rhamnolipid is one of the most popular glycolipids (Mnif and Ghribi 2016) and is made up of one or two (*L*)-rhamnose molecules with a glycosidic linkage. One or two *L*-rhamnose molecules are hydrophobic group which contains one or two β-hydroxyl fatty acids. Microorganisms can produce different types of rhamnolipid congeners during fermentation. Rhamnolipid congeners vary in chain length and are different in numbers of rhamnose molecules and unsaturation for the fatty acid chain potential (Chong and Li 2017).

Rhamnolipid consists of rhamnose units which are composed of β-glycosidic bond that can assist rhamnose units to link to 3-hydroxyl fatty unit/units. *O*-Glycosidic bonds help rhamnose units to link to each other. Thus, an ester bond also allows

3-hydroxyl fatty acids to link to each other (Reddy et al. 2018; Twigg et al. 2018). Rhamnolipid is a viscous sticky oily yellowish-brown liquid with fruity odor produced by *Pseudomonas aeruginosa* (Irorere et al. 2017). The rhamnolipid production by different microbes mainly in area such as soil/water samples or industrial facilities (Irorere et al. 2017), though *P. aeruginosa* has been confirmed as the best-known representative of organisms that produce rhamnolipids (Kaczorek et al. 2018).

Rhamnolipids can be used for degradation of hydrocarbon in a site contaminated with the compound or in the petroleum industry (Irorere et al. 2017; Müller et al. 2012), and that is why they are good candidate for environmental application (Shao et al. 2017). Zhong investigated the effect of low concentration mono-rhamnolipid using *Pseudomonas aeruginosa* ATCC 9027 which was properly prepared and grown on glucose or hexadecane to glass beads. It comprises of hydrophobic or hydrophilic surface and was conducted via batch adsorption experiments. The result obtained reveals that there is a hydrophobic interaction on the bacteria cells during adsorption to the surfaces. This results to the reduction of bacterial adsorption with much implication on cell surface hydrophobicity (Zhong et al. 2015). It confirms that it enhances an effective degradation of hydrophobic compounds which begins with solubilization and alters cell affinity to hydrophobic compounds by microorganism (Zeng et al. 2018). Application of rhamnolipid biosurfactant such as degradation of hydrophobic compounds by means of solubilization, water treatment, and soil and waste treatment is referred as promising approach on environmental freedom (Catherine 2013), although there are challenges with rhamnolipids as they negatively play a role in the biodegradation of relatively volatile hydrocarbons such as *n*-alkanes with short chains (Chen et al. 2013). Rhamnolipid surfactants are capable of reducing from 72 to 28 mN/m of the surface tension of water and 43 to <1 mN/m of interfacial tension of water-oil (W/O) (Zhou et al. 2019). It was referred as promising target to the environmental application because it can enhance the pollutants uptake by increasing the bacterial membrane to targeted compounds/candidate that aids bioremediation processes (Kaczorek et al. 2018). It is also used for agricultural control of plant diseases and protecting stored products (Mnif and Ghribi 2016).

Cellobiose Lipids

Cellobiose lipids (CLs) are biosurfactants and a member of glycolipid which is made up of cellobiose moiety comprising hydrophilic parts and acetyl groups and fatty acids referred as hydrophobic part (Morita et al. 2013). CLs are produced extracellularly by yeast and mycelia fungi. The CLs with a cellobiose moiety can be used is fungicidal application, e.g., yeast which can help to preserve food from fungal attack (Morital et al. 2011). The cellobiose lipids produced by *Cryptococcus humicola have a high surface activity (*López et al. 2006*)*.

They are surface-active compounds owing to its ability to lower the tension of water solutions. Its specific function depends on pH and temperature stability making the cellobiose promising compounds in the application of agricultural use such as fumigation. And that is why the study of cellobiose lipids by biochemists and genetics and its ecological role are relevant (Morital et al. 2011; Trilisenko et al. 2012). The ability of microorganism to degrade cellobiose via cellulolytic biocatalysts helps in bioprocessing (Guo et al. 2015).

Sophorolipids

Sophorolipids (SLPs) are a family of glycolipid biosurfactant which are made up of disaccharide sophorose (2′-O-β-D-glucopyranosyl-β-glycoside) and have become the most promising glycolipid biosurfactant. SLPs have various characteristics that make them superior to synthetic surfactants such as temperature, salinity, and stability in the wide range of pH (Oliveira et al. 2015). *Candida bombicola* produces SLPs which have become the most studied SLPs producing yeast sophorolipids (Oliveira et al. 2015).

SLPs can also be used as biosurfactants instead of using classical chemistry-derived surfactants in food, petroleum, cleanings, and cosmetic industries. Reports showed that SLPs exhibit medical features including anticancer, anti-inflammatory, anti-HIV, and antiviral activities which are useful in medical application (Ivancic et al. 2018). Currently, SLPs are the first microbiological biosurfactants on the market (Müller et al. 2012). SLP biosurfactant can be applied for biofuel, drug delivery, detergent, and cleaners (Nguyen and Sabatini 2011). Sophorolipid biosurfactants are biodegradable and less toxic, and they have been approved by the FDA (Vasudevan and Prabhune 2018).

Trehalolipids

Trehalolipids occur naturally and consist of three isomers (α, α-; α, β-; and β, β-). It is also called α-D-glucopyranosyl-(1 → 1)-α-D-glucopyranoside (Reis et al. 2018). The structures of trehalose-containing glycolipids are made of hydrophobic moiety with long complex fatty acid (Kuyukina and Ivshina 2019). Trehalolipids are characterized with high surfactant activity. The chemical diversity of trehalolipids is vast which includes monomycolates, trehalose dimycolates, and trimycolates. Others are nonionic acylated and anionic trehalose tetraesters and succinoyl trehalolipids (White et al. 2013).

Trehalolipids are produced by various organisms, and the trehalolipids produced contain different number of atoms, size, structural pattern of mycolic acid, and degree of unsaturation (Desai and Banat 1997). *Rhodococcus* sp. is the most notable bacteria that produce trehalolipid biosurfactant using hydrophobic substrates (sunflower oil). They are excellent emulsifying compounds and can be used in the microbial application for oil recovery and degradation of oil-contaminated environment (Sen et al. 2017).

Mannosylerythritol Lipid

Mannosylerythritol lipids (MELs) are a glycolipid class of biosurfactants. MELs are active compound of glycolipid biosurfactant with excellent interfacial biochemical properties (Souayeh et al. 2014). Thus, MELs possess hydrophilic and hydrophobic parts (Morita 2013).

They are mainly produced by *Ustilago* and *Pseudozyma* on n-alkanes. For the fact that they are active surface compound, it can be applied in pharmaceutical, food, and cosmetic industries due to their excellent surface activities and other peculiar bioactivities (Yu et al. 2015). MELs are regarded as promising biosurfactants because they are environmentally friendly and have structural diversity, mild production versatile biochemical functions, and self-assembling properties with high yield (Souayeh et al. 2014; Niu et al. 2019).

6.4.1.2 Lipopeptides

Lipopeptide is a low molecular weight (LMW) biosurfactant derived from amino acids. Lipopeptide is obtained via the mixture of cyclic lipopeptides mainly from hydroxy fatty acid chains and heptapeptides (Gudiña et al. 2013). Lipopeptides can lower surface tension and are referred as an important bioactive compound (Baltz 2018; Kubicki et al. 2019). It is produced by various clades of microorganisms such as the bacterial genera *Lactobacillus*, *Bacillus*, *Pseudomonas*, *Streptomyces*, *and Serratia (*Kubicki et al. 2019*;* Park et al. 2019*).* Lipopeptide possesses antifungal activity (Toral et al. 2018). Generally, lipopeptides are promising microbial surfactants applied in the environment for oil recovery. *Bacillus subtilis* can produce cyclic lipopeptide surfactin which can lower 72 to 27.9 mN/m of the surface tension at 0.005% concentration (Arun et al. 2011; Souayeh et al. 2014).

6.4.1.3 Phospholipids

Phospholipids (PLs) are surfactants which play a vital function during cell growth in plant and animal (Willem et al. 2015). They are amphiphilic molecules with surface-active compounds made up of head called polar and lipophilic tail. Thus, amphiphilic features of PL deemed it fit to be used as emulsifier, solubilizer, and wetting agent (Jing et al. 2015). It can stablize emulsions due to its good emulsifying features, though phospholipids enhance the hydrophilic and hydrophobic properties due to their surface-active wetting features (Rui et al. 2013).

Bacteria and yeasts are organisms that can produce phospholipids mostly on n-alkanes when isolated, and they are very useful in the application of various industries such as food, pharmaceutical, and cosmetic industries (Vikbjerg et al. 2005). Meanwhile, they are also useful in the environmental application (Murínová and Dercová 2014).

6.4.2 Polymeric Biosurfactants

Polymeric surfactants are macromolecules with hydrophilic and hydrophobic parts in their structure. They are also called polymers with surfactant characteristics. Polymeric microbial surfactant is produced by microorganisms with a combination of many chemical types, and their chemical structures are exploited for commercial purposes (Bustamante et al. 2012). The polymeric surfactants are high molecular weight biosurfactants (Kapadia et al. 2013). They are environmentally friendly and low toxicity, and its biodegradability can be controlled during extreme condition such as pH, temperature, and salinity (Almeida et al. 2016). Polysaccharide protein consists of polysaccharide, generally produced from the surface coat of bacteria, linked to protein carriers. Examples of polysaccharide protein complex are lipopolysaccharides (Gautam and Tyagi 2006; Floris et al. 2018). Lipopolysaccharides (LPS) are the primary component of the outer membrane of Gram-negative bacteria; it is composed of high molecular weight component associated with phospholipid and protein. LPS are useful in environmental application (Zhang et al. 2016a: Saddler 1978).

Liposan, lipomannam, alasan, emulsan and mannoprotein are polysaccharide protein complexes which are the most studied polymeric biosurfactants. Reports reveal that emulsan consists of 80%(w/w) lipopolysaccharide and high molecular weight exopolysaccharide 20% (w/w). Lipopolysaccharide is made up of fatty acid, charge, and solution behavior, whereas exopolysaccharide is cationic in nature for the formation of the emulsan complex by electrostatic binding mechanism. The exopolysaccharide enhances emulsion stabilization (Pandey et al. 2015) and are produced by *Acinetobacter venetianus*. Emulsan can be used as emulsifier for hydrocarbons in water at 0.001% to 0.01% concentration. Emulsan possesses Newtonian flow features and therefore undergoes conformational changes at W/O interface. Thus, it can be applied in the bioremediation processes for oil removal. It can also be used in the preparation of cosmetics. Emulsan helps to prevent bacteria from adhering to buccal epithelial cells and has become a promising candidate for cosmetic application such as toothpaste production (Saddler 1978; Mercaldi et al. 2008). Liposan can acts as emulsifier in the extracellular water-soluble, and can be synthesized by *Candida lipolytica*. They are made up of carbohydrate (83%) and protein (17%) which can be employed to different industries such as food, cosmetics (Santos et al. 2016), and oil industries where it has many adverse processing conditions (de Cássia et al. 2014). Mannoproteins are produced by *Saccharomyces cerevisiae*; it is referred as having excellent emulsifier activity toward several oils, organic solvents, and alkane.

6.5 Enhanced Biodegradation of PAHs by Biosurfactant

The existence of a hydrophobic organic compound in the soil has an impact on the environmental-related problems (Cheng et al. 2018). PAHs are highly lipophilic and soluble in organic solvent (Kafilzadeh 2015), and the low aqueous solubility of PAHs decreases upon molecular weight increase enhancing accumulation in the bottom sediments because of its ability to settle out of the water (Kafilzadeh 2015). For the fact that PAHs are nonpolar in water, an increase in their molecular weight decreases its hydrophobicity. Normally, PAHs bind to particles in the soil and are absorbed. Therefore, the persistence of hydrophobicity of hydrocarbon in the contaminated site (soil) affect the degradation potentials of the participating microbes as it can lower water solubility thereby increasing their sorption to soil particle thereby limiting their biodegradability (Barkay et al. 1999). There is an overwhelming setback in biodegradation of PAHs because of their hydrophobicity coupled with low aqueous solubility of different PAHs (Kaczorek et al. 2018; Hou et al. 2018).

6.5.1 Biodegradation in Micelles

It is very necessary to increase the apparent solubility of hydrophobic hydrocarbon via addition of bioemulsifier alasan (surfactants) so that biodegradation will be enhanced which can transfer PAHs to water and PAHs in micelles or emulsion for microbial action (Kafilzadeh 2015; Barkay et al. 1999; Díaz et al. 2001). The hydrophobic and hydrophilic nature of surfactant possesses a different degree of hydrogen bonding and polarity (Olivera et al. 2008). Micelles could be initiated by biosurfactant and emulsifiers, thereby enhancing biodegradation of solid or liquid substances because they help to maintain direct cellular contact with several compounds (Al-Turki et al. 2009). The biosurfactant released from microorganisms is referred as detergent molecules. Detergent molecules like structure are composed of hydrophilic head and lipophilic tail and have the ability to form spherical micelles especially once the micelle concentration and compound are greater than surfactant concentration (Reis et al. 2013). The hydrophobic and hydrophilic moiety features can possibly lower surface tension and interfacial tension among molecules at the surface and interface (Gautam and Tyagi 2006; Karlapudi et al. 2018) and are easily biodegradable by microorganisms. The nature of their molecular weight determines its ability to reduce interfacial surface tension such as low molecular weight, whereas high molecular weights are stabilizing agents (Park et al. 2019; Santos et al. 2016). It can be applied in many industries including food, pharmaceutical, and cosmetic industries (Santos et al. 2016). They function with varying temperature, salinity, pH, greater selectivity range (Zhang et al. 2016b), biodegradability, environmental compatibility, and its ability to adopt to extreme temperature and low toxicity (Saddler et al. 1979).

The use of surfactant to increase the bioavailability of poor carbon such as hydrophobic compounds to allow free mass transfer to a contaminated soil so as to lower the interfacial tension thereby enhancing an increase in mass transfer of the contaminants (Norman et al. 2002; Szczepaniak et al. 2016). The surfactant-mediated biodegradation is a process involved when PAHs in the soil are partitioned into the hydrophobic core of surfactant micelle solubilization using surfactant at concentration above critical micelle concentration (CMC) values. Thus, the quantity and types of the surfactant, time, surfactant-soil interactions, and hydrophobicity of surfactant determine the efficiency of degradation. Time plays a significant role because the time at which hydrophobic compounds, e.g., PAHs, get in contact with the soil is very important during the process of degradation (Szczepaniak et al. 2016).

Enhancing the biodegradation of hydrocarbons using surfactant via solubilization aids hydrocarbon uptake by microorganisms. Biodegradation can be enhanced in micellar solution increasing the solubility and bioavailability of substrate to bacteria under the influence of surfactant which further increases the interfacial area.

Importantly, the presence of surfactant enables the bacteria to get in contact at the hydrocarbon-water interface. Also, the presence of surfactant can influence contact between cells' nonaqueous liquid phase and reduce diffusion path length between the site of adsorption and site of microorganism uptake for cell adsorption to hydrocarbon in soil particles. Thus, the effectiveness of the surfactant in enhancing biodegradation of PAHs is specific interactions dependent between bacteria and surfactant (Barkay et al. 1999). The biosurfactant-enhanced solubility of pollutants is a good strategy for a potential application in biodegradation.

For the fact that low water solubility affects the hydrophobicity degradation of PAHs, the addition of biosurfactant will enhance biodegradation, (Barkay et al. 1999) which can overcome the problems associated with PAH low aqueous solubility (Ukiwe et al. 2013). Application of biosurfactant in biodegradation enhances the bioavailability of hydrophobic compounds. Meanwhile, the microbial growth also increases the biosurfactant released into environment and also enhances bioavailability of pollutants (Araújo et al. 2019; Borah and Yadav 2017). Finally, the effect of degradation of hydrocarbon also depends on the growth factor and media components involved (Ibrahim 2016). The production of biosurfactant enhances the mechanism needed for biodegradation of water-immiscible pollutants (Schmid et al. 1998).

6.5.2 Biosurfactant Acting as Bioemulsifier

An aqueous-organic solvent two-phase system usually an organic water-immiscible solvent and an aqueous solution is used as a model for hydrophobic compound/ product synthesis (Oberbremer 1990). The transportation rate of the lipophilic substrates from the organic phase to the cells is relatively and may affect the growth in the systems. There are possible steps to improve this limitation by increasing the volume fraction of the dispersed organic phase and also increasing the organic-aqueous interfacial area using surfactants (Sifour et al. 2007). Biosurfactant is a surface-active compound and can lower the surface and interfacial tension between different phases such as liquid to liquid with a low CMC and can form stable emulsions (Satpute 2010).

Bioemulsifiers are surface-active compounds with profound characteristics such as biodegradability, foaming, biocompatibilities, nontoxicity, low concentrations, pH, salinity, and temperatures (Alizadeh-Sani et al. 2018; Satpute et al. 2010). It is a bioactive compound that currently draws much research interest because of its function and structural diversity in the degradation of hydrocarbons. High emulsification helps to facilitate the bioavailability of organic compound for faster hydrocarbon degradation by the participating bacteria. The ability of the marine organism to deliver bioemulsifiers can reduce overdependence on synthetic surfactant due to their influence in the use of environmental biodegradation surface-active molecule (Amodu et al. 2014).

Biosurfactant produced from *Pseudomonas aeruginosa* has the ability to emulsify two liquid phases. The two liquid phases can be hydrocarbon, hydrocarbon mixtures, and vegetable oil and capable of forming stable emulsions. The emulsification characteristics of rhamnolipid to hydrocarbons and vegetable oils occur simply via lowering surface tension of the culture media (Satpute 2010). The surface-active agent functions effectively by lowering the surface tension of the medium, thereby enhancing the formation of two phases of emulsions and thus enhancing the bioavailability of hydrophobic compounds (Suryanti et al. 2017). Biosurfactant such as phospholipid and glycolipid can emulsify the hydrocarbon substrate via synthetization, enhancing transport into the cell. For example, application of biosurfactant can lower the surface tension of water and benzene used in the study of the emulsification of biosurfactant by observing the emulsion stability which suggests that biosurfactants are good emulsifiers. It also established that biosurfactant emulsion is water in oil (W/O) emulsion (Uzoigwe et al. 2015). The stability of emulsion decreases when the temperature is high, and it will affect the features of oil, interfacial film, water, and solubility of surfactant in the oil and water phases (Suryanti et al. 2017).

Bioemulsifiers are well known for its ability to emulsify liquids, thereby having no reduction effect in surface/interfacial tension of their growth medium or between different phases (Suryanti et al. 2017). Meanwhile, reduction effect on surface tension reduction is very important because surface tension reduction should be less than 35 mN/m even though many reports have recorded biosurfactant containing high emulsification capacity of hydrophobic organic compounds where medium's surface tensions are above 35 mN/m. In addition, emulsion stability remains the basic tool used in the environmental application of biosurfactants. Importantly, pH, soil variation, salinity, and temperature are environmental factors that can lead to de-emulsification. De-emulsification occurs as a result of stimulation and ionizations of acid constituents of interfacial films (Satpute 2010; Suryanti et al. 2017).

6.6 Conclusions

PAHs are common causes of environmental pollution, and they have characteristic features such as mutagenic, carcinogenic, genotoxic, and toxic. Occupational exposures to benzo[α]pyrene through inhalation are dangerous to health. Thus, the persistent exposures to PAHs contributed to an increased rate of cancer and other related diseases to human. Generally, PAHs are classified into HMW and LMW, which are based on the different aromatic rings present. HMW PAHs are less degradable by native microbes, whereas LMW PAHs are less carcinogenic when compared to HMW PAHs. Utilization of hydrocarbon (fluoranthene, pyrene, and naphthalene) by microorganisms (*Pseudomonas* species) as their energy source via enzymatic pathway is effective. Biosurfactants such as phospholipid and glycolipid are apparently synthesized by the PAH-degraded microorganism to solubilize via micelles or on other hand emulsify the hydrocarbon substrate enhancing its transport into the cells. Meanwhile, the mass transport in emulsion is very essentials for the movement of oil molecules from emulsion droplets to the surrounding aqueous medium via surfactant.

Conflict of Interest All authors declare no conflict of interest.

References

Abdel-Mawgoud AM, Stephanopoulos G (2017) Simple glycolipids of microbes: chemistry, biological activity and metabolic engineering. Synth Syst Biotechnol 3(1):3–19. https://doi.org/10.1016/j.synbio.2017.12.001

Abdel-Shafy HI, Mansour MSM (2016) A review on polycyclic aromatic hydrocarbons: source, environmental impact, effect on human health and remediation. Egypt J Pet 25(1):107–123. https://doi.org/10.1016/j.ejar.2015.06.004

Abramsson-Zetterberg L, Maurer BM (2015) Fluoranthene and phenanthrene, two predominant PAHs in heat-prepared food, do not influence the frequency of micronucleated mouse erythrocytes induced by other PAHs. Toxicol Rep 2:1057–1063. https://doi.org/10.1016/j.toxrep.2015.07.016

Akhtar S, Husain Q (2006) Potential applications of immobilized bitter gourd (*Momordica charantia*) peroxidase in the removal of phenols from polluted water. Chemosphere 65 (7):1228–1235. https://doi.org/10.1016/j.chemosphere.2006.04.049

Alizadeh-Sani M et al (2018) Bioemulsifiers derived from microorganisms: applications in the drug and food industry. Adv Pharmaceut Bull 8(2):191–199. https://doi.org/10.15171/apb.2018.023

Almeida D et al (2016) Biosurfactants: promising molecules for petroleum biotechnology advances. Front Microbiol 7:1718. https://doi.org/10.3389/fmicb.2016.01718

Al-Turki AI et al (2009) Microbial polycyclic aromatic hydrocarbons degradation in soil. Res J Environ Toxicol 3:1–8. https://doi.org/10.3923/rjet.2009.1.8

Amodu OS, Ntwampe SK, Ojumu TV (2014) Emulsification of hydrocarbons by biosurfactant: exclusive use of agrowaste. Bioresources 9(2):3508–3525

Araújo HW et al (2019) Sustainable biosurfactant produced by Serratia marcescens UCP 1549 and its suitability for agricultural and marine bioremediation applications. Microb Cell Fact 18 (12):2. https://doi.org/10.1186/s12934-018-1046-0

Armstrong B et al (2004) Lung cancer risk after exposure to polycyclic aromatic hydrocarbons: a review and meta-analysis. Environ Health Perspect 112(9):970–978. https://doi.org/10.1289/ehp.6895

Arun K et al (2011) Crude oil PAH constitution, degradation pathway and associated bioremediation microflora: an overview. Int J Environ Sci 1:7. https://doi.org/10.6088/ijessi.00107020004

Balcıoğlu EB (2016) Potential effects of polycyclic aromatic hydrocarbons (PAHs) in marine foods on human health: a critical review. Toxin Rev 35(3–4):98–105. https://doi.org/10.1080/15569543.2016.1201513

Baltz R (2018) Synthetic biology, genome mining, and combinatorial biosynthesis of NRPS-derived antibiotics: a perspective. J Ind Microbiol Biotechnol 45(7):635–649. https://doi.org/10.1007/s10295-017-1999-8

Bamforth SM, Singleton L (2005) Bioremediation of polycyclic aromatic hydrocarbons: current knowledge and future directions. J Chem Technol Biotechnol 80(7):723–736. https://doi.org/10.1002/jctb.1276

Barkay T et al (1999) Enhancement of solubilization and biodegradation of polyaromatic hydrocarbons by the bioemulsifier alasan. Appl Environ Microbiol 65(6):2697–2702

Bhattacharya S et al (2014) Mycoremediation of benzo[a]pyrene by *pleurotus ostreatus* in the presence of heavy metals and mediators. 3 Biotech 4(2):205–211. https://doi.org/10.1007/s13205-013-0148-y

Bisht S et al (2015) Bioremediation of polyaromatic hydrocarbons (PAHs) using rhizosphere technology. Braz J Microbiol 46(1):7–21. https://doi.org/10.1590/S1517-838246120131354

Boada LD et al (2015) Exposure to polycyclic aromatic hydrocarbons (PAHs) and bladder cancer: evaluation from a gene-environment perspective in a hospital-based case-control study in the Canary Islands (Spain). Int J Occup Environ Health 21(1):23–30. https://doi.org/10.1179/2049396714Y.0000000085

Borah D, Yadav RNS (2017) Bioremediation of petroleum based contaminants with biosurfactant produced by a newly isolated petroleum oil degrading bacterial strain. Egypt J Pet 26(1):181. https://doi.org/10.1016/j.ejpe.2016.02.005

Bustamante M et al (2012) Biosurfactants are useful tools for the bioremediation of contaminated soil: a review. J Soil Sci Plant Nutr 12(4):667–687. dx.doi.org. https://doi.org/10.4067/S0718-95162012005000024

Cadet J et al (2012) Oxidatively generated complex DNA damage: tandem and clustered lesions. Cancer Lett 327(1–2):5–15. https://doi.org/10.1016/j.canlet.2012.04.005

Calfee MW et al (2005) Solubility and bioactivity of the *pseudomonas* quinolone signal are increased by a *pseudomonas aeruginosa*-produced surfactant. Infect Immun 73(2):878–882. https://doi.org/10.1128/IAI.73.2.878-882.2005

Capó X et al (2015) Oxidative status assessment of the endemic bivalve pinna nobilis affected by the oil spill from the sinking of the Don Pedro. Mar Environ Res 110:19–24. https://doi.org/10.1016/j.marenvres.2015.07.013

Catherine NM (2013) Chapter 15. Rhamnolipid biosurfactants: solubility and environmental issues. In: Thermodynamics, solubility and environmental issues, pp 279–298. https://doi.org/10.1016/B978-044452707-3/50017-3

Cavallotti C et al (1999) 1, 2-Dihydro-1, 2-dihydroxynaphthalene dehydrogenase containing recombinant strains: preparation, isolation and characterisation of 1, 2-dihydroxynaphthalenes and 1, 2-naphthoquinones. Tetrahedron 55:4467–4480. https://doi.org/10.1016/S0040-4020(99)00135-0

Cenci G et al (1999) Dioxygenase activity and relative behaviour of pseudomonas strains from soil in the presence of different aromatic compounds. World J Microbiol Biotechnol 15:41. https://doi.org/10.1023/A:1008868124715

Chen Q et al (2013) Rhamnolipids enhance marine oil spill bioremediation in laboratory system. Mar Pollut Bull 71(1–2):269–275. https://doi.org/10.1016/j.marpolbul.2013.01.037

Cheng M et al (2018) Tween 80 surfactant-enhanced bioremediation: toward a solution to the soil contamination by hydrophobic organic compounds. Crit Rev Biotechnol 38(1):17–30. https://doi.org/10.1080/07388551.2017.1311296

Chong H, Li Q (2017) Microbial production of rhamnolipids: opportunities, challenges and strategies. Microb Cell Fact 16(1):137. https://doi.org/10.1186/s12934-017-0753-2

Cui X et al (2011) Biodegradation of pyrene in sand, silt and clay fractions of sediment. Biodegradation 22:297–307. https://doi.org/10.1007/s10532-010-9399-z

Daugulis AJ, McCracken CM (2003) Microbial degradation of high and low molecular weight polyaromatic hydrocarbons in a two-phase partitioning bioreactor by two strains of *Sphingomonas* sp. Biotechnol Lett 25:1441. https://doi.org/10.1023/A:1025007729355

de Cássia FSSR et al (2014) Applications of biosurfactants in the petroleum industry and the remediation of oil spills. Int J Mol Sci 15(7):12523–12542. https://doi.org/10.3390/ijms150712523

Dean-Ross D et al (2002) Utilization of mixtures of polycyclic aromatic hydrocarbons by bacteria isolated from contaminated sediment. FEMS Microbiol Ecol 41:1–7. https://doi.org/10.1111/j.1574-6941.2002.tb00960.x

Desai JD, Banat IM (1997) Microbial production of surfactants and their commercial potential. Microbiol Mol Biol Rev 61(1):47–64

Díaz E et al (2001) Biodegradation of aromatic compounds by *escherichia coli*. Microbiol Mol Biol Rev 65(4):523–569. https://doi.org/10.1128/MMBR.65.4.523-569.2001

Díaz E et al (2012) Aerobic degradation of aromatic compounds. Curr Opin Biotechnol 24(3):431–442. https://doi.org/10.1016/j.copbio.2012.10.010

Dizdaroglu M, Jaruga P (2012) Mechanisms of free radical-induced damage to DNA. Free Radic Res 46(4):382–419. https://doi.org/10.3109/10715762.2011.653969

Dizdaroglu M et al (2015) Measurement of oxidatively induced DNA damage and its repair, by mass spectrometric techniques. Free Radic Res 49(5):525–548. https://doi.org/10.3109/10715762.2015.1014814

Eaton RW, Chapman PJ (1992) Bacterial metabolism of naphthalene: construction and use of recombinant bacteria to study ring cleavage of 1, 2-dihydroxynaphthalene and subsequent reactions. J Bacteriol 174(23):7542–7554. https://doi.org/10.1128/jb.174.23.7542-7554.1992

Evans MD et al (2004) Oxidative DNA damage and disease: induction, repair and significance. Mutat Res 567(1):1–61. https://doi.org/10.1016/j.mrrev.2003.11.001

Ferraro DJ et al (2017) One enzyme, many reactions: structural basis for the various reactions catalyzed by naphthalene 1, 2-dioxygenase. IUCrJ 4(5):648–656. https://doi.org/10.1107/S2052252517008223

Floris R et al (2018) Biosurfactants from marine microorganisms. In: Bacteriology, vol 27. IntechOpen, London, pp 1263–1272

Friedberg Errol C et al (2006) DNA repair and mutagenesis, vol 26, 2nd edn. ASM Press, Washington, DC, p 221. https://doi.org/10.1097/01.shk.0000232588.61871.ff

Gagnaire B et al (2010) In vivo indirect measurement of cytochrome P450-associated activities in freshwater gastropod molluscs. Environ Toxicol 25(6):545–553. https://doi.org/10.1002/tox.20515

Garrido-Sanz D et al (2019) Metagenomic insights into the bacterial functions of a diesel-degrading consortium for the rhizoremediation of diesel-polluted soil. Genes 10(6):456. https://doi.org/10.3390/genes10060456

Gautam KK, Tyagi VK (2006) Microbial surfactants. J Oleo Sci 55(4):155–166. https://doi.org/10.5650/jos.55.155

Gerischer U et al (1995) Spontaneous mutations in pcaH and -G, structural genes for protocatechuate 3,4-dioxygenase in acinetobacter calcoaceticus. J Bacteriol 177 (5):1336–1347. https://doi.org/10.1128/jb.177.5.1336-1347.1995

Ghosal D et al (2016a) Corrigendum: current state of knowledge in microbial degradation of polycyclic aromatic hydrocarbons (PAHs): a review. Front Microbiol 7:1837. https://doi.org/10.3389/fmicb.2016.01837

Ghosal D et al (2016b) Current state of knowledge in microbial degradation of polycyclic aromatic hydrocarbons (PAHs): a review. Front Microbiol 7:1369. https://doi.org/10.3389/fmicb.2016.01369

Ghosh I et al (2014) Biodegradation of pyrene by a *pseudomonas aeruginosa* strain RS1 isolated from refinery sludge. Bioresour Technol 166:548–558. https://doi.org/10.1016/j.biortech.2014.05.074

Gudiña EJ et al (2013) Potential therapeutic applications of biosurfactants. Trends Pharmacol Sci 34 (12):667–675. https://doi.org/10.1016/j.tips.2013.10.002

Guo Z et al (2015) Development of cellobiose-degrading ability in Yarrowia lipolytica strain by overexpression of endogenous genes. Biotechnol Biofuels 8:109. https://doi.org/10.1186/s13068-015-0289-9

Guo L et al (2019) Detection of BPDE-DNA adducts in human umbilical cord blood by LC-MS/MS analysis. J Food Drug Anal 27(2):518–525. https://doi.org/10.1016/j.jfda.2019.03.001

Haristash AK, Kaushik CP (2009) Biodegradation aspects of polycyclic aromatic hydrocarbons (PAHs): a review. Hazard Mater 169(1–3):1–15. https://doi.org/10.1016/j.jhazmat

Heipieper HJ et al (2010) Surface properties and cellular energetics of bacteria in response to the presence of hydrocarbons. In: Handbook of hydrocarbon and lipid microbiology. Springer, Berlin. https://doi.org/10.1007/978-3-540-77587-4-113

Herwijnen RV et al (2003) Influence of phenanthrene and fluoranthene on the degradation of fluorene and glucose by *sphingomonas sp.* strain LB126 in chemostat cultures. FEMS Microbiol Ecol 46(1):105–111. https://doi.org/10.1016/S0168-6496(03)00202-2

Hossain M, Salehuddin SM (2012) Polycyclic aromatic hydrocarbons (PAHs) in edible oils by gas chromatography coupled with mass spectroscopy. Arab J Chem 5:391–396. https://doi.org/10.1016/j.arabjc.2010.09.012

Hou N et al (2018) Biodegradation of phenanthrene by biodemulsifier-producing strain *achromobacter sp.* LH-1 and the study on its metabolisms and fermentation kinetics. Ecotoxicol Environ Saf 163:205–214. https://doi.org/10.1016/j.ecoenv.2018.07.064

Husain S (2008a) Effect of surfactants on pyrene degradation by *pseudomonas fluorescens* 29L. World J Microbiol Biotechnol 24(11):2411–2419. https://doi.org/10.1007/s11274-008-9756-9
Husain S (2008b) Identification of pyrene degradation pathways: bench-scale studies using *pseudomonas fluorescens* 29L. Remediat J 18(3):119–142
Ibrahim HM (2016) Biodegradation of used engine oil by novel strains of *Ochrobactrum anthropi* HM-1 and *citrobacter freundii* HM-2 isolated from oil-contaminated soil. 3 Biotech 6(2):226. https://doi.org/10.1007/s13205-016-0540-5
Imura T et al (2014) Aqueous gel formation from sodium salts of cellobiose lipids. J Oleo Sci 63 (10):1005–1010. https://doi.org/10.5650/jos.ess14104
Irorere UV et al (2017) Microbial rhamnolipid production: a critical re-evaluation of published data and suggested future publication criteria. Appl Microbiol Biotechnol 101(10):3941–3951. https://doi.org/10.1007/s00253-017-8262-0
Ivancic M et al (2018) Production of microbial lipids from lignocellulosic biomass. In: Advances in biofuels and bioenergy. IntechOpen, London, pp 137–164. https://doi.org/10.5772/intechopen.74013
Jelena M et al (2015) Initial microbial degradation of polycyclic aromatic hydrocarbons. Chem Ind Chem Eng Q 22(00):43–43. https://doi.org/10.2298/CICEQ150606043M
Jerina DM et al (1976) Dihydrodiols from anthracene and phenanthrene. J Am Chem Soc 98 (19):5988–5996. https://doi.org/10.1021/ja00435a035
Jing L et al (2015) A review on phospholipids and their main applications in drug delivery systems. Asian J Pharmaceut Sci 10(2):81–98. https://doi.org/10.1016/j.ajps.2014.09.004
Jõesaar M et al (2017) Strategy of *pseudomonas pseudoalcaligenes* C70 for effective degradation of phenol and salicylate. PLoS One 12(3):e0173180. https://doi.org/10.1371/journal.pone.0173180
Juhasz AL et al (1997) Degradation of benzo[*a*]pyrene, dibenz[*a,h*]anthracene and coronene by *burkholderia cepacia*. Water Sci Technol 36(10):45–51. https://doi.org/10.1016/S0273-1223(97)00641-0
Kaczorek E et al (2018) The impact of biosurfactants on microbial cell properties leading to hydrocarbon bioavailability increase. Colloids Interfaces 2(3):35. https://doi.org/10.3390/colloids2030035
Kafilzadeh F (2015) Distribution and sources of polycyclic aromatic hydrocarbons in water and sediments of the Soltan Abad River, Iran. Egypt J Aquat Res 41(3):227–231. https://doi.org/10.1016/j.ejar.2015.06.004
Kanaly RA, Harayama S (2000) Biodegradation of high-molecular-weight polycyclic aromatic hydrocarbons by bacteria. J Bacteriol 182(8):2059–2067. https://doi.org/10.1128/jb.182.8.2059-2067.2000
Kapadia SG et al (2013) Current trend and potential for microbial biosurfactants. Asian J Exp Biol Sci 4:1–8
Karimi B et al (2015) Biodegradation of naphthalene using Pseudomonas aeruginosa by up flow anoxic-aerobic continuous flow combined bioreactor. J Environ Health Sci Eng 13:26. https://doi.org/10.1186/s40201-015-0175-1
Karlapudi AP et al (2018) Role of biosurfactants in bioremediation of oil pollution—a review. Petroleum 4(3):241–249. https://doi.org/10.1016/j.petlm.2018.03.007
Kim S et al (2007) Complete and integrated pyrene degradation pathway in *mycobacterium vanbaalenii* PYR-1 based on systems biology. J Bacteriol 189(2):464–472
Kim KH et al (2013) A review of airborne polycyclic aromatic hydrocarbons (PAHs) and their human health effects. Environ Int 60:71–80. https://doi.org/10.1016/j.envint.2013.07.019
Kubicki S et al (2019) Marine biosurfactants: biosynthesis, structural diversity and biotechnological applications. Mar Drugs 17(7):408. https://doi.org/10.3390/md17070408
Kuyukina MS, Ivshina IB (2019) Production of trehalolipid biosurfactants by *rhodococcus*. In: Alvarez H (ed) Biology of *rhodococcus*. Microbiology monographs, vol 16. Springer, Cham
Kweon O et al (2011) Polycyclic aromatic hydrocarbon metabolic network in Mycobacterium vanbaalenii PYR-1. J Bacteriol 193(17):4326–4337. https://doi.org/10.1128/JB.00215-11

Lily MK et al (2009) Degradation of benzo [a] Pyrene by a novel strain *bacillus subtilis* BMT4i (MTCC 9447). Braz J Microbiol 40(4):884–892. https://doi.org/10.1590/S1517-838220090004000020
Loeb LA (2011) Human cancers express mutator phenotypes: origin, consequences and targeting. Nat Rev Cancer 11(6):450–457. https://doi.org/10.1038/nrc3063
López Z et al (2006) Metabolism of fluoranthene by mycobacterium sp. strain AP1. *Appl Microbiol Biotechnol* 70(6):747–756. https://doi.org/10.1007/s00253-005-0120-9
Luch A, Baird WM (2010) Carcinogenic polycyclic aromatic hydrocarbons. Comprehens Toxicol 38(2):85–123
Ma J et al (2013) Characterization of pyrene degradation by Pseudomonas sp. strain Jpyr-1 isolated from active sewage sludge. Bioresour Technol 140:15–21. https://doi.org/10.1016/j.biortech.2013.03.184
Mercaldi MP et al (2008) Discovery of the dual polysaccharide composition of emulsan and the isolation of the emulsion stabilizing component. Biomacromolecules 9(7):1988–1996
Michel C, Vincent-Hubert F (2015) DNA oxidation and DNA repair in gills of zebra mussels exposed to cadmium and benzo(a)pyrene. Ecotoxicology 24(9):2009–2016. https://doi.org/10.1007/s10646-015-1536-3
Mishra S, Singh SN (2014) Biodegradation of benzo(a)pyrene mediated by catabolic enzymes of bacteria. Int J Environ Sci Technol 11:1571–1580. https://doi.org/10.1007/s13762-013-0300-6
Mnif I, Ghribi D (2016) Glycolipid biosurfactants: main properties and potential applications in agriculture and food industry. *J Sci Food Agric* 96(13):4310–4320. https://doi.org/10.1002/jsfa.7759
Montaño-Soto MT et al (2014) Health effects due to exposure to polycyclic aromatic hydrocarbons from the petroleum refining industry. Med Univ 16(64):136–140
Morita T (2013) Production of mannosylerythritol lipids and their application in cosmetics. Appl Microbiol Biotechnol 97(11):4691–4700. https://doi.org/10.1007/s00253-013-4858-1
Morita T et al (2013) Accumulation of cellobiose lipids under nitrogen-limiting conditions by two *ustilaginomycetous* yeasts, *pseudozyma aphidis* and *pseudozyma hubeiensis*. FEMS Yeast Res 13(1):44–49. https://doi.org/10.1111/1567-1364.12005
Morital T et al (2011) Production of glycolipid biosurfactants, cellobiose lipids, by *cryptococcus humicola* JCM 1461 and their interfacial properties. Biosci Biotechnol Agrochem 75 (8):1597–1599
Müller MM et al (2012) Rhamnolipids—next generation surfactants. J Biotechnol 162(4):366–380. https://doi.org/10.1016/j.jbiotec.2012.05.022
Murínová S, Dercová K (2014) Response mechanisms of bacterial degraders to environmental contaminants on the level of cell walls and cytoplasmic membrane. Int J Microbiol 2014:873081. https://doi.org/10.1155/2014/873081
Nguyen TT, Sabatini DA (2011) Characterization and emulsification properties of rhamnolipid and sophorolipid biosurfactants and their applications. Int J Mol Sci 12(2):1232–1244. https://doi.org/10.3390/ijms12021232
Niu Y et al (2019) Production and characterization of a new glycolipid, mannosylerythritol lipid, from waste cooking oil biotransformation by *pseudozyma aphidis* ZJUDM34. Food Sci Nutr 7 (3):937–948. https://doi.org/10.1002/fsn3.880
Nogales J et al (2017) Degradation of aromatic compounds in *pseudomonas*: a systems biology view. In: Rojo F (ed) Aerobic utilization of hydrocarbons, oils and lipids. Handbook of hydrocarbon and lipid microbiology. Springer, Cham
Norman R et al (2002) Variability in *pseudomonas aeruginosa* lipopolysaccharide expression during crude oil degradation. Appl Environ Microbiol 68(10):5096–5103. https://doi.org/10.1128/aem.68.10.5096-5103.2002
Nwaichi EO, Ntorgbo SA (2016) Assessment of PAHs levels in some fish and seafood from different coastal waters in the Niger Delta. Toxicol Rep 3:167–172. https://doi.org/10.1016/j.toxrep.2016.01.005

Obayori OS et al (2008) Pyrene-degradation potentials of *pseudomonas* species isolated from polluted tropical soils. World J Microbiol Miotechnol 24:2639. https://doi.org/10.1007/s11274-008-9790-7

Oberbremer A (1990) Effect of the addition of microbial surfactants on hydrocarbon degradation in a soil population in a stirred reactor. Appl Microbiol Biotechnol 32(4):485–489. https://doi.org/10.1007/BF00903788

Oliveira MR et al (2015) Review: sophorolipids a promising biosurfactant and it's applications. Int J Adv Biotechnol Res 6(2):161–174

Olivera NL et al (2008) Isolation and characterization of biosurfactant-producing *alcanivorax* strains: hydrocarbon accession strategies and alkane hydroxylase gene analysis. Microbiol Res 160(1):19–26. https://doi.org/10.1016/j.resmic.2008.09.011

Pandey A et al (2015) Industrial biorefineries and white biotechnology, 1st edn. Elsevier, Amsterdam

Park JH et al (2005) Formation of 8-oxo-7,8-dihydro-2′-deoxyguanosine (8-oxo-dGuo) by PAH o-quinones: involvement of reactive oxygen species and copper(II)/copper(I) redox cycling. Chem Res Toxicol 18(6):1026–1037. https://doi.org/10.1021/tx050001a

Park T et al (2019) Modification of interfacial tension and wettability in oil–brine–quartz system by in situ bacterial biosurfactant production at reservoir conditions: implications for microbial enhanced oil recovery. Energy Fuel 33(6):4909–4920. https://doi.org/10.1021/acs.energyfuels.9b00545

Patel TR, Barnsley EA (1980) Naphthalene metabolism by pseudomonads: purification and properties of 1, 2-dihydroxynaphthalene oxygenase. J Bacteriol 143(2):668–673

Patel PA et al (2014) Draft genome sequence of petroleum hydrocarbon-degrading *pseudomonas aeruginosa* strain PK6, isolated from the saurashtra region of Gujarat, India. Genome Announc 2(1):e00002–e00014. https://doi.org/10.1128/genomeA.00002-14

Paulino BN et al (2016) Current status in biotechnological production and applications of glycolipid biosurfactants. Appl Microbiol Biotechnol 24:10265–10293. https://doi.org/10.1007/s00253-016-7980-z

Phale PS et al (2019) Variability in assembly of degradation operons for naphthalene and its derivative, carbaryl, suggests mobilization through horizontal gene transfer. Genes 10(8):569. https://doi.org/10.3390/genes10080569

Potin O et al (2004) Biodegradation of polycyclic aromatic hydrocarbons (PAHs) by *Cladosporium sphaerospermum* isolated from an aged PAH contaminated soil. FEMS Microbiol Ecol 51 (1):71–78. https://doi.org/10.1016/j.femsec.2004.07.013

Priyangshu MS et al (2004) Degradation of polycyclic aromatic hydrocarbons by a newly discovered enteric bacterium, Leclercia adecarboxylata. Appl Environ Microbiol 70(5):3163–3166. https://doi.org/10.1128/aem.70.5.3163-3166.2004

Rand GM, Petrocelli SR (1985) Fundamentals of aquatic toxicology: methods and applications, vol 37. Hemisphere Publishing Corporation, Washington, DC, pp 181–182

Raquel S et al (2013) Biodegradation of high-molecular-weight polycyclic aromatic hydrocarbons by a wood-degrading consortium at low temperatures. FEMS Microbiol Ecol 83(2):438–449. https://doi.org/10.1111/1574-6941.12006

Reddy PV et al (2018) Biodegradation of fluoranthene by *Paenibacillus* sp. strain PRNK-6: a pathway for complete mineralization. Arch Microbiol 200:171. https://doi.org/10.1007/s00203-017-1431-9

Rehmann K et al (2001) Fluoranthene metabolism in mycobacterium sp. strain KR20: identity of pathway intermediates during degradation and growth. Microbiology 147:2783–2794

Reis RS et al (2013) In: Chamy R, Rosenkran F (eds) Biosurfactants: production and applications, biodegradation—life of science. IntechOpen, London. https://doi.org/10.5772/56144

Reis CBLD et al (2018) First report of the production of a potent biosurfactant with α,β-trehalose by *fusarium fujikuroi* under optimized conditions of submerged fermentation. Braz J Microbiol 49:185–192. https://doi.org/10.1016/j.bjm.2018.04.004

Reizer E et al (2019) Formation mechanism of benzo(a)pyrene: one of the most carcinogenic polycyclic aromatic hydrocarbons (PAH). Molecules 24(6):1040. https://doi.org/10.3390/molecules24061040

Resnick S (1996) Diverse reactions catalyzed by naphthalene dioxygenase from *Pseudomonas* sp strain. J Ind Microbiol Biotechnol 17:438. https://doi.org/10.1007/BF01574775

Rosenberg E, Ron E (1999) High- and low-molecular-mass microbial surfactants. Appl Microbiol Biotechnol 52:154. https://doi.org/10.1007/s002530051502

Rui Y et al (2013) Role of phospholipids and copolymers in enhancing stability and controlling degradation of intravenous lipid emulsions. Colloids Surf A Physicochem Eng Asp 436:434–442

Saddler JN (1978) Biodegradation of bacterial lipopolysaccharides. PhD thesis, University of Glasgow

Saddler JN et al (1979) Degradation of bacterial lipopolysaccharide by the slime mould physarum polycephalum. Can J Microbiol 25(2):124–129. https://doi.org/10.1139/m79-020

Samanta SK et al (2002) Polycyclic aromatic hydrocarbons: environmental pollution and bioremediation. Trends Biotechnol 20(6):243–248. https://doi.org/10.1016/s0167-7799(02)01943-1

Santos DKF et al (2016) Biosurfactants: multifunctional biomolecules of the 21st century. Int J Mol Sci 17(3):401. https://doi.org/10.3390/ijms17030401

Satpute SK (2010) Methods for investigating biosurfactants and bioemulsifiers: a review. Crit Rev Biotechnol 30(2):127–144. https://doi.org/10.3109/07388550903427280

Satpute SK et al (2010) Methods for investigating biosurfactants and bioemulsifiers: a review. Crit Rev Biotechnol 30(2):127–144. https://doi.org/10.3109/07388550903427280

Saunders CR (2003) Fluoranthene-induced neurobehavioral toxicity in F-344. *Int J Toxicol* 22 (4):263–276

Schmid A et al (1998) Effects of biosurfactant and emulsification on two-liquid phase *pseudomonas oleovorans* cultures and cell-free emulsions containing n-decane. Enzyme Microb Technol 22 (6):487–493

Selifonov SA et al (1996) Oxidation of naphthenoaromatic and methyl-substituted aromatic compounds by naphthalene 1, 2-dioxygenase. Appl Environ Microbiol 62(2):507–514

Sen S et al (2017) Production, characterization, and antifungal activity of a biosurfactant produced by *rhodotorula babjevae* YS3. Microb Cell Fact 16(1):95. https://doi.org/10.1186/s12934-017-0711-z

Seo JS et al (2009) Bacterial degradation of aromatic compounds. Int J Environ Res Public Health 6 (1):278–309. https://doi.org/10.3390/ijerph6010278

Shao B et al (2017) Effects of rhamnolipids on microorganism characteristics and applications in composting: a review. Microbiol Res 200:33–44. https://doi.org/10.1016/j.micres.2017.04.005

Shusheng L et al (2014) Pyrene degradation accelerated by constructed consortium of bacterium and microalga: effects of degradation products on the microalgal growth. Environ Sci Technol 48(23):13917–13924. https://doi.org/10.1021/es503761j

Sifour M, Al-Jilawi MH, Aziz GM (2007) Emulsification properties of biosurfactant produced from *pseudomonas aeruginosa* RB 28. Pak J Biol Sci 10(8):1331–1335. https://doi.org/10.3923/pjbs.2007.1331.1335

Simarro R et al (2013) Assessment of the efficiency of *in situ* bioremediation techniques in a creosote polluted soil: change in bacterial community. J Hazard Mater 262(15):158–167

Singh B, Sharma N (2008) Mechanistic implications of plastic degradation. Polym Degrad Stab 93 (3):561–584. https://doi.org/10.1016/j.polymdegradstab.2007.11.008

Snežana M (2013) Petroleum hydrocarbon biodegradability in soil—implications for bioremediation, hydrocarbon. IntechOpen, London. https://doi.org/10.5772/50108

Soberón-Chávez G, Maier RM (2011) Biosurfactants: from genes to applications, book review, vol 20. Springer, Berlin. https://doi.org/10.1007/978-3-642-14490-5-1

Soberón-Chávez G et al (2005) Production of rhamnolipids by *pseudomonas aeruginosa*. Appl Microbiol Biotechnol 68:718. https://doi.org/10.1007/s00253-005-0150-3

Somtrakoon K et al (2008) Phenanthrene stimulates the degradation of pyrene and fluoranthene by *burkholderia* sp. VUN10013. World J Microbiol Biotechnol 24:523–531. https://doi.org/10.1007/s11274-007-9503-7

Sosa BS et al (2017) Human health risk due to variations in PM_{10}-$PM_{2.5}$ and associated PAHs levels. Atmos Environ 160:27–35. https://doi.org/10.1016/j.atmosenv.2017.04.004

Souayeh M et al (2014) Optimization of a low-concentration *bacillus subtilis* strain biosurfactant toward microbial enhanced oil recovery. Energy Fuel 28(9):5606–5611. https://doi.org/10.1021/ef500954u

Sousa JC et al (2017) A review on environmental monitoring of water organic pollutants identified by EU guidelines. J Hazard Mater 344:146–162. https://doi.org/10.1016/j.jhazmat

Stegeman JJ, Lech JJ (1991) Cytochrome P-450 monooxygenase systems in aquatic species: carcinogen metabolism and biomarkers for carcinogen and pollutant exposure. Environ Health Perspect 90:101–109. https://doi.org/10.1289/ehp.90-1519513

Sugimoto K et al (1999) Crystal structure of an aromatic ring opening dioxygenase LigAB, a protocatechuate 4, 5-dioxygenase, under aerobic conditions. Structure 7(8):953–965

Suryanti V et al (2017) Physicochemical properties of biosurfactant produced by *pseudomonas fluorescens* grown on whey tofu. IOP Conf Ser Mater Sci Eng 176(1):012003. https://doi.org/10.1088/1757-899x/176/1/012003

Szczepaniak Z et al (2016) Influence of soil contamination with PAH on microbial community dynamics and expression level of genes responsible for biodegradation of PAH and production of rhamnolipids. Environ Sci Pollut Res Int 23(22):23043–23056. https://doi.org/10.1007/s11356-016-7500-9

Tersagh I et al (2016) Aerobic degradation of fluoranthene, benzo(b)fluoranthene and benzo(k)-fluoranthene by aerobic heterotrophic bacteria- cyanobacteria interaction in brackish water of bodo creek. J Pet Environ Biotechnol 7:292. https://doi.org/10.4172/2157-7463.1000292

Tomás-Gallardo L et al (2009) Molecular and biochemical characterization of the tetralin degradation pathway in *rhodococcus* sp. strain TFB. J Microbial Biotechnol 2:262–273. https://doi.org/10.1111/j.1751-7915.2009.00086.x

Tomás-Gallardo L et al (2014) Combination of degradation pathways for naphthalene utilization in *rhodococcus sp.* strain TFB. J Microbial Biotechnol 7(2):100–113. https://doi.org/10.1111/1751-7915.12096

Toral L et al (2018) Antifungal activity of lipopeptides from *bacillus* XT1 CECT 8661 against *botrytis cinerea*. Front Microbiol 9:1315. https://doi.org/10.3389/fmicb.2018.01315

Toräng L (2004) Biodegradation rates of chemicals in surface water and groundwater assessed in batch simulation tests. In: Environment & resources DTU. Technical University of Denmark, Lyngby, p 69

Trilisenko LV et al (2012) The antifungal effect of cellobiose lipid on the cells of *saccharomyces cerevisiae* depends on carbon source. SpringerPlus 1:18. https://doi.org/10.1186/2193-1801-1-18

Twigg MS et al (2018) Identification and characterisation of short chain rhamnolipid production in a previously uninvestigated, non-pathogenic marine pseudomonad. *Appl Microbiol Biotechnol* 102(19):8537–8549. https://doi.org/10.1007/s00253-018-9202-3

Ukiwe LN et al (2013) Polycyclic aromatic hydrocarbons degradation techniques: a review. Int J Chem 5:4. https://doi.org/10.5539/ijc.v5n4p43

Uzoigwe C et al (1999) Bioemulsifiers are not biosurfactants and require different screening approaches. Front Microbiol 6:245. https://doi.org/10.3389/fmicb.2015.00245

Uzoigwe C et al (2015) Bioemulsifiers are not biosurfactants and require different screening approaches. Front Microbiol 6:245. https://doi.org/10.3389/fmicb.2015.00245

Vaidya S et al (2017) Metabolism of pyrene through phthalic acid pathway by enriched bacterial consortium composed of *pseudomonas*, *burkholderia*, and *rhodococcus* (PBR). 3 Biotech 7 (1):29. https://doi.org/10.1007/s13205-017-0598-8

Vaidya S et al (2018) Degradation of chrysene by enriched bacterial consortium. Front Microbiol 9:1333. https://doi.org/10.3389/fmicb.2018.01333

Vasudevan S, Prabhune AA (2018) Photophysical studies on curcumin-sophorolipid nanostructures: applications in quorum quenching and imaging. R Soc Open Sci 5(2):170865. https://doi.org/10.1098/rsos.170865

Vikbjerg AF et al (2005) Continuous production of structured phospholipids in a packed bed reactor with lipase from *thermomyces lanuginosa*. J Am Oil Chem Soc 82(4):237–242. https://doi.org/10.1007/s11746-005-1061-x

Von SC (2006) Free-radical-induced DNA damage and its repair: a chemical perspective. Springer-Verlag, Berlin. https://doi.org/10.1007/3-540-30592-0

Wang S et al (2007) Light-induced cytotoxicity of 16 polycyclic aromatic hydrocarbons on the US EPA priority pollutant list in human skin HaCaT keratinocytes: relationship between phototoxicity and excited state properties. Environ Toxicol 22(3):318–327. https://doi.org/10.1002/tox.20241

Wang J et al (2012) Endogenous formation and repair of oxidatively induced G[8-5 m]T intrastrand cross-link lesion. Nucleic Acids Res 40(15):7368–7374. https://doi.org/10.1093/nar/gks357

Weissenfels WD et al (1990) Degradation of phenanthrene, fluorene and fluoranthene by pure bacterial cultures. Appl Microbiol Biotechnol 32(4):479–484. https://doi.org/10.1007/bf00903787

White DA et al (2013) Production and characterization of a trehalolipid biosurfactant produced by the novel marine bacterium Rhodococcus sp., strain PML026. J Appl Microbiol 115 (3):744–755. https://doi.org/10.1111/jam.12287

Willem N et al (2015) Production and utilization of natural phospholipids, in polar lipids biology, chemistry, and technology. In: Polar lipids, pp 245–276. https://doi.org/10.1016/B978-1-63067-044-3.50013-3

Wu L et al (2018) Degradation of phenol via ortho-pathway by *kocuria sp*. strain TIBETAN4 isolated from the soils around Qinghai Lake in China. PLoS One 13(6):e0199572. https://doi.org/10.1371/journal.pone.0199572

Xing-Fang L et al (1996) Microbial degradation of pyrene and characterization of a metabolite. Sci Total Environ 177(1–3, 5):17–29

Yamamoto S et al (2012) The moisturizing effects of glycolipid biosurfactants, mannosylerythritol lipids, on human skin. J Oleo Sci 61(7):407–412. https://doi.org/10.5650/jos.61.407

Yamanashi T et al (2015) *In vitro* reconstitution of the catabolic reactions catalyzed by PcaHG, PcaB, and PcaL: the protocatechuate branch of the β-ketoadipate pathway in *rhodococcus jostii* RHA1. Biosci Biotechnol Biochem 79(5):830–835. https://doi.org/10.1080/09168451.2014.993915

Yan S, Wu G (2017) Reorganization of gene network for degradation of polycyclic aromatic hydrocarbons (PAHs) in *pseudomonas aeruginosa* PAO1 under several conditions. J Appl Genet 58(4):545–563. https://doi.org/10.1007/s13353-017-0402-9

Yen K-M et al (1988) Genetics of naphthalene catabolism in *pseudomonads*. CRC Crit Rev Microbiol 15(3):247–268. https://doi.org/10.3109/10408418809104459

Yu H (2002) Environmental carcinogenic polycyclic aromatic hydrocarbons: photochemistry and phototoxicity. Journal of environmental science and health. Part C. Environ Carcinog Ecotoxicol Rev 20(2):149–183. https://doi.org/10.1081/GNC-120016203

Yu M et al (2015) Characteristics of mannosylerythritol lipids and their environmental potential. Carbohydr Res 30(407):63–72. https://doi.org/10.1016/j.carres.2014.12.012

Zeng Z et al (2018) Mechanisms for rhamnolipids-mediated biodegradation of hydrophobic organic compounds. *Sci Total Environ* 634:1–11. https://doi.org/10.1016/j.scitotenv.2018.03.349

Zhang J et al (2016a) Production of lipopeptide biosurfactants by *bacillus atrophaeus* 5-2a and their potential use in microbial enhanced oil recovery. Microb Cell Fact 15(1):168. https://doi.org/10.1186/s12934-016-0574-8

Zhang YL et al (2016b) Isolation and characterization of a novel paraffin wax-degrading bacterium, *pseudomonas sp* strain PW-1, from petroleum-contaminated sites. Genet Mol Res J 15(2):gmr8021. https://doi.org/10.4238/gmr.15028021

Zhong H et al (2015) Effect of low-concentration rhamnolipid on adsorption of *pseudomonas aeruginosa* ATCC 9027 on hydrophilic and hydrophobic surfaces. J Hazard Mater 285:383–388. https://doi.org/10.1016/j.jhazmat.2014.11.050

Zhou J et al (2019) High di-rhamnolipid production using *pseudomonas aeruginosa* KT1115, separation of mono/di-rhamnolipids, and evaluation of their properties. Front Bioeng Biotechnol 7:245. https://doi.org/10.3389/fbioe.2019.00245

Zhuo S et al (2017) Source-oriented risk assessment of inhalation exposure to ambient polycyclic aromatic hydrocarbons and contributions of non-priority isomers in urban Nanjing, a megacity located in Yangtze River Delta, China. Environ Pollut 224:796–809. https://doi.org/10.1016/j.envpol.2017.01.039

Surfactin: A Biosurfactant Against Breast Cancer

7

Pimpi Sahu, Kumar Kritartha Kaushik, Tingli Lu, and Kai Dong

Abstract

Surfactin is a biosurfactant produced by different species of the genus *Bacillus*. It poses anticancer activity against breast and other different cancers. Surfactin suppresses cancerous growth by cell cycle arrest and programed cell death and also captures metastasis. As a result of the fantastic impact, surfactin is widely studied. Here the synthesis, structure, and properties of surfactin, along with its effectiveness against breast cancer, are briefly discussed.

Keywords

Biosurfactant · Surfactin · Breast cancer · Apoptosis · ROS

7.1 Introduction

Globally women have breast cancer, which is uncontrollable malice (Lukong 2017; Torre et al. 2015). Because of the tremendous frequency of the disease, it is the secondary conduit inference for the cost of life in ladies, and it is a distant imperative to ratify vigorous remedial strategy. Breast cancer cells catalog to unveil nonintervention in different molecular ways (Singhal et al. 2016). The applicability of

P. Sahu (✉) · T. Lu · K. Dong
Key Laboratory for Space Bioscience and Biotechnology, School of Life Sciences, Northwestern Polytechnical University, Xi'an, Shaanxi, China
e-mail: pimpisahu@mail.nwpu.edu.cn; lutinglixinxin@nwpu.edu.cn; dongkai120@nwpu.edu.cn

K. K. Kaushik
Department of Zoology, Sipajhar College, Darrang, Assam, India

Department of Molecular Biology and Biotechnology, Tezpur University, Napaam, Sonitpur, Assam, India
e-mail: kritartha@tezu.ernet.in

Inamuddin et al. (eds.), *Microbial Biosurfactants*, Environmental and Microbial Biotechnology, https://doi.org/10.1007/978-981-15-6607-3_7

nanotherapeutics provides the help of raising surfactin distribution for augmented anticancer remedy (Wu et al. 2017). Biosurfactants have straightaway spurt as a propitious particle for its constitutional innovation, skillfulness, and plentiful possessions that are possibly constructive for curative operation. Explicitly, considering their surface action, those molecules merge with cellular membranes of diverse organisms and with the encompassing surroundings and hence in remedial treatment. Some microbial surfactants comprise lipopeptides and glycolipids, which were confirmed to restrain the propagation of cancer cells collectively and to agitate cell membranes, which lead to their breakdown by programed cell death. Furthermore, biosurfactants as drug carrier wagon provide economically luring and experimental unique programs (Gudiña et al. 2013).

7.2 Biosurfactants and Its Types

Biosurfactants consist of different amphipathic particles with wonderful synthetic systems yielded by numerous microorganisms. These molecules, which can be specially fashioned as other metabolites, show vital aspects within the continuation in their generation of microorganisms by way of expediting nutriment conveyor, intrusive in a host and microbe synergy, and quorum sensing contrivance or by using as biocide operators (Marchant and Banat 2012). Their diagnosed capacity and organic nature have stimulated numerous studies on their available healing programs (Fracchia et al. 2012; Rodrigues 2011). Those composites are principal to manufactured surfactants, due to their microbe-starting place, environment friendliness, and shallow toxicant (Marchant and Banat 2012). Due to those reasons, it has been studied for foodstuff and demulcent commercial enterprise, better grease restoration, and bioprocessing (Marchant and Banat 2012). Glycolipids are low molecular weight biosurfactants, whereas polysaccharides, proteins, and others are high molecular weight biosurfactants. Biosurfactants with low molecular weight contain striking surface features (e.g., Figs. 7.1 and 7.2). Biosurfactants can influence the attachment of microorganisms as they screen at the junction of the liquid state with awesome polarity and hydrogen bonding. Adjustments inside the substantial membrane framework or adjustments in protein arrangement arise, for this reason changing good-sized membrane capabilities that incorporate transportation and effective production (Rodrigues 2011; Van Hamme et al. 2006; Sodagari et al. 2013). Additionally, these admixture can muddle cell membranes that results in breaking by using extend permeability of membrane and, in the long run, to leakage of metabolites (Bharali et al. 2013). Antibacterial, antifungal, and antiviral are some of its features, additional to their antiadhesive quality in opposition to pathogenicity, and probiotic attributes are the highest appropriate for the fitness-associated application (Marchant and Banat 2012; Fracchia et al. 2012; Rodrigues 2011; Raaijmakers et al. 2010). Drugs synthesized from biosurfactants can be used as remedial measures. Biosurfactant is an antiadhesive coating for biological materials and for lung immune therapy (Rodrigues et al. 2006). Lately, it has come out that biosurfactants can act on cancer cells. As an example, lipopeptide surfactin was

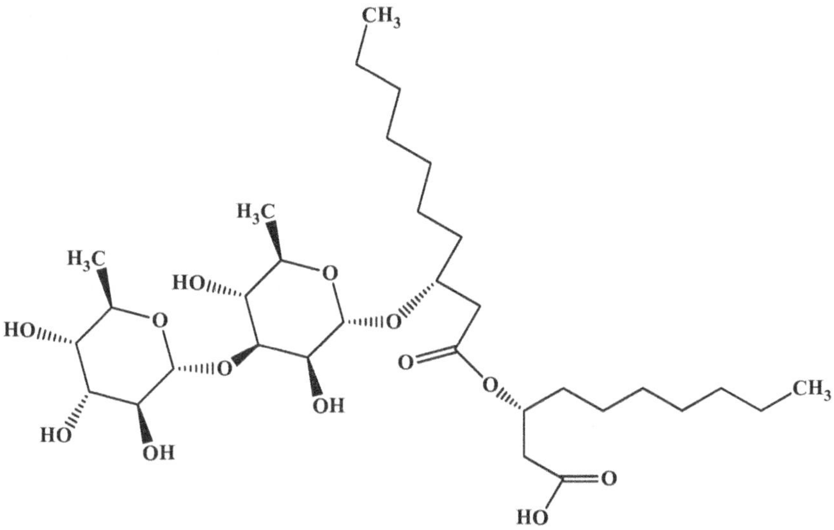

Fig. 7.1 Rhamnolipid

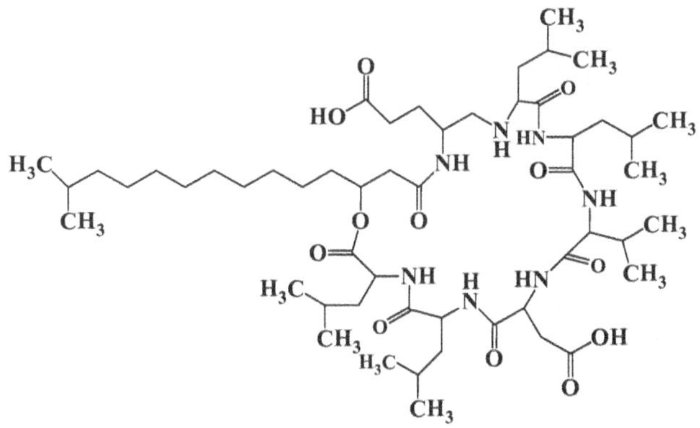

Fig. 7.2 Surfactin

located to set off programed cell death in breast cancer cells (Cao et al. 2010). Mannosylerythritol and succinoyl lipids were worried about boom detention and programed cell death of tumor cells (Zhao et al. 2000; Isoda and Nakahara 1997; Isoda et al. 1995). Many other healing applications were suggested for biosurfactants together with new uses in nanotechnology, in particular primarily occupying on their adjustable self-assembly (Kitamoto et al. 2009). For the therapeutic remedy for herpes simplex virus, b-sitosterol-b-D-glucoside, complex DNA, which is a liposomal vector, is in practice (Maitani et al. 2006). Nanovectors involving biosurfactant raise the potentiality in genetic engineering (Nakanishi et al. 2009). Though it appears that biosurfactants are used for multipurpose activity and effective particles for remedial measures, some may be life-risking factors for human. For example, nosocomial infection occurs by *Pseudomonas aeruginosa*, which produces glycolipids for numerous pharmaceutical-associated packages (Hošková et al. 2013; Abbasi et al. 2012).

7.2.1 Glycolipids

It is a saccharide with a polar head and hydrocarbon tail (Veenanadig et al. 2000; Chen et al. 2007).

7.2.1.1 Rhamnolipids

It includes rhamnose (Lang and Wullbrandt 1999; Rahman et al. 2002). These were formed with the aid of bacteria *Pseudomonas* sp. (Neto et al. 2008) and *P. putida*. (Cha et al. 2008). *P. fluorescens* that forms rhamnolipids is a disaccharide with methyl pentose and lipids. However, ester and carboxyl groups are present in the lipid segment (Rahman and Gakpe 2008).

7.2.1.2 Sophorolipids

These are made up of sophorose (Asmer et al. 1988). They are yielded via *Torulopsis* sp. or *Candida* sp. (Felse et al. 2007), *T. petrophilum* (Cooper and Goldenberg 1987), *T. apicola* (Tulloch et al. 1967), and *Candida bogoriensis* (Cutler and Light 1979). *Wickerhamiella domercqiae* was additionally suggested to supply sophorolipids (Chen et al. 2006).

Non-lactonic sophorolipids that have hydroxyl fatty acids contain carboxylic acid. Lactonic SLs form a lactone ring (Hu and Ju 2001).

7.2.1.3 Trehalolipids

These are made up of disaccharide trehalose (Desai and Banat 1997). The microorganisms which produce trehalolipids are *Mycobacterium*, *Nocardia*, and *Corynebacterium* (Rahman and Gakpe 2008). Trehalose lipids constituted of *Rhodococcus erythropolis*, and *Arthrobacter* sp. was additionally pronounced (Li et al. 1984).

7.2.2 Lipopeptides

Surfactin is produced from *Bacillus* sp. (Kakinuma et al. 1969). Microorganism, *Alcaligenes* sp., produces lipopeptide containing ester group (Huang et al. 2009). *B. subtilis* produces surfactin along with bacillomycin L and plipastatin (Roongsawang et al. 2002). Biosurfactants are also produced from *Klebsiella* sp. (Lee et al. 2008). Fengycin (Deleu et al. 2008) is a lipopeptide formed by different types of *B. subtilis* (Vanittanakom et al. 1986; Jacques et al. 1999).

7.2.3 Fatty Acids

They are made up of carboxyl groups, for example, corynomycolic acids (Mulligan 2005). Aerobically *Penicillium spiculisporum* forms spiculisporic acid (Ban et al. 1998). Alkanes that form fatty acid can be regarded as surfactin (Rehm and Reiff 1981).

7.2.4 Phospholipids

Aspergillus sp. (Käppeli et al. 1984) and *Thiobacillus thiooxidans* (Beebe and Umbreit 1971) form phospholipids. Phosphatidylethanolamine wealthy sacs were formed by using *Acinetobacter* sp. developed on hexadecane (Käppeli and Finnerty 1979). Phosphatidylethanolamine additionally is formed by using *Rhodococcus* sp. grown on n-alkane (Rahman and Gakpe 2008).

7.2.5 Polymeric Biosurfactant

This may be carbohydrate or protein on the basis of lipids attached to it. It is of high molecular weight (Rosenberg et al. 1979). *Acinetobacter calcoaceticus* A2 can also form biodispersan, a heteropolysaccharide. Alasan, another heteropolysaccharide protein biosurfactant, can be isolated from *Acinetobacter radioresistens* KA-53. Liposan is formed by *Candida lipolytica* made up of 83% and 17% protein ((Navon-Venezia et al. 1995; Desai and Banat 1997; Rahman and Gakpe 2008). *Saccharomyces cerevisiae* became incredible in generating big amounts of mannoprotein. It confirmed notable emulsifier activity in the direction of numerous oils, alkanes, and natural solvents (Cameron et al. 1988).

7.3 Surfactin: Structure, Membrane Interaction, Biosynthesis, and Regulation

Surfactin (SUR) is formed by *Bacillus subtilis* (Hwang et al. 2005). It also has antibacterial (Lee et al. 2016; Loiseau et al. 2015), antiviral (Kracht et al. 1999), anti-inflammatory (Zhang et al. 2015; Gan et al. 2016), antiproliferative (Kim et al. 2007), adjuvant for immunization, and antitumor properties (Vollenbroich et al. 1997; Cao et al. 2009a, b, 2010, 2011; Christova et al. 2013). Surfactin constrains the propagation of MCF-7 cells through activating programed cell death (Cao et al. 2011). Surfactin can combine together to form nanoparticles (Straus and Hancock 2006). Combining the anticancer activity of SUR and the traits of nanoparticles consisting of enhanced permeability and retention effect results and multidrug resistance retraction; it enhances chemotherapy utilizing surfactant to lade anticancer drugs.

7.3.1 Structure

It is formed by heptapeptide (Abdel-Mawgoud et al. 2008).

7.3.2 Membrane Interaction

Through hydrophobic interactions, surfactin invades automatically into lipid membranes. Peptide communicates with polar heads in cancerous cells (Wu et al. 2010).

7.3.3 Biosynthesis

Non-ribosomal process using surfactin synthetase leads to the biosynthesis of surfactin (Kluge et al. 1988; Shaligram and Singhal 2010). In the peptide chain, phosphopantetheinyl cofactors act as an acceptor of the developing peptide chain. Adenylation of ATP occurs to reinforce seven amino acids in surfactin (Nakano and Zuber 1990; Vater et al. 1997).

7.3.4 Regulation

Genetic evaluation performs an essential part in governing biosynthesis of surfactin (Sen 2010). Surfactin synthetase is meditated through chromosomal genes (Gabriela and Jaroslava 2008). Srf and sfp control the synthesis of surfactin (Sen 2010). Inducible operon, srfA, open reading frame encodes surfactin subunits (Fernandez-Abalos et al. 2003).

7.4 Surfactin and Breast Cancer

B. subtilis-derived surfactin inhibits T47D and MDAMB-231 in a time- and dose-dependent way (Duarte et al. 2014). The programed cell death effect of Bcap-37 cells is due to the enhanced membrane fluidization (Liu et al. 2010).

Surfactin, a flexible bioactive molecule, is comprised of antitumor activity (Sachdev and Cameotra 2013). Surfactin has been said to show antitumor interest in opposition to murine mammary carcinoma cells (Sivapathasekaran et al. 2010). JNK and ERK1/2 regulate apoptosis (Cao et al. 2010).

Oxidative stress is a consequence in the induction of apoptosis inside the cell. DNA nicking also leads to programed cell death. DNA nicking depends on the number of lipopolysaccharides. Some biomolecules are responsible for programed cell death, regulation of transcription, and DNA repair (Alonso et al. 2003). Peripheral chromatin condensation occurs when incubated HeLa recombine with AIF (apoptosis-inducing factor) (Cregan et al. 2004). Apoptosis-inducing factor launched from mitochondria and subsequent cell demise changed into proven to be precipitated by using immoderate cellular calcium inflow (Bröker et al. 2005).

Doxorubicin (DOX)-SUR results in potent cytotoxicity against human breast cancer. The difference in the structure of biosurfactants may be beneficial for interdependent drug transportation (Akiyode et al. 2016).

7.5 Conclusion

There have been numerous research at the anticancer consequences of surfactin in opposition to most cancer cellular traces that suggest its selective cytotoxicity. Surfactin isolated from *B. subtilis* 573 has a cytotoxic effect on ordinary MCT-3T3-E1 fibroblast cells (Duarte et al. 2014). Cytotoxicity of surfactin C is examined via oral dose to rats (HWANG et al. 2009). Surfactin, drawback as an anticancer promoter due to hemolytic action above 0.05 g/l (Dehghan-Noudeh et al. 2005). To conquer this hemolytic nature, some surfactin has been synthesized, which is nonhemolytic in nature (Gabriela and Jaroslava 2008). An alternative way to deliver surfactin is nanoformulation to decrease toxic effects and more desirable anticancer effects (Wu et al. 2017).

References

Abbasi H, Noghabi KA, Ortiz A (2012) Interaction of a bacterial monorhamnolipid secreted by Pseudomonas aeruginosa MA01 with phosphatidylcholine model membranes. Chem Phys Lipids 165:745–752

Abdel-Mawgoud AM, Aboulwafa MM, Hassouna NA-H (2008) Characterization of surfactin produced by Bacillus subtilis isolate BS5. Appl Biochem Biotechnol 150:289–303

Akiyode O, George D, Getti G, Boateng J (2016) Systematic comparison of the functional physico-chemical characteristics and biocidal activity of microbial derived biosurfactants on blood-derived and breast cancer cells. J Colloid Interface Sci 479:221–233

Alonso M, Tamasdan C, Miller DC, Newcomb EW (2003) Flavopiridol induces apoptosis in Glioma cell lines independent of retinoblastoma and p53 tumor suppressor pathway alterations by a Caspase-independent Pathway1. Mol Cancer Ther 2:139–150

Asmer H-J, Lang S, Wagner F, Wray V (1988) Microbial production, structure elucidation and bioconversion of sophorose lipids. J Am Oil Chem Soc 65:1460–1466

Ban T, Sato T, Yen TF (1998) Interfacial activity of n-alkyl amines from microbially produced spiculisporic acid. J Petrol Sci Eng 21:223–238

Beebe JL, Umbreit W (1971) Extracellular lipid of Thiobacillus thiooxidans. J Bacteriol 108:612

Bharali P, Saikia J, Ray A, Konwar B (2013) Rhamnolipid (RL) from Pseudomonas aeruginosa OBP1: a novel chemotaxis and antibacterial agent. Colloids Surf B Biointerfaces 103:502–509

Bröker LE, Kruyt FA, Giaccone G (2005) Cell death independent of caspases: a review. Clin Cancer Res 11:3155–3162

Cameron DR, Cooper DG, Neufeld R (1988) The mannoprotein of Saccharomyces cerevisiae is an effective bioemulsifier. Appl Environ Microbiol 54:1420–1425

Cao X-H, Liao Z-Y, Wang C-L, Yang W-Y, Lu M-F (2009a) Evaluation of a lipopeptide biosurfactant from Bacillus natto TK-1 as a potential source of anti-adhesive, antimicrobial and antitumor activities. Braz J Microbiol 40:373–379

Cao XH, Liao ZY, Wang CL, Cai P, Yang WY, Lu MF, Huang GW (2009b) Purification and antitumour activity of a lipopeptide biosurfactant produced by Bacillus natto TK-1. Biotechnol Appl Biochem 52:97–106

Cao X-H, Wang A-H, Wang C-L, Mao D-Z, Lu M-F, Cui Y-Q, Jiao R-Z (2010) Surfactin induces apoptosis in human breast cancer MCF-7 cells through a ROS/JNK-mediated mitochondrial/caspase pathway. Chem Biol Interact 183:357–362

Cao X-H, Zhao S-S, LIU D-Y, Wang Z, Niu L-L, Hou L-H, Wang C-L (2011) ROS-Ca2+ is associated with mitochondria permeability transition pore involved in surfactin-induced MCF-7 cells apoptosis. Chem Biol Interact 190:16–27

Cha M, Lee N, Kim M, Kim M, Lee S (2008) Heterologous production of Pseudomonas aeruginosa EMS1 biosurfactant in Pseudomonas putida. Bioresour Technol 99:2192–2199

Chen J, Song X, Zhang H, Qu Y (2006) Production, structure elucidation and anticancer properties of sophorolipid from Wickerhamiella domercqiae. Enzyme Microb Technol 39:501–506

Chen S-Y, Wei Y-H, Chang J-S (2007) Repeated pH-stat fed-batch fermentation for rhamnolipid production with indigenous Pseudomonas aeruginosa S2. Appl Microbiol Biotechnol 76:67–74

Christova N, Tuleva B, Kril A, Georgieva M, Konstantinov S, Terziyski I, Nikolova B, Stoineva I (2013) Chemical structure and in vitro antitumor activity of rhamnolipids from Pseudomonas aeruginosa BN10. Appl Biochem Biotechnol 170:676–689

Cooper DG, Goldenberg BG (1987) Surface-active agents from two Bacillus species. Appl Environ Microbiol 53:224–229

Cregan SP, Dawson VL, Slack RS (2004) Role of AIF in caspase-dependent and caspase-independent cell death. Oncogene 23:2785–2796

Cutler AJ, Light RJ (1979) Regulation of hydroxydocosanoic acid sophoroside production in Candida bogoriensis by the levels of glucose and yeast extract in the growth medium. J Biol Chem 254:1944–1950

Dehghan-Noudeh G, Housaindokht M, Bazzaz BSF (2005) Isolation, characterization, and investigation of surface and hemolytic activities of a lipopeptide biosurfactant produced by Bacillus subtilis ATCC 6633. J Microbiol 43:272–276

Deleu M, Paquot M, Nylander T (2008) Effect of fengycin, a lipopeptide produced by Bacillus subtilis, on model biomembranes. Biophys J 94:2667–2679

Desai JD, Banat IM (1997) Microbial production of surfactants and their commercial potential. Microbiol Mol Biol Rev 61:47–64

Duarte C, Gudiña EJ, Lima CF, Rodrigues LR (2014) Effects of biosurfactants on the viability and proliferation of human breast cancer cells. AMB Express 4:40

Felse PA, Shah V, Chan J, Rao KJ, Gross RA (2007) Sophorolipid biosynthesis by Candida bombicola from industrial fatty acid residues. Enzyme Microb Technol 40:316–323

Fernandez-Abalos JM, Reviejo V, Díaz M, Rodríguez S, Leal F, Santamaría RI (2003) Posttranslational processing of the xylanase Xys1L from Streptomyces halstedii JM8 is carried out by secreted serine proteases. Microbiology 149:1623–1632

Fracchia L, Cavallo M, Martinotti MG, Banat IM (2012) Biosurfactants and bioemulsifiers biomedical and related applications—present status and future potentials. Biomed Sci Eng Technol 14:326–335

Gabriela S, Jaroslava S (2008) Review of surfactin chemical properties and potential applications. Cent Eur J Med 3:123–133

Gan P, Jin D, Zhao X, Gao Z, Wang S, Du P, Qi G (2016) Bacillus-produced surfactin attenuates chronic inflammation in atherosclerotic lesions of ApoE−/− mice. Int Immunopharmacol 35:226–234

Gudiña EJ, Rangarajan V, Sen R, Rodrigues LR (2013) Potential therapeutic applications of biosurfactants. Trends Pharmacol Sci 34:667–675

Hošková M, Schreiberová O, Ježdík R, Chudoba J, Masák J, Sigler K, Řezanka T (2013) Characterization of rhamnolipids produced by non-pathogenic Acinetobacter and Enterobacter bacteria. Bioresour Technol 130:510–516

Hu Y, Ju L-K (2001) Purification of lactonic sophorolipids by crystallization. J Biotechnol 87:263–272
Huang X-F, Liu J, Lu L-J, Wen Y, Xu J-C, Yang D-H, ZHOU Q (2009) Evaluation of screening methods for demulsifying bacteria and characterization of lipopeptide bio-demulsifier produced by Alcaligenes sp. Bioresour Technol 100:1358–1365
Hwang M-H, Lim J-H, Yun H-I, Rhee M-H, Cho J-Y, HSU WH, Park S-C (2005) Surfactin C inhibits the lipopolysaccharide-induced transcription of interleukin-1β and inducible nitric oxide synthase and nitric oxide production in murine RAW 264.7 cells. Biotechnol Lett 27:1605–1608
HWANG Y-H, KIM M-S, SONG I-B, PARK B-K, LIM J-H, PARK S-C, YUN H-I (2009) Subacute (28 day) toxicity of surfactin C, a lipopeptide produced by Bacillus subtilis, in rats. J Health Sci 55:351–355
Isoda H, Nakahara T (1997) Mannosylerythritol lipid induces granulocytic differentiation and inhibits the tyrosine phosphorylation of human myelogenous leukemia cell line K562. Cytotechnology 25:191–195
Isoda H, Shinmoto H, Matsumura M, Nakahara T (1995) Succinoyl trehalose lipid induced differentiation of human monocytoid leukemic cell line U937 into monocyte-macrophages. Cytotechnology 19:79–88
Jacques P, Hbid C, Destain J, Razafindralambo H, Paquot M, De Pauw E, Thonart P (1999) Optimization of biosurfactant lipopeptide production from Bacillus subtilis S499 by Plackett-Burman design. In: Twentieth symposium on biotechnology for fuels and chemicals. Springer, Berlin, pp 223–233
Kakinuma A, Ouchida A, Shima T, Sugino H, Isono M, Tamura G, Arima K (1969) Confirmation of the structure of surfactin by mass spectrometry. Agric Biol Chem 33:1669–1671
Käppeli O, Finnerty W (1979) Partition of alkane by an extracellular vesicle derived from hexadecane-grown Acinetobacter. J Bacteriol 140:707–712
Käppeli O, Walther P, Mueller M, Fiechter A (1984) Structure of the cell surface of the yeast Candida tropicalis and its relation to hydrocarbon transport. Arch Microbiol 138:279–282
Kim S-Y, Kim JY, Kim S-H, Bae HJ, Yi H, Yoon SH, Koo BS, Kwon M, Cho JY, Lee C-E (2007) Surfactin from Bacillus subtilis displays anti-proliferative effect via apoptosis induction, cell cycle arrest and survival signaling suppression. FEBS Lett 581:865–871
Kitamoto D, Morita T, Fukuoka T, Konishi M-A, Imura T (2009) Self-assembling properties of glycolipid biosurfactants and their potential applications. Curr Opin Colloid Interface Sci 14:315–328
Kluge B, Vater J, Salnikow J, Eckart K (1988) Studies on the biosynthesis of surfactin, a lipopeptide antibiotic from Bacillus subtilis ATCC 21332. FEBS Lett 231:107–110
Kracht M, Rokos H, Özel M, Kowall M, Pauli G, Vater J (1999) Antiviral and hemolytic activities of surfactin isoforms and their methyl ester derivatives. J Antibiot 52:613–619
Lang S, Wullbrandt D (1999) Rhamnose lipids–biosynthesis, microbial production and application potential. Appl Microbiol Biotechnol 51:22–32
Lee S-C, Lee S-J, Kim S-H, Park I-H, Lee Y-S, Chung S-Y, Choi Y-L (2008) Characterization of new biosurfactant produced by Klebsiella sp. Y6-1 isolated from waste soybean oil. Bioresour Technol 99:2288–2292
Lee M-H, Lee J, Nam Y-D, Lee JS, Seo M-J, Yi S-H (2016) Characterization of antimicrobial lipopeptides produced by Bacillus sp. LM7 isolated from chungkookjang, a Korean traditional fermented soybean food. Int J Food Microbiol 221:12–18
Li ZY, Lang S, Wagner F, Witte I, Wray V (1984) Formation and identification of interfacial-active glycolipids from resting microbial cells. Appl Environ Microbiol 48:610–617
Liu X, Tao X, Zou A, Yang S, Zhang L, Mu B (2010) Effect of the microbial lipopeptide on tumor cell lines: apoptosis induced by disturbing the fatty acid composition of cell membrane. Protein Cell 1:584–594
Loiseau C, Schlusselhuber M, Bigot R, Bertaux J, Berjeaud J-M, Verdon J (2015) Surfactin from Bacillus subtilis displays an unexpected anti-legionella activity. Appl Microbiol Biotechnol 99:5083–5093

Lukong KE (2017) Understanding breast cancer—the long and winding road. BBA Clin 7:64–77
Maitani Y, Yano S, Hattori Y, Furuhata M, Hayashi K (2006) Liposome vector containing biosurfactant-complexed DNA as herpes simplex virus thymidine kinase gene delivery system. J Liposome Res 16:359–372
Marchant R, Banat IM (2012) Microbial biosurfactants: challenges and opportunities for future exploitation. Trends Biotechnol 30:558–565
Mulligan CN (2005) Environmental applications for biosurfactants. Environ Pollut 133:183–198
Nakanishi M, Inoh Y, Kitamoto D, FURUNO T (2009) Nano vectors with a biosurfactant for gene transfection and drug delivery. J Drug Deliv Sci Technol 19:165–169
Nakano MM, Zuber P (1990) Molecular biology of antibiotic production in Bacillus. Crit Rev Biotechnol 10:223–240
Navon-Venezia S, Zosim Z, Gottlieb A, Legmann R, Carmeli S, Ron E, Rosenberg E (1995) Alasan, a new bioemulsifier from Acinetobacter radioresistens. Appl Environ Microbiol 61:3240–3244
Neto DC, Meira JA, De Araújo JM, Mitchell DA, Krieger N (2008) Optimization of the production of rhamnolipids by Pseudomonas aeruginosa UFPEDA 614 in solid-state culture. Appl Microbiol Biotechnol 81:441
Raaijmakers JM, De Bruijn I, Nybroe O, Ongena M (2010) Natural functions of lipopeptides from Bacillus and Pseudomonas: more than surfactants and antibiotics. FEMS Microbiol Rev 34:1037–1062
Rahman P, Gakpe E (2008) Production, characterisation and applications of biosurfactants-review. Biotechnology 7:360–370
Rahman K, Rahman TJ, McClean S, Marchant R, Banat IM (2002) Rhamnolipid biosurfactant production by strains of Pseudomonas aeruginosa using low-cost raw materials. Biotechnol Prog 18:1277–1281
Rehm H, Reiff I (1981) Mechanisms and occurrence of microbial oxidation of long-chain alkanes. In: Reactors and reactions. Springer, Berlin
Rodrigues LR (2011) Inhibition of bacterial adhesion on medical devices. In: Bacterial adhesion. Springer, Berlin
Rodrigues L, Banat IM, Teixeira J, Oliveira R (2006) Biosurfactants: potential applications in medicine. J Antimicrob Chemother 57:609–618
Roongsawang N, Thaniyavarn J, Thaniyavarn S, Kameyama T, Haruki M, Imanaka T, Morikawa M, Kanaya S (2002) Isolation and characterization of a halotolerant Bacillus subtilis BBK-1 which produces three kinds of lipopeptides: bacillomycin L, plipastatin, and surfactin. Extremophiles 6:499–506
Rosenberg E, Zuckerberg A, Rubinovitz C, Gutnick D (1979) Emulsifier of Arthrobacter RAG-1: isolation and emulsifying properties. Appl Environ Microbiol 37:402–408
Sachdev DP, Cameotra SS (2013) Biosurfactants in agriculture. Appl Microbiol Biotechnol 97:1005–1016
Sen R (2010) Surfactin: biosynthesis, genetics and potential applications. In: Biosurfactants. Springer, Berlin
Shaligram NS, Singhal RS (2010) Surfactin—a review on biosynthesis, fermentation, purification and applications. Food Technol Biotechnol 48:119–134
Singhal SK, Usmani N, Michiels S, Metzger-Filho O, Saini KS, Kovalchuk O, Parliament M (2016) Towards understanding the breast cancer epigenome: a comparison of genome-wide DNA methylation and gene expression data. Oncotarget 7:3002
Sivapathasekaran C, Das P, Mukherjee S, Saravanakumar J, Mandal M, SEN R (2010) Marine bacterium derived lipopeptides: characterization and cytotoxic activity against cancer cell lines. Int J Pept Res Ther 16:215–222
Sodagari M, Wang H, Newby B-MZ, Ju L-K (2013) Effect of rhamnolipids on initial attachment of bacteria on glass and octadecyltrichlorosilane-modified glass. Colloids Surf B Biointerfaces 103:121–128

Straus SK, Hancock RE (2006) Mode of action of the new antibiotic for gram-positive pathogens daptomycin: comparison with cationic antimicrobial peptides and lipopeptides. Biochim Biophys Acta Biomembr 1758:1215–1223

Torre LA, Bray F, Siegel RL, Ferlay J, Lortet-Tieulent J, Jemal A (2015) Global cancer statistics, 2012. CA Cancer J Clin 65:87–108

Tulloch A, Hill A, Spencer J (1967) A new type of macrocyclic lactone from Torulopsis apicola. Chem Commun (Camb) 7:584–586

Van Hamme JD, Singh A, Ward OP (2006) Physiological aspects: part 1 in a series of papers devoted to surfactants in microbiology and biotechnology. Biotechnol Adv 24:604–620

Vanittanakom N, Loeffler W, Koch U, Jung G (1986) Fengycin-a novel antifungal lipopeptide antibiotic produced by Bacillus subtilis F-29-3. J Antibiot 39:888–901

Vater J, Stein T, Vollenbroich D, Kruft V, Wittmann-Liebold B, Franke P, Liu L, Zuber P (1997) The modular organization of multifunctional peptide synthetases. J Protein Chem 16:557–564

Veenanadig N, Gowthaman M, Karanth N (2000) Scale up studies for the production of biosurfactant in packed column bioreactor. Bioprocess Eng 22:95–99

Vollenbroich D, Özel M, Vater J, Kamp RM, Pauli G (1997) Mechanism of inactivation of enveloped viruses by the biosurfactant surfactin from Bacillus subtilis. Biologicals 25:289–297

Wu Y, Lou Z, Miao Y, Yu Y, Dong H, Peng W, Bartlam M, Li X, Rao Z (2010) Structures of EV71 RNA-dependent RNA polymerase in complex with substrate and analogue provide a drug target against the hand-foot-and-mouth disease pandemic in China. Protein Cell 1:491–500

Wu Y-S, Ngai S-C, Goh B-H, Chan K-G, Lee L-H, Chuah L-H (2017) Anticancer activities of surfactin and potential application of nanotechnology assisted surfactin delivery. Front Pharmacol 8:761

Zhang Y, Liu C, Dong B, Ma X, Hou L, Cao X, Wang C (2015) Anti-inflammatory activity and mechanism of surfactin in lipopolysaccharide-activated macrophages. Inflammation 38:756–764

Zhao X, Geltinger C, Kishikawa S, Ohshima K, Murata T, Nomura N, Nakahara T, Yokoyama KK (2000) Treatment of mouse melanoma cells with phorbol 12-myristate 13-acetate counteracts mannosylerythritol lipid-induced growth arrest and apoptosis. Cytotechnology 33:123–130

Anti-Cancer Biosurfactants

8

Flávia F. Magalhães, João C. F. Nunes, Marília T. Araújo, Ana M. Ferreira, Mafalda R. Almeida, Mara G. Freire, and Ana P. M. Tavares

Abstract

Biosurfactants produced by different types of microorganisms are amphiphilic biomolecules with active surface properties. Recently, biosurfactants have emerged as promising agents for cancer therapy since a high diversity of these molecules has shown the ability to induce cytotoxicity against many cancer cell lines, thus regulating cancer progression processes. In this sense, microbial biosurfactants are a potential alternative to current cancer therapeutics and as drug delivery systems of anti-cancer drugs. This chapter covers the current knowledge of microbial surfactants with anti-cancer potential, providing information on their production, structure, and research on diverse cancer cell lines, namely, breast and lungs cancer, leukemia, melanoma, and colon cancer. The potential application of biosurfactants in drug delivery is also reviewed.

Keywords

Microbial surfactants · Structure · Production · Anti-cancer activity · Drug delivery

F. F. Magalhães · J. C. F. Nunes · A. M. Ferreira · M. R. Almeida · M. G. Freire · A. P. M. Tavares (✉)
Department of Chemistry, CICECO—Aveiro Institute of Materials, University of Aveiro, Aveiro, Portugal
e-mail: aptavares@ua.pt

M. T. Araújo
Faculdade de Medicina, Universidade do Porto, Porto, Portugal

Inamuddin et al. (eds.), *Microbial Biosurfactants*, Environmental and Microbial Biotechnology, https://doi.org/10.1007/978-981-15-6607-3_8

8.1 Introduction

Cancer is one of the main cause of death worldwide. The World Health Organization (WHO) presented in 2018 data estimating 9.6 million deaths and new cases growing to 18.1 million (WHO 2018). Chemotherapy remains one of the major options in cancer treatment, despite having known limitations and undesirable side effects (Feinberg et al. 2019). Chemotherapeutic drugs target nonspecific, highly proliferative cells, presenting toxicity to normal tissues, and interfering with the life quality of cancer patients (Sak 2012). Thus, many efforts have been made to find new drugs for cancer treatment focusing on novel agents that selectively target cancer cells (Gudiña et al. 2016). Recently, due to their ability to control some functions on mammalian cells, biosurfactants have demonstrated capability in the treatment of cancer (Gudiña et al. 2013).

Biosurfactants, amphiphilic molecules with emulsifying and surface activity, are obtained from microorganisms. With a wide-ranging of chemical structures, these compounds show diverse properties and physiological functions (Rodrigues and Teixeira 2010). Some of the natural roles of these compounds include a growing of the surface area, ability for regulating the removal and binding of microorganisms from surfaces, bioavailability of hydrophobic subtracts, heavy metal binding, biofilm formation, antibacterial pathogenesis, antimicrobial activity, and quorum sensing (Ron and Rosenberg 2001). Furthermore, biosurfactants usually have lower toxicity and higher biodegradability than synthetic surfactants (Rodrigues and Teixeira 2010; Marchant and Banat 2012). Some biosurfactants have also a potential as biologically active compounds, being suitable therapeutic alternatives to synthetic drugs (Banat et al. 2000; Singh and Cameotra 2004). Aditionally, biosurfactants have been explored in gene delivery (Igarashi et al. 2006), drug delivery (Ag Seleci et al. 2016), as adjuvants in immunology (Cameotra and Makkar 2004), as antimicrobial (Ndlovu et al. 2017), antifungal (Sen et al. 2017) and antiviral (Kracht et al. 1999) agents, and as anti-cancer therapeutics (Gudiña et al. 2016).

The anti-cancer activity of biosurfactants is related to their capacity to inhibit cancer cells growth (Sivapathasekaran et al. 2010), apoptosis (cell death) induction (Chen et al. 2006), activity on differentiation (Isoda and Nakahara 1997), necrosis (Fu et al. 2008), and cell cycle arrest (Chen et al. 2006). These biosurfactants comprise mannosylerythritol lipids (MELs) (Shu et al. 2019; Coelho et al. 2020), succinoyl trehalose lipids (STLs) (Sudo et al. 2000), sophorolipids (Ribeiro et al. 2015), rhamnolipids (RLs), surfactin (Wu et al. 2017), serrawettins (Perez Tomas et al. 2005; Clements et al. 2019), and monoolein (Chiewpattanakul et al. 2010). In addition to the high potential of biosurfactants as anti-cancer therapeutics, these molecules can also be incorporated or used as vehicles or drug delivery systems (DDS) of anti-cancer drugs (Wu et al. 2017).

This chapter discusses the current research and knowledge of microbial biosurfactants as anti-cancer drugs, with emphasis on their structure, production, DDS, and potential anti-cancer activity in the treatment of breast and lung cancer, leukemia, melanoma, and colon cancer.

8.2 Biosurfactants Classification and Structure

Biosurfactants are mainly categorized in high- and low-molecular-weight compounds. The first class encompasses phospholipids, fatty acids, lipoproteins, lipopeptides, and glycolipids, while the second group comprises polymeric biosurfactants (Bajpai Tripathy and Mishra 2016). In this chapter, only biosurfactants with anti-cancer properties, namely glycolipids and lipopetides, are reviewed.

8.2.1 Mannosylerythritol Lipids (MELs)

Mannosylerythritol lipids (MELs) were first characterized in 1970 by Bhattacharjee et al. (1970). MELs are glycolipids with a acylated derivative of 4-*O*-β-D-mannopyranosyl-D-erythritol, and fatty acids representing the hydrophobic groups (Bhattacharjee et al. 1970). The structural variants of MELs occur due to: (1) the number and position of the acetyls group on erythritol and mannose; (2) acyl groups in erythritol and mannose amount; and (3) fatty acid chain length and saturation (Yu et al. 2015; Arutchelvi et al. 2008). Based on the level of acetylation in mannopyranosyl, namely, at C-6′ and C-4′ position, MELs are categorized as MEL-A, MEL-B, MEL-C, and MEL-D (Fig. 8.1). MEL-A is diacetylated, while MEL-B is monoacetylated at C-6′ and MEL-C at C-4′. The completely deacetylated compound is known as MEL-D (Yu et al. 2015; Arutchelvi et al. 2008). Novel types of MELs have been reported and identified as tri-acetylated and mono-acylated, with C-6′, C-4′, and C-2′ of mannopyranosyl connected with OH or OAc (Yu et al. 2015). Fukuoka et al. (2008) found a MEL-B diastereomer type with a sugar moiety, named as 1-*O*-β-D-mannopyranosyl-erythritol, but distinct from the 4-*O*-β-D-mannopyranosyl-erythritol of traditional MELs, in a stereochemical way. Morita et al. (2009) reported a distinct MEL containing mannitol as the hydrophilic part instead of erythritol, and a mannosyl-mannitol lipid. The hydrophobic part of MELs contains C18:1, C18:0, C16:1, C16:0, C14:1, C14:0, C12:0 and C2:0 fatty acids. MELs fatty acid profiles are very diverse according to the species (same genus), with

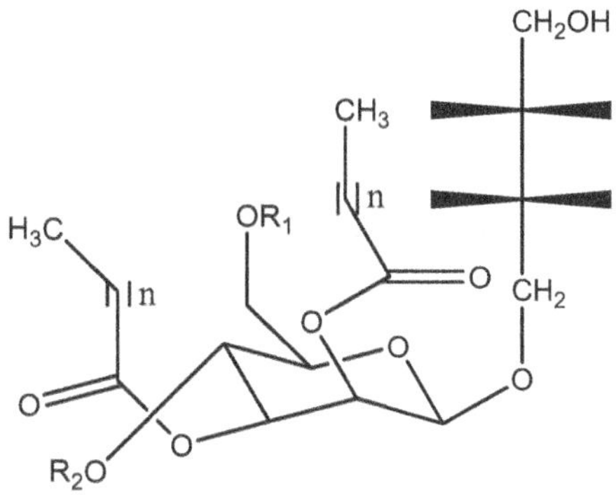

Fig. 8.1 Structure of mannosylerythritol lipids (MELs): MEL A: $R_1 = R_2 =$ Ac; MEL-B $R_1 =$ Ac, $R_2 =$ H; MEL-C: $R_1 =$ H, $R_2 =$ Ac; MEL-D: $R_1 =$ H, $R_2 =$ H. $n = 6–10$. (Adapted from Arutchelvi et al. (2008))

one MEL produced as a main product (Yu et al. 2015). For example, MEL-A was the principal compound produced by *Candida pseudozyma* sp. SY16, with C14:1, C14:0, C12:0, and C6:0 (Kim et al. 1999), while MEL-C was the major MEL produced by *Pseudozyma hubeiensis* KM-59 with C16:2, C12:0, C10:0, and C6:0 (Morita et al. 2007).

8.2.2 Succinoyl Trehalose Lipids (STLs)

Succinoyl trehalose lipids (STLs) are glycolipids synthesized from *n*-alkanes and the most exciting form of trehalose lipids (Inaba et al. 2013). The chemical structure of STLs have two or three fatty acids and one or two succinic acids linked to a trehalose portion (Inaba et al. 2013; Tokumoto et al. 2009). The main STL-1 portion is described as 3,4-di-*O*-palmitoyl- 2,2′-di-*O*-succinoyl-α,α-trehalose, whilst STL-2 and STL-3 are known as 2,3,4-di-*O*-alkanoyl-*O*-succinoyl-α,α-trehalose and 2,3,4,2′-mono-*O*-succinoyl-tri-*O*-alkanoyl-trehalose, respectively (Fig. 8.2) (Jana et al. 2017). The acyl chains proper position in STL-2 and STL-3 is not validated, but concerning the STL-1 structure, it is assumed that succinic acid is in the O2

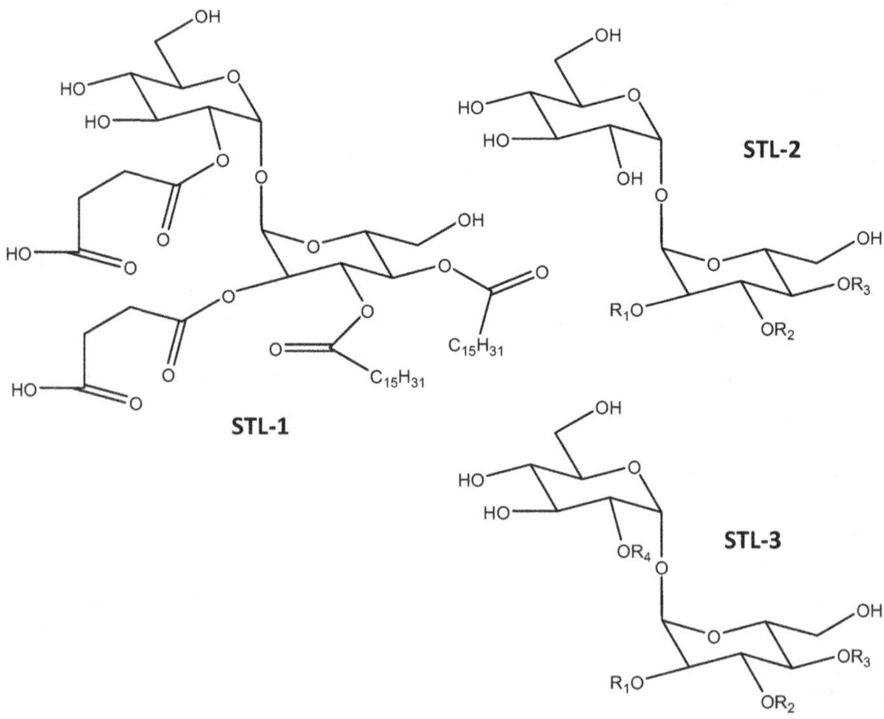

Fig. 8.2 Structures of succinoyl trehalose lipids (STLs): STL-2: R1 − R3 = 1×succionyl + 2×alkanoyl; STL-3: R1 − R4 = 1×succionyl + 3×alkanoyl. (Adapted from Jana et al. (2017))

position (Jana et al. 2017). Uchida et al. (1989) reported the synthesis of two principal forms of STL homologues from *Rhodococcus* sp. strain SD-74, STL-1 and STL-2. These compounds have hydrophobic acyl groups and present the identical carbon chain length used by *n*-alkane as substrate. Tokumoto et al. (2009) reported the structural characterization of STL-1 (Fig. 8.2), with trehalose lipid comprising a hexadecanoyl residue and two succinoyl residues as described by Uchida et al. (1989). The purified STL-1 and its fatty acid structure was assessed by gas chromatography–mass spectrometry, with C16 (63.5%) being the major fatty acid found, while C14 (26.6%) and C12 (9.9%) were found in less amounts.

8.2.3 Sophorolipids

Sophorolipids are biosurfactants composed of a residue of sophorose (acetylated 2-*O*-β-D-glucopyranosyl-D-glucopyranose), a disaccharide of glucose residues linked by a β-1,2′ bond, and a hydroxy fatty acid with a long-chain (Fig. 8.3a) (Price et al. 2012). These glycolipids can be different depending on the position and number of acetate groups as *O*-substituents in the carbohydrate residue and the fatty acid residues (Kulakovskaya and Kulakovskaya 2014). Sophorose may be acetylated in positions 6′- and/or 6, and one terminal or subterminal of the hydroxylated fatty acid is β-glycosidically linked to the sophorose molecule (Price et al. 2012). The hydroxy fatty acid residue is generally C18 or C16 and may have one or more unsaturated bonds (Fig. 8.3a) (Price et al. 2012). Moreover, the carboxylic group of fatty acid is free (open form) or esterified (lactonic form) (Fig. 8.3b) (Kulakovskaya and Kulakovskaya 2014). Sophorolipids can exist as lactones, and as monomeric or dimeric forms containing C22 fatty acid residues (Nunez et al. 2004). Sophorolipids from *Starmerella bombicola* and *Candida batistae* are different in the hydroxylic group position in the fatty acid residue: in sophorolipids from *S. bombicola* the fatty acid residues are hydroxylated mostly in the ω-1 position, whereas those from *C. batistae* are hydroxylated mostly in the ω-position (Konishi et al. 2008). *Candida apicola* is able to synthesize non-acetyl, mono-*O*-acetyl, and di-*O*-acetyl sophorolipids (Price et al. 2012). *Rhodotorula bogoriensis* produces sophorolipids containing a C22 fatty acid residue as an aglycone (Fig. 8.3b) (Nunez et al. 2004).

8.2.4 Rhamnolipids (RLs)

Rhamnolipids (RLs) are glycolipid biosurfactants produced by several bacteria (Abdel-Mawgoud et al. 2010). RLs (Fig. 8.4) were discovered in 1946 by Bergström et al. (1946a, b) and being synthesized from *Pseudomonas pyocyanea*. Later, in 1965, Edwards and Hayashi (1965) found that between the two rhamnose fractions there is an α-1,2-glycosidic bond, due to methylation and periodate oxidation. Therefore, the authors chemically identified these RLs as 2-*O*-α-1,2-L-rhamnopyranosyl-α-L-rhamnopyranosyl-β-hydroxydecanoyl-β-hydroxydecanoate.

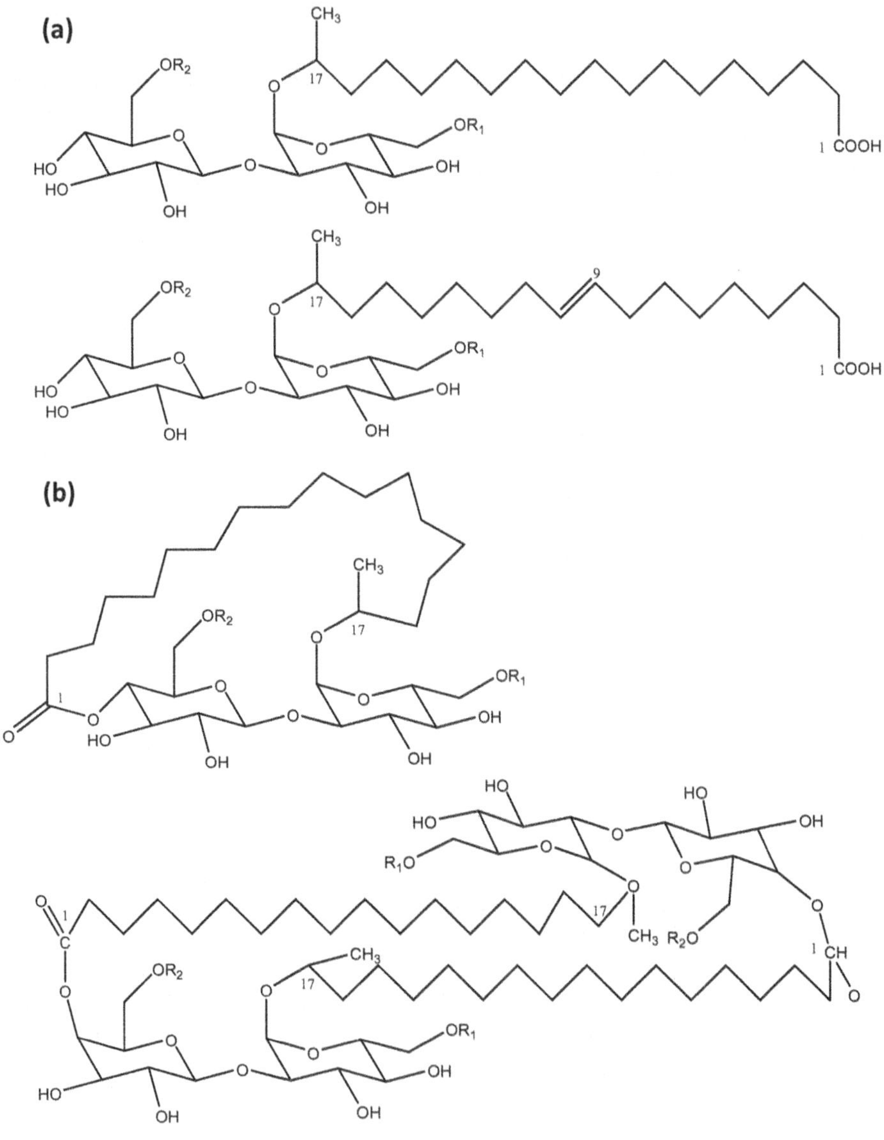

Fig. 8.3 Structures of sophorolipids in acid form (**a**) and in lactone form (**b**). (Adapted from Kulakovskaya and Kulakovskaya (2014))

Overall, eight RLs congeners were found until the mid-1980s (Abdel-Mawgoud et al. 2010).

At the end of the last century, a significant number of new RLs were identified, and novel analogues were frequently described. Therefore, and regarding the structures, RLs can be defined as glycosides with rhamnose and lipidic fractions,

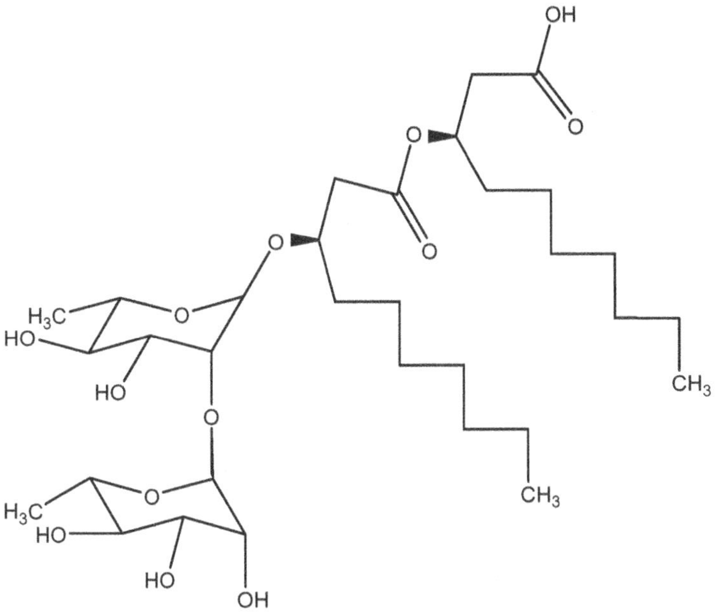

Fig. 8.4 Chemical structure of the first identified rhamnolipid; simply named as α-L-rhamnopyranosyl-α-L-rhamnopyranosyl-β-hydroxydecanoyl-β-hydroxydecanoate and symbolized as Rha-Rha-C10-C10. (Adapted from Abdel-Mawgoud et al. (2010))

both connected by an *O*-glycosidic linkage. The rhamnose part has one (mono-RL) or two (di-RL) rhamnose parts linked by an α-1,2-glycosidic linkage. The lipidic part, nevertheless, is composed of one or two (occasionally three) β-hydroxy fatty acid chains (polyunsaturated, mono-, or saturated, from C16 to C8) connected by an ester bond established among the β-hydroxyl group of the distal chain with the carboxyl group of the proximal chain (Fig. 8.4). In the majority of the cases, the carboxyl group of the distal β-hydroxy fatty acid chain remains free; nevertheless, some analogues have this group esterified with a short alkyl group (Abdel-Mawgoud et al. 2010). Likewise, the distal 2-hydroxyl group (in relation to the glycosidic bond) rhamnose group stays mainly free, even though in some unique homologues it may be acylated with long chain alkenoic acid (Abdel-Mawgoud et al. 2010).

Differences in the chemical structures of RLs lead to several RLs homologues. The variations between these homologues are in the rhamnose and/or the lipidic fractions, contributing to the biodiversity of RLs (Abdel-Mawgoud et al. 2010).

8.2.5 Myrmekiosides

Myrmekiosides are glycolipids biosurfactants with a structure containing mono-*O*-alkyldiglycosylglycerol. These types of biosurfactants are produced from *Myrmekioderma sp.* and *Trikentrion loeve*. These compounds consist of a glycerol

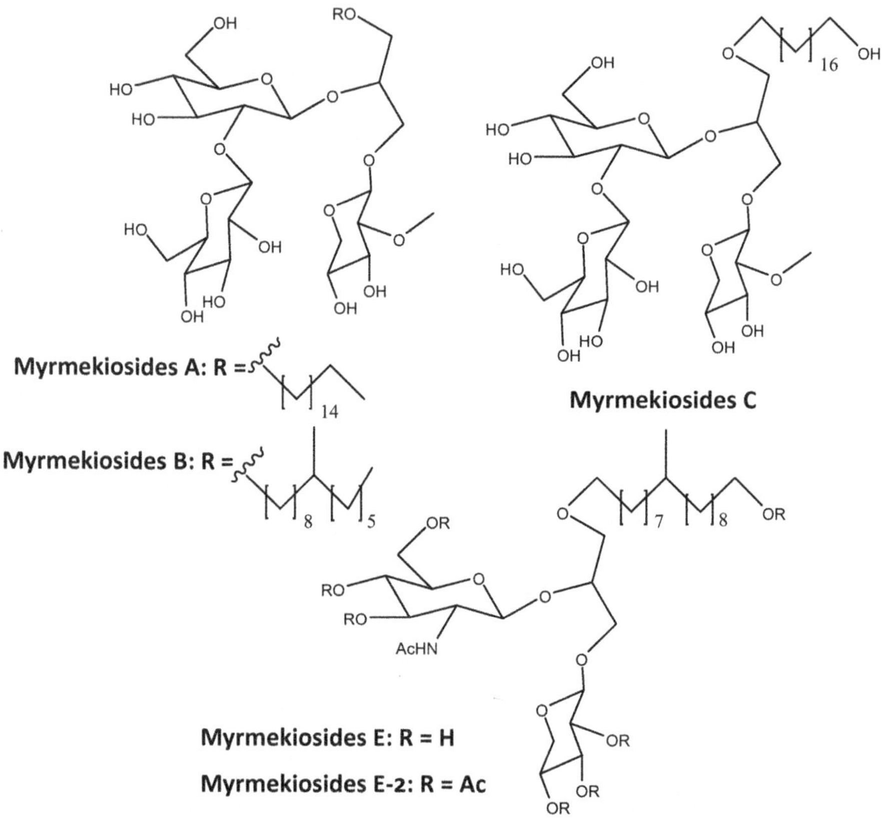

Fig. 8.5 Structure of myrmekiosides. (Adapted from Zhang et al. (2015))

backbone with a mono- or diglucosyl unit or a monoglucosamine residue attached to C-2′, and a xylose and an alkyl chain at the terminal hydroxyl positions (Fig. 8.5). Myrmekiosides A–C, are constituted by similar sugar fractions, however, presenting distinct *O*-alkyl chains. Myrmekioside E-2 is a myrmekioside peracetylated derivative with *N*-acetylglucosamine and xylose, and a terminal alcohol group (Zhang et al. 2015).

8.2.6 Cyclic Lipopeptides (CLPs)

Cyclic lipopeptides (CLPs) are synthesized through different bacteria (Raaijmakers et al. 2006). Structurally, CLPs comprise an oligopeptide peptidically linked to a fatty acid. The lipid chain length (from C6 to C18) as the oxidation degree can differ (Götze et al. 2017). The oligopeptide C-terminus (up to 25 amino acids) creates a lactone with an amino, phenol or hydroxyl functional group from the side chains of the peptide or the lipid fraction, thus providing different cycles of variable sizes (typically from 4 to 16 amino acids) (Götze et al. 2017).

8.2.6.1 Amphisin, Tolaasin, and Syringomycin CLPs

The chemical structure analysis of all CLPs from *Pseudomonas* sp. might be classified into two main groups, and different subgroups (amphisin, tolaasin, and syringomycin groups) (Raaijmakers et al. 2006). The main groups differ in the amino acids content in the cyclic peptide fraction, whereas the subgroups can be different due to specific substitutions in the peptide fraction, and more specifically in the amino acids (Nielsen et al. 2002). Amphisin CLPs, including amphisin and tensin, contain a peptide portion with eleven amino acids, linked to 3-hydroxyoctanoic acid. Both compounds contain helical structures, with a cyclic peptide covering a water molecule. Contrarily, tolaasin group CLPs are distinct because of the several differences in the length and composition of lipid tail (3-hydroxyoctanoic acid) and the peptide chain. The peptide part of tolaasin includes several rare amino acids, such as homoserine and 2,3-dihydro-2-aminobutyric acid. The cyclic fraction of the peptide part comprises from five to eight amino acids while the lactone ring is established among the C-terminal amino acid and the allo-Thr residue. Finally, CLPs in the syringomycin group contain also rare amino acids, including the C-terminal 4-chlorothreonine and 2,4-diamino butyric acid. Moreover, the lactone ring is formed between the C-terminal Thr(4-Cl) and the N-terminal Ser. The CLPs fatty acid tail in the syringomycin group may contain a 3,4-dihydroxy or 3-hydroxy fatty acid (10–14 carbon atoms) (Raaijmakers et al. 2006).

The structure of other CLPs from *Pseudomonas* sp. has been elucidated in the last years. For example, a new compound was recognized as CLP (pseudofactin II) containing an N-terminal group of eight amino acids in the peptide fraction linked to a palmitic acid. The carboxylic group of the final amino acid forms a lactone with the hydroxyl of Thr3 (Janek et al. 2012).

8.2.6.2 Iturin and fengycin CLPs

CLPs produced by other bacteria have been also reported. Iturin (Fig. 8.6a) from *Bacillus subtillis* is a cyclic peptide containing seven amino acids connected to a

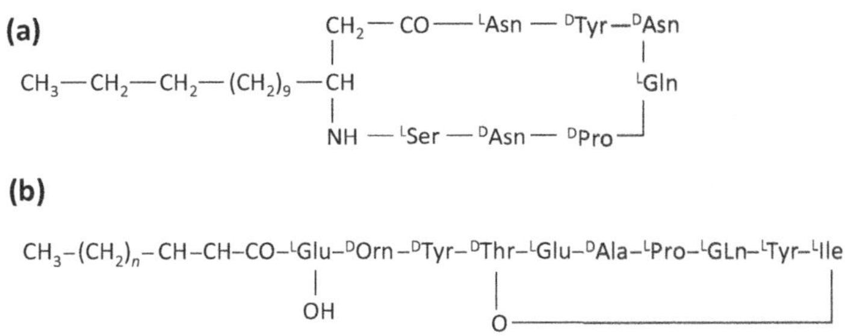

Fig. 8.6 Primary structure of iturin (**a**); primary structure of fengycin, $n = 14–17$ (**b**). (Adapted from Meena and Kanwar (2015))

β-amino fatty acid chain which differ in the number of carbon (C14–C17) (Meena and Kanwar 2015). Mixirins A–C (three cyclic acylpeptides) included in the iturin class have been also synthesized by *Bacillus* sp. (Mondol et al. 2013).

Fengycin (Fig. 8.6b), also produced by *B. subtillis*, contains saturated or unsaturated bonds in the β-hydroxy fatty acid of the lactone ring (Meena and Kanwar 2015). The structure of fengycin contains a fatty acid chain varying from C14 to C17 carbon atoms linked to a peptide chain of ten amino acids, with isomers and different homologous compounds giving rise (Fig. 8.6b). Moreover, fengycin compounds display heterogeneity in the peptide fraction (6th position) and in chain length of the β-hydroxy fatty acid, being classified as fengycin A (comprises Ala at position 6) and fengycin B (contains Val at position 6) (Meena and Kanwar 2015).

8.2.6.3 Surfactin CLP

The surfactin biosurfactant, produced by different strains of *B. subtilis*, is the most studied CLP. Surfactin is constituted by a heptapeptide with a chiral sequence connected to the chain of β-hydroxy fatty acid, which forms a closed cyclic lactone ring structure (Fig. 8.7) (Wu et al. 2017; Tsan et al. 2007). Surfactin hydrophobic amino acids are found at the 2′, 3′, 4′, 6′, and 7′ positions. On the other hand, hydrophilic aspartyl and glutamyl residues are placed in positions 5′ and 1′, giving the molecule two negative charges (Wu et al. 2017). Two conformations of surfactin, S1 and S2, have been found by Bonmatin et al. (1994). Surfactins S1 and S2 exhibit a conformation with two charged side chains gathered on the same side. Surfactin conformations form a "claw" and offer a hydrophilic head and opposite to a hydrophobic domain (Tsan et al. 2007). Baumgart et al. (1991) described that *B. subtilis* OKB 105 and ATCC 21332 produced three types of surfactin compounds. Among them, two are specifically different in the Leu amino acid, since it is substituted by Val and Ile. In another work, Liu et al. (2007) reported that *B. subtilis* produced three distinct surfactins in the peptide fractions: (1) N-Asp-Leu-Leu-Val-Glu-Leu-Leu-C sequence; (2) N-Glu-Leu-Leu-Val-Asp-Leu-Leu-C sequence; (3) peptide chain methyl esterified. Moreover, these biosurfactants contain different fatty acid fractions, such as iso C12, iso C13, anteiso C13, iso C14, n C14, iso C15, anteiso C15, n C15, anteiso C16, and anteiso C17 beta-hydroxy fatty acids (Liu et al. 2007).

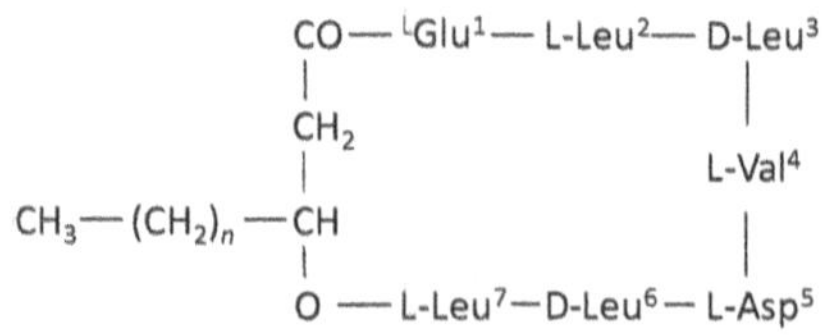

Fig. 8.7 Primary structure of surfactin, $n = 9–11$. (Adapted from Wu et al. (2017))

8.2.7 Rakicidns and Apratoxins

Rakicidins contain a 3-hydroxyfatty acid and three amino acids. So far, four different compounds (Fig. 8.8), namely, rakicidins A and B produced by *Micromonospora* sp. and rakicidins C and D by *Streptomyces* sp., are described (Igarashi et al. 2010). All rakicidins share mutual amino acid constituents, namely, glycine, 4-amino-2,4-pentadienoate and hydroxyaspargine (or glutamine in rakicidin C), and an unit of 3-hydroxyfatty acid differing in the methylation pattern and chain length (Valliappan et al. 2015).

Marine cyanobacteria produce several secondary metabolites with fascinating biological characteristics and molecular constructions, namely, apratoxins. Apratoxins are cyclodepsipeptides with *N*-methylated amino acids, a proline residue, a dehydroxylated fatty acid fraction, and a modified cysteine residue (Fig. 8.9) (Masuda et al. 2014). Apratoxins A to G have been categorized and several modifications, comprising the nonexistence of a C- or *N*-methyl group at different

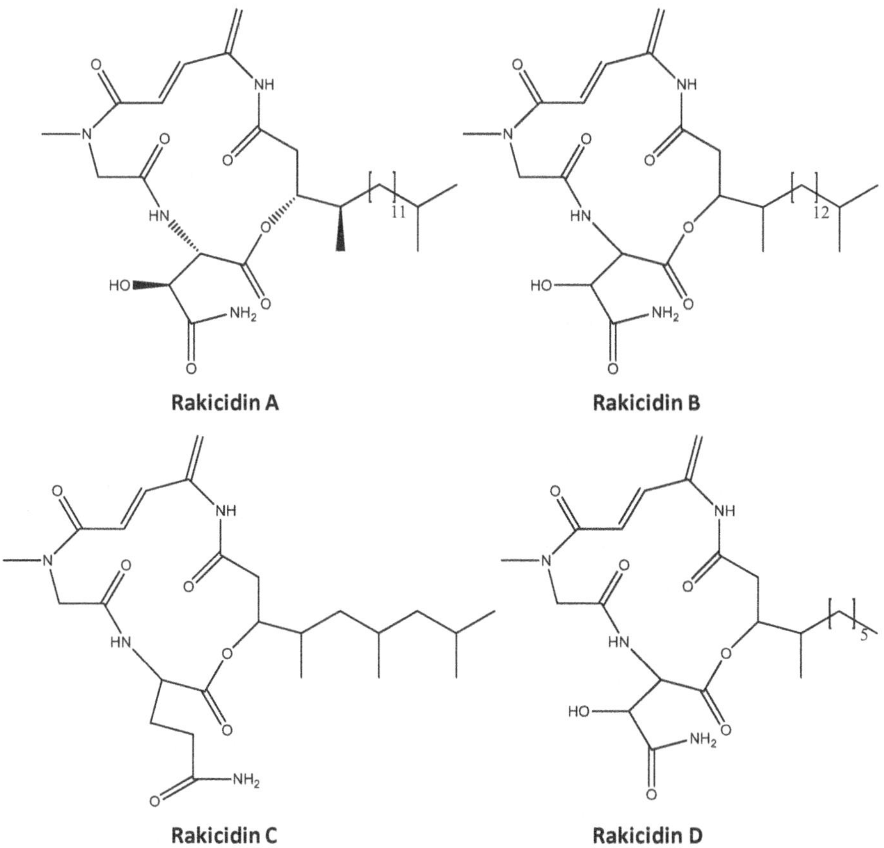

Fig. 8.8 Structure of rakicidins A, B, C, and D. (Adapted from Sang et al. (2016))

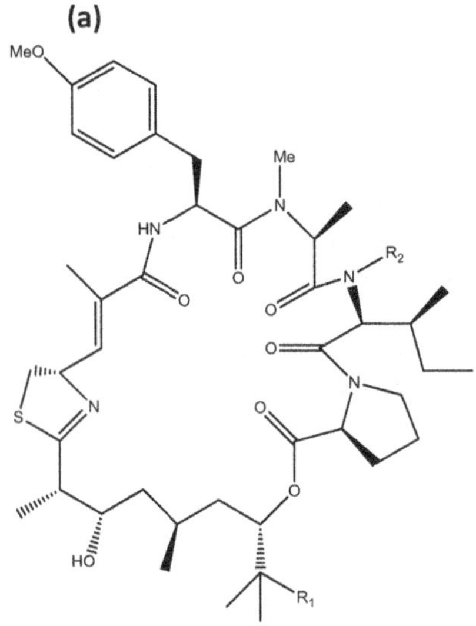

Fig. 8.9 Chemical structures of apratoxins: (**a**) apratoxin A: $R_1 = Me$, $R_2 = Me$; apratoxin B: $R_1 = Me$, $R_2 = H$; apratoxin C: $R_1 = H$, $R_2 = Me$; (**b**) apratoxin D. (Adapted from Masuda et al. (2014))

sites, a replacement of a terminating proline residue with an *N*-methyl alanine, and an additional polyketide synthases module early in the biosynthetic sequence, were observed (Nunnery et al. 2010).

8.2.8 Serrawettins

Serrawettins are produced by different *Serratia* genera and were first isolated in 1985. Specifically, serrawettin W1 and serrawettin W2 are biosynthesized by *S. marcescens* and *S. surfactantfaciens* strains, respectively (Clements et al. 2019).

Serrawettin W1 (or serratamolide A (Fig. 8.10)), includes in its structure a symmetrical dilactone with two L-serine amino acids connected to two β-hydroxy-based fatty acids (Eckelmann et al. 2018). However, different homologues have been described, namely, serratamolides B to G, related to differences in the chain length of the fatty acid (from C8 to C14) and the absence or presence of double bonds in the structure of serrawettin W1 (serratamolide A) (Clements et al. 2019).

Otherwise, serrawettin W2 comprises five amino acids (D-leucine/isoleucine-L-serine-L-threonine-D-phenylalanine-L-isoleucine/ leucine) connected to the fatty acid fraction of β-hydroxy, and variations in the chain length of the fatty acid (C8 or C10) or the 1st, 2nd, or 5th amino acid positions result in serrawettin W2 equivalents (Clements et al. 2019).

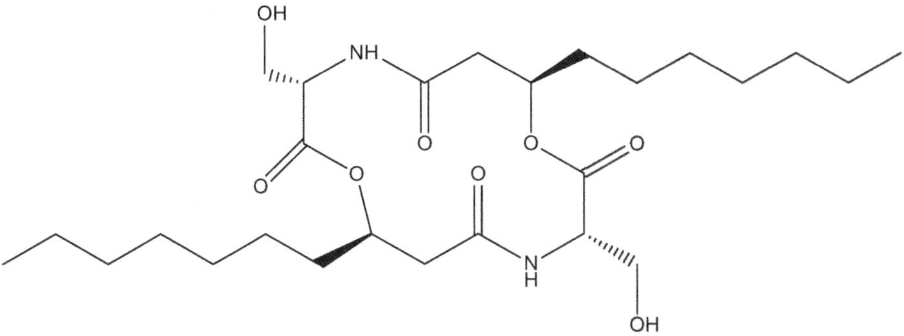

Fig. 8.10 Structure of serrawettin W1 (serratamolide A). (Adapted from Shanks et al. (2012))

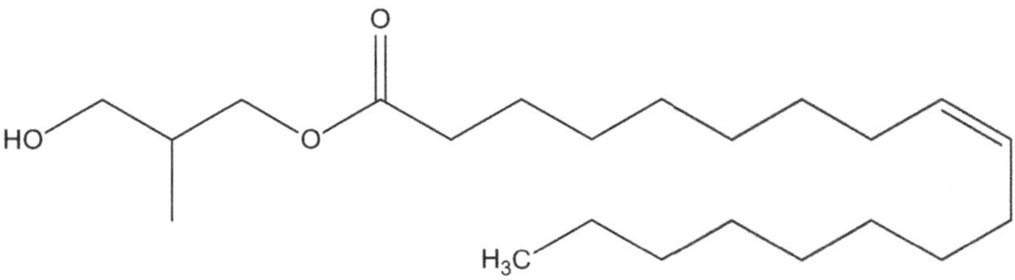

Fig. 8.11 Structure of monnolein. (Adapted from Kulkarni et al. (2011))

8.2.9 Monoolein

Monoolein (1-Monoolein (1-(cis-9-Octadecenoyl)-rac-glycerol)) includes a chain of hydrocarbon (oleic acid) connected, by an ester bond, to the glycerol backbone (Fig. 8.11). The other two hydroxyl groups from the glycerol fraction give hydrophilic characteristics to this part of the compound (Kulkarni et al. 2011; Ganem et al. 2000). In an aqueous environment, the glycerol part can establish hydrogen bonds with water, and is commonly referred to the head group (Ganem et al. 2000). The strongly hydrophobic hydrocarbon chain includes a *cis* double bond at the nine and tenth positions (Kulkarni et al. 2011; Ganem et al. 2000), making monoolein a molecule with amphiphilic characters.

8.2.10 Fellutamides

Fellutamides are a class of lipopeptide biosurfactants containing a (3R)-β-hydroxy alkanoate tail and a C-terminal aldehyde (Fig. 8.12). A, C, and D types of fellutamides include a β-L-threo-hydroxy-glutamine amino acid, whereas fellutamides A and B contain a fatty chain based on β-hydroxylated amide, which is originated from (3R)-hydroxy lauric acid (Giltrap et al. 2013). However, in the literature other distinct products were called fellutamide C, and therefore, to solve

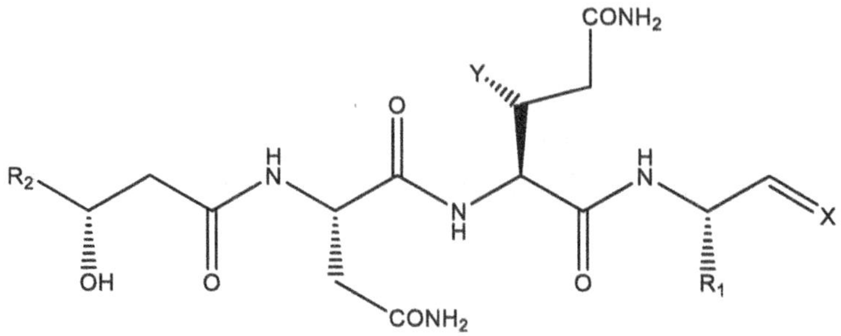

Fellutamide	R_1	R_2	X	Y
A	iBu	C_9H_{19}	O	OH
B	iBu	C_9H_{19}	O	H
C	iBu	C_9H_{19}	H, OH	H
D	iBu	$C_{11}H_{23}$	O	OH
E	iPr	$C_{11}H_{23}$	O	OH
F	iAmyl	C_9H_{19}	OH, OH	H

Fig. 8.12 Structure of fellutamides. (Adapted from Pirrung et al. (2016))

the similarity between them, Singh's fellutamide C has been renamed as fellutamide E (Pirrung et al. 2016).

8.3 Biosurfactants Production

The manufacture of biosurfactants should be performed by applying secure and nonpathogenic microorganisms to prevent problems with pathogenicity (Ghasemi et al. 2019). Furthermore, most biosurfactants are considered secondary metabolites, playing crucial roles for the survival of the producing microorganisms, since they facilitate nutrient transport, promote microbe-host interactions, and act as natural green biocides (Van Hamme et al. 2006). Accordingly, many biosurfactant applications substantially depend on whether they can be economically produced. Thus, extensive efforts have been made in process optimization. Biosurfactants production from low-cost raw materials and cheap substrates can decrease production cost (Mukherjee et al. 2006). In this sense, substrates such as olive oil mill effluents, corn steep liquor, vegetable cooking oil waste, animal fat, soap stock, and dairy industry waste, among others, have been deeply investigated (Santos et al. 2016). The selection of waste biomass for biosurfactant production should guarantee a balanced supply of nutrients to ensure a proper microbial growth (Santos et al.

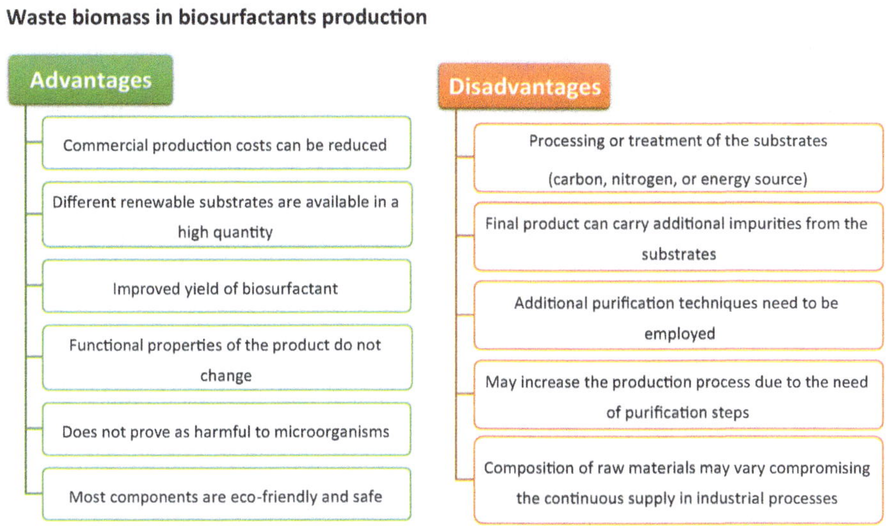

Fig. 8.13 Advantages and disadvantages associated with the use of waste biomass in biosurfactants production. (Adapted from Banat et al. (2014))

2016). Nevertheless, it should take into account that the use of substrates with reduced prices for biosurfactants production does not bring only advantages, but also some disadvantages, as summarized in Fig. 8.13. Banat et al. (2014) discussed this topic (renewable sources) and the cost-effectiveness of the related processes in their review.

In summary, the development of economically competitive biosurfactant production processes is urgent, and must include the culture conditions optimization and the development of cost-effective recovery processes to improve the yield and quality of biosurfactants.

8.3.1 Factors Involved in Biosurfactants Production

The production of biosurfactants (type, quality and quantity) is usually affected by the sources of nitrogen and carbon, and by the concentration of different ions (phosphorous, magnesium, ferric, and manganese) present in the broth medium (Bajpai Tripathy and Mishra 2016). Moreover, the growth conditions of the culture can be affected by the biosurfactant synthesis procedure, i.e., agitation speed, temperature, pH, aeration, and dilution rate (Bajpai Tripathy and Mishra 2016). During their production, different preliminary techniques have been used to find and choose the more effective factors/conditions to achieve a high yield and quality of biosurfactants (Bajpai Tripathy and Mishra 2016). These factors are discussed below.

8.3.1.1 Source of Carbon

The growth of microorganisms and biosurfactants production are closely related to the carbon source. Usually, the sources of carbon applied in biosurfactant production are carbohydrates, hydrocarbons, and vegetable oils (Kaskatepe and Yildiz 2016). Among them, aqueous soluble carbon sources like dextrose, sucrose, fructose, and glycerin are normally applied (Bajpai Tripathy and Mishra 2016). Reports from the literature indicated that the application of carbon mixtures in biosurfactants production can be more efficient than the use of one source. For instance, Cooper and Paddock (1984) demonstrated that the use of a carbon source composed only of glucose or vegetable oil does not change the biosurfactant yield (1 g/L) by *T. bombicola.* However, if both sources were supplied at the same time, the yield improved up to 70 g/L (Cooper and Paddock 1984). A similar behavior was found for the sophorolipids production from *C. bombicola*, with the best production yield (120 g/L) obtained using a mixture of carbon sources (sugar and oil) (Casas et al. 1997). Furthermore, when an industrial waste (soap stock) was used for the production of sophorolipids by *C. apicola* and *Candida antarctica*, yields of 7.3 g/L and 13.4 were obtained, respectively (Bednarski et al. 2004). Additionally, *Pseudozyma* (*C. antarctica*) converted C12 to C18 *n*-alkanes into MELs with a yield of 140 g/L applying the following carbon source: soybean oil (Kitamoto et al. 2001).

8.3.1.2 Source of Nitrogen

Nitrogen source is a crucial parameter in the production of biosurfactants by microorganisms, since it is essential for growth and the regulation of protein synthesis (Bajpai Tripathy and Mishra 2016). Distinct sources of inorganic and organic nitrogen, such as sodium nitrate, ammonium sulfate, urea, and ammonium nitrate have been used in the production of biosurfactants (Bajpai Tripathy and Mishra 2016). Usually, a high carbon/nitrogen (C/N) ratio conducts to a reduction in cell growth and a rise in the cell metabolism (Santos et al. 2016). On the other hand, a low C/N ratio leads to a reduction in cell growth and an increase in cell metabolism (Santos et al. 2016). For instance, biosurfactant production using *Arthrobacter paraffineus* was enhanced when nitrogen sources such as ammonium salts and urea were used, while by *Pseudomonas aeruginosa* and *Rhodococcus* sp. the highest yield was obtained when nitrates were used (Bajpai Tripathy and Mishra 2016). In a similar work, Mulligan and Gibbs (1989) reported biosurfactant production by *P. aeruginosa* using ammonium, amino acids, and nitrates as nitrogen sources. In general, when compared to ammonium, the assimilation of nitrate is usually lower, simulating nitrogen restriction, being one advantage in the production of a biosurfactant (Santos et al. 2016). As an example, the overproduction of RLs biosurfactants by *Pseudomonas* sp. strain DSM-2874 was achieved by nitrogen restriction at the beginning of the stationary phase growing (Mouafo et al. 2018). Amani et al. (2013) also showed maximum RLs production after nitrogen limitation (120 mg/L). As reported by Hommel et al. (1987), nitrogen concentration appears to be a mandatory factor for a high biosurfactant yield.

8.3.1.3 Effect of Ions

The addition of multivalent cations such as magnesium, ferric, manganese, among others, into the culture media has been found to affect the production of biosurfactants (Bajpai Tripathy and Mishra 2016). More specifically, the limitation of multivalent cations is reported to enhance the production of biosurfactants (Bajpai Tripathy and Mishra 2016). For instance, better yields of RLs produced by *B. subtilis* were achieved by reducing the concentrations of calcium, sodium, potassium, and magnesium salts and residue elements (Bajpai Tripathy and Mishra 2016). Furthermore, the addition of other chemicals like ethylenediaminetetraacetic acid (EDTA), ethambutol, chloramphenicol, and penicillin influence the production of biosurfactants (Karanth et al. 1999). The biosurfactant production control, using these type of compounds, is produced by their impact on the solubilization of apolar substrates or through their impact on the production of polar substrates.

8.3.1.4 Physical Factors

Growth culture conditions, namely, pH, temperature, time, and agitation speed, also affect biosurfactants production (Bajpai Tripathy and Mishra 2016). For example, at a pH range between 6.0 and 6.5, the production of RLs by *Pseudomonas* sp. was highest, decreasing considerably at pH values above 7.0. On the other hand, the production of surfactin by *B. subtilis* was favored at neutral pH (Abdel-Mawgoud et al. 2008), whereas the production of sophorolipids by *Candida batistae* was maximized at pH 6.0 (Konishi et al. 2008).

Overall, the most favorable temperature during biosurfactants production is around 30 °C, as verified for several species of *Candida* (Santos et al. 2016). The same temperature conditions were verified during surfactin production by *B. subtilis* (Hmidet et al. 2017). However, the highest RLs concentration (1892 mg/mL) was reached when the fermentation of *P. aeruginosa* was carried out at 42 °C (Eraqi et al. 2016). In relation to the incubation time, it was found that after 5 days the maximum production of RLs by *P. aeruginosa* was achieved (Kaskatepe and Yildiz 2016), whereas the incubation periods for *C. bombicola* in the production of sophorolipids were from 7 to 11 days (Mouafo et al. 2018; Felse et al. 2007). Finally, the agitation is also an important factor since it is responsible for an efficient oxygen transfer during the biosurfactant production from the phase of gas to the phase of liquid. Oliveira et al. (2009) found the rise in agitation favored the production of RLs by *Pseudomonas alcaligenes*. Wei et al. (2005) studied the agitation speeds (between 50 and 250 rpm) in the production of RLs by *P. aeruginosa* and observed a better result at 200 rpm. However, other studies have concluded that high agitation speed (>500 rpm) had a negative effect on surfactin production by *B. subtilis* (Ha et al. 2018).

8.4 Anti-Cancer Activity of Biosurfactants

Biosurfactants have been investigated as anti-cancer compounds due to their promising intercellular recognition steps, which comprise selective block cancer cells proliferation by signal transduction (Rodrigues et al. 2006). The first approach in cancer treatment includes induction of terminal differentiation and apoptosis (the main method of programmed cell death) pathways of cancer cells (Rodrigues et al. 2006; Reed 2003). Diverse mechanisms have been proposed to describe biosurfactants' anti-cancer capacity, such as: (1) cell evolution delay; (2) inhibition of signaling pathways; (3) apoptosis induction by death receptors in cancer cells; (4) natural killer T (NKT) cells stimulation; and (5) angiogenesis decrease (Fig. 8.14) (Duarte et al. 2014). Moreover, biosurfactants are able to disrupt cell membranes by lysis and increasing membrane permeability (Gudiña et al. 2013).

Hannun and Bell (1989) reported a review article regarding the discovery of glycosphingolipids and lysosphingolipids as active compounds in cell differentiation and in cell proliferation modulation in oncogenesis (Hannun and Bell 1989). Since then, a diversity of compounds, such as polar compounds, glucocorticoids, fatty acids with short-chain and retinoids, were noticed to stimulate differentiation in cell

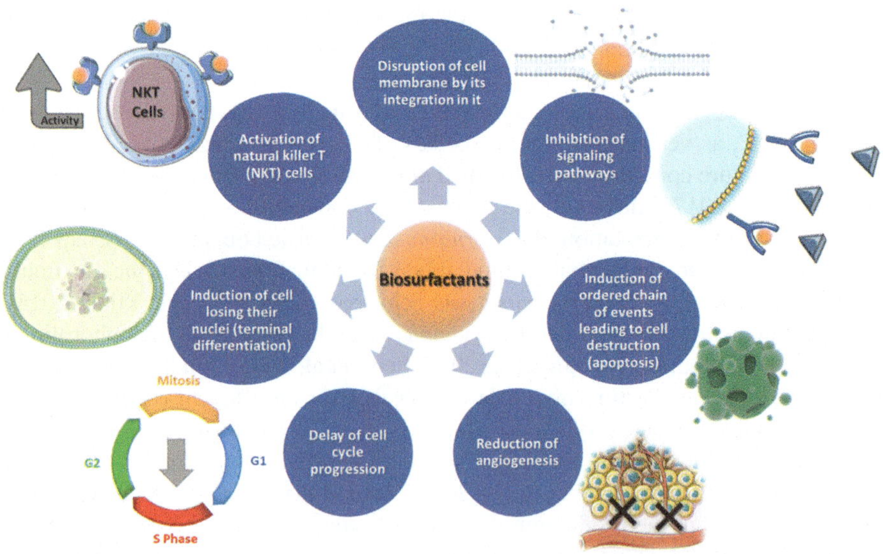

Fig. 8.14 Mechanisms to exemplify the anti-cancer activity of microbial biosurfactants

lines and can trigger apoptotic events (Wakamatsu et al. 2001; Bursch et al. 1992). The current state of biosurfactant as potential therapeutic applications of cancer (breast and lungs cancer, leukemia, melanoma, and colon cancer) are presented in Table 8.1 and discussed in this section.

8.4.1 Breast Cancer

Breast cancer is a common type of cancer affecting one in eight women, with more than one million additional cases per year (Chu and Lu 2008; Davis et al. 2020). Early detection of breast cancer can be an effective strategy to reduce the number of cases; however, in an advanced stage, conventional cancer treatments are required.

Despite improvements in early detection and treatment, around 50% of patients will either not succeed in chemotherapy or will develop resistance to chemotherapeutic drugs (Jana et al. 2020). Biosurfactants have been reported as a potential therapeutic alternative to combat breast cancer. Cao et al. (2009, 2010) showed the apoptosis of MCF-7 human breast cells by surfactin. This biosurfactant produced reactive oxygen species, indicating the involvement of reactive oxygen species generation in surfactin-induced cell death (Cao et al. 2010). Duarte et al. (2014) also studied the anti-tumor capacity of surfactin from *B. subtilis* 573 for both cell lines T47D and MDA-MB-231. The results show that surfactin leads to a decrease in cell viability and proliferation by induced cell cycle arrest at G1 phase after a contact of 48 h, without negatively affecting normal fibroblasts (Duarte et al. 2014). In an additional work, a surfactin from *Micromonospora marina* was tested on MCF-7 breast cancer cell lines (Ramalingam et al. 2019). This compound presented cytotoxicity against cancer cells by inducing apoptosis and cleaving mitochondrial membrane, while did not affecting normal cells (Ramalingam et al. 2019).

Other types of biosurfactants studied against breast cancer cell lines are iturins (lipopeptides), produced by *Bacillus* strains. Iturins were found to significantly lead the initiation of apoptosis in breast cancer cells (Dey et al. 2015; Trischman et al. 1994; Zhang et al. 2004). In vitro and in vivo tests of iturin A were carried out in human breast cancer. The results revealed that apoptosis occurred by proliferation inhibition of the cell lines MCF-7 and MDA-MB-231 (Dey et al. 2015). Moreover, iturin A was able to decrease tumor growth with lowered expressions of proteins based on P-MAPK, CD-31, Ki-67, P-GSK3β, P-FoxO3a, and P-Akt (Dey et al. 2015).

S. bombicola was able to produce biosurfactants, namely sophorolipids. These compounds were evaluated regarding to their cytotoxicity in breast cancer MDA-MB-231 cells (Ribeiro et al. 2015). High cytotoxic effect at the Critical Micelle Concentration (CMC) of the sophorolipids was observed in cancer cells. The higher cytotoxicity was obtained in sophorolipids (C18:0 and C18:1), with compound C18:1 being able to intracellularly increase reactive oxygen species (ROS) involved in cancer cell death, and inhibit cells migration without cellular damage. This finding has a potential in the treatment of tumor growth in initial phases (Ribeiro et al. 2015).

Table 8.1 Anti-tumor activity of biosurfactants against cancer cells

Cancer type	Biosurfactant	Activity	References
Breast cancer	Surfactin	Induces apoptosis, inhibits proliferation, reduces cell viability and induces cell cycle arrest at G1 phase	Cao et al. (2009, 2010) and Ramalingam et al. (2019)
	BioEG from *Lactobacillus paracasei*	*Induces cell cycle arrest at G1 phase*	Duarte et al. (2014)
	Iturin from *Bacillus* sp.	Leads the apoptosis induction and inhibits tumor growth	Dey et al. (2015), Trischman et al. (1994) and Zhang et al. (2004)
	Sophorolipids from *Starmerella bombicola*	Interferes with cell migration and intracellular ROS increase	Ribeiro et al. (2015)
	RLs from *Pseudomonas aeruginosa*	Induces p53 gene	Rahimi et al. (2019)
	Rakicidins from the *Micromonospora*	Interferes with the invasiveness	Poulsen (2011)
Lungs cancer	Surfactin	*Induces cell cycle arrest at G0/G1 phase and induces* apoptosis	Routhu et al. (2019)
	Somocystinamide A from *Lyngbya majuscula*	Induces apoptosis	Wrasidlo et al. (2008)
	Glycolipoprotein from *Acinetobacter M6*	*Decreases cell viability and induces cell cycle arrest at G1 phase*	Karlapudi et al. (2020)
	Fellutamides from *Aspergillus versicolor*	Cytotoxic effects	Lee et al. (2010, 2011)
	Fengycin from *Bacillus subtilis*	Induces cell cycle arrest at the G0/G1 phase and promotes apoptosis	Yin et al. (2013)
	Rakicidin B from *Micromonospora chalcea*	Induces apoptosis	Xie et al. (2011)
	Apratoxins from *Moorea*	Cytotoxic effects	Thornburg et al. (2013)
	Myrmekioside from *Myrmekioderma dendyi*	Cytotoxic effects	Farokhi et al. (2013)
	Dolyemycins A and B	Anti-proliferative effects	Liu et al. (2018)

(continued)

Table 8.1 (continued)

Cancer type	Biosurfactant	Activity	References
Leukemia	STL from *Rhodococcus erythropolis*	Induces cell differentiation, inhibit growth and induce morphological changes	Isoda et al. (1997b)
	MEL from *Candida Antarctica*	Induces cell differentiation	Isoda et al. (1997b)
	Cyclic lipopeptide from *Bacillus natto*	Inhibits cell growth by inducing apoptosis	Wang et al. (2007)
	RLs	Antiproliferation	Shen et al. (2020)
	Iturin from *Bacillus subtilis*	Induces paraptosis and apoptosis, and inhibits autophagy	Zhao et al. (2018)
	Serratamolide from *Serratia marcenses*	Induces apoptosis	Perez Tomas et al. (2005)
	Monoolein from *Exophiala dermatitidis* SK80	Morphological cell changes such as cell shrinkage, membrane blebbing, and DNA fragmentation	Chiewpattanakul et al. (2010)
Melanoma	MEL	Inhibits cell growth and induce apoptosis	Zhao et al. (1999)
	PFII from *Pseudomonas fluorescens*	Induces apoptosis	Janek et al. (2013)
	MEL from *Candida Antarctica*	Induces cell differentiation by promoting apoptosis by the condensation of chromatin, DNA fragmentation, and sub-G1 arrest	Sudo et al. (2000)
	Mixirins from *Bacillus* sp.	Cytotoxic effects and inhibits cancer growth	Zhang et al. (2004)
Colon cancer	Surfactin and fengycin from *B. circulans*	Selective anti-proliferative activity	Sivapathasekaran et al. (2010)
	Marine lipopeptide	Anti-proliferative activity	Das et al. (2015)
	Rakicidins	Cytotoxic effects	
	Serrawettin W2 from *Serratia surfactantfaciens*	Selective cancer cell lines growth suppresion	Su et al. (2016)
	New molecule from *Sphingobacterium detergens*	Anti-proliferative effects and apoptosis activity	Burgos-Díaz et al. (2013)
	Surfactin from *Bacillus subtilis*	Apoptosis induction, cell cycle arrest and survival signaling suppression	Kim et al. (2007)

(continued)

Table 8.1 (continued)

Cancer type	Biosurfactant	Activity	References
	Viscosin from *Pseudomonas libanensis*	Inhibits migration of metastatic cells	Saini et al. (2008)

RLs, the most popular glycolipid biosurfactants, were studied by Rahimi et al. (2019). In this work, the cytotoxic effect of mono and di-RLs produced by *P. aeruginosa* on MCF-7 breast cancer cells was explored. Both compounds studied were found to induce the expression of the p53 gene (Rahimi et al. 2019). A new glycolipid biosurfactant produced by *Planococcus maritimus* was evaluated as anti-cancer agent in MCF-7 cell line with a high cytotoxicity effect (Waghmode et al. 2019, 2020). The anti-cancer capacity was associated with hydrophobic and Van der Waal interactions; however, the mechanism involved was not presented (Waghmode et al. 2019, 2020).

Rakicidins from *Micromonospora* are also known as breast anti-cancer agents. Even though no cytotoxicity of the derivatives C and D of rakicidin (containing small chain of lipids) was found, derivative E inhibited the cancer cell lines (Banat and Thavasi 2019).

8.4.2 Lung Cancer

Lung cancer is a typical and frequent cancer following breast cancer (females) and prostate cancer (males). Recently, the prevalence this type of cancer exceeded diseases related to the heart, the primary cause of mortality being due to smoking. Unfortunately, most lung cancer patients are diagnosed in an advanced stage. Thus, researchers are continuously looking for improved diagnosis and better alternatives of lung cancer treatment (Huq et al. 2009), which include the use of biosurfactants as suitable alternative therapeutic agents. A surfactin produced by *B. atrophaeus was studied by Routhu et al. (*2019). This biosurfactant presented cytotoxicity against A549 lung carcinoma cell line (Routhu et al. 2019*). The anti-cancer activity occurred due to cancer cell inhibition, cycle progression in G_0/G_1, and induced* apoptosis via ROS accumulation (Routhu et al. 2019).

Wrasidlo et al. (2008) studied somocystinamide A production by *Lyngbya majuscula*. Somocystinamide A is a lipopeptide type of biosurfactant displaying considerable cytotoxicity for cell line A549. The anti-proliferative activity was due to the induction of programmed cell death. Moreover, caspase 8 colocalization and ceramide aggregation in treated cells was found (Wrasidlo et al. 2008).

A biosurfactant of the glycolipoprotein class produced by *Acinetobacter M6* strain demonstrated anti-cancer activity against A549 cancer cells. The overall results showed that cell viability decreased with increasing biosurfactant concentrations and incubation time (Karlapudi et al. 2020).

Fellutamides biosurfactants produced by the fungus *Aspergillus versicolor* presented cytotoxic properties for A549 cell lines (Lee et al. 2010, 2011). Fengycin produced by *B. subtilis* was used in mice to combat the growth cell line 95D (Yin et al. 2013). This biosurfactant can inhibit the proliferation of cancer cells due to apoptosis via mitochondrial pathway and at G0/G1 phase, the arrest of cell cycle (Yin et al. 2013).

Rakicidin B (FW523-3) derivative from *Micromonospora chalcea* has also been described as a functional compound for A549 and 95D cell lines (Xie et al. 2011). This biosurfactant promoted apoptosis in cell lines of lung cancer while blocking the pathways signs of MAPK and JNK/p38. Moreover, this rakicidin inhibited cell growth of cancer cells while at the same time induced apoptosis of the cancer cell through the MAPK and mitochondrial routes (Xie et al. 2011).

Another type of biosurfactants for lung cancer are apratoxins analogues. For example, apratoxins produced by *Moorea producens*, namely, apratoxin H and apratoxin A sulfoxide, were studied against NCI-H460 human cell line (Thornburg et al. 2013).

Myrmekioside, an *o*-alkyl-diglycosylglycerol, is a glycolipid biosurfactant produced by *Myrmekioderma dendyi*. Myrmekioside derivatives were used to combat cell lines A549 and NSCLC-N6 (Farokhi et al. 2012). It was concluded that cytotoxicity was not similar among myrmekiosides E-1, E-2, and E-3 due to their different polarities. Peracetylated myrmekioside E-2 has a liposoluble character. Thus, this compound can easily diffuse into lipid bilayer and cross quickly the cell membrane (Farokhi et al. 2012). Liu et al. (2018) isolated two new cyclopepetides, dolyemycins A and B, from *Streptomyces griseus*. These molecules presented action in combating the spread of A549 cell lines. The mechanism of action of these molecules on lung cancer cells was yet not presented (Liu et al. 2018).

8.4.3 Leukemia

Leukemia is a group of highly heterogeneous cancer types of the blood, characterized by excessive cell proliferation of lymphoid or myeloid origin in the bone marrow and secondary blood (Al Ageeli 2020). Leukemia can be subdivided into different varieties, such as: (1) acute myeloid leukemia; (2) acute lymphoblastic leukemia; (3) chronic myeloid leukemia; and (4) chronic lymphocytic. Eight percent of cancers are from these four types of leukemia. Additionally, leukemia is the main type of cancer in infants (Kouhpeikar et al. 2019). Besides a number of alternatives for the therapy of leukemia, such as radiation, chemotherapy, and transplantation of bone marrow, novel methodologies applying biosurfactants have also been explored.

STLs produced by *Rhodococcus erythropolis* and MELs produced by *Candida antarctica* T-34, both glycolipid biosurfactants, were evaluated in human HL-60 promyelocytic leukemia cells (Isoda et al. 1997a). In this work, MELs and STLs have shown to markedly induce HL-60 cell differentiation towards granulocytes instead of cell proliferation, suggesting that STLs and MELs differentiation capacity are not by the surfactant effect (Isoda et al. 1997a). In another work from the same research group, STL-1 from *R. erythropolis* SD-74 significantly inhibited U937 human monocytoid leukemia cell line growth, and also promoted variations in the morphology (Isoda et al. 1995). STL-3 containing saturated carbon chains with even or odd numbers induced HL-60 cell differentiation (Sudo et al. 2000). The results proved that STL-3 on HL-60 was dependent on the of STL-3 structure (Sudo et al. 2000).

The cyclic lipopeptide biosurfactant from *Bacillus natto* inhibited human K562 cells progression by inducing apoptosis (Wang et al. 2007). Surfactin cyclic lipopeptide stimulated apoptosis in leukemia K562 cancer cells by regulating the activation of Ca^{2+} extracellular-related protein kinase (Wang et al. 2009).

The anti-cancer potential of RLs was studied in human chronic myeloid leukemia K562 cells (Shen et al. 2020). The study showed an antiproliferation activity of cancer cells, without affecting healthy blood cells. This phenomenon was related to the stiffness of the cells, since K562 cells are characterized by a greater corticol membrane tension than healthy blood cells (Sudo et al. 2000).

The lipopeptide iturin produced by *B. subtilis* was evaluated for the treatment of chronic myelogenous leukemia, using K562 cells (Zhao et al. 2018). The biosurfactant showed an anti-proliferative activity against cancer cells and acted via three pathways. The existence of a caspase as an inhibitor, stimulated by iturin paraptosis, prevented the autophagy progress, and also induced apoptosis by causing ROS burst (Zhao et al. 2018).

The application of serratamolide (serrawettin W1) as a chemotherapeutic agent against various cancer types was patented in 2005 by Tomas et al. (2005). The studies were based on acute human T cell leukemia cells (jurkat clone E6-1) and peripheral blood acute human lymphoblastic leukemia (Molt-4) (Perez Tomas et al. 2005). This biosurfactant was found to induce apoptosis, reducing cancer cell viability, with no negative effects on healthy cell lines (Perez Tomas et al. 2005).

Chiewpattanakul et al. (2010) discovered monoolein produced by *Exophiala dermatitidis*. In U937 leukemia cell lines, this molecule presents an anti-proliferative activity, in addition to not showing toxic effects on healthy cells (Chiewpattanakul et al. 2010). Monoolein acts by morphologically modifying the cell and its DNA, including cell shrinkage, membrane blebbing, and DNA fragmentation (Chiewpattanakul et al. 2010).

8.4.4 Melanoma

Melanoma is a destructive type of skin cancer with no fast improvements in novel treatments. This concern raises the need to discover new therapeutic agents. Some studies from the literature have shown that biosurfactants induce apoptosis and growth arrest in melanoma tumor cells. The work of Zhao et al. (1999) corresponds to the first evidence that MELs significantly reduce the growth and apoptosis of B16 melanoma cells. In the following work of Zhao et al. (2000), MELs were used as a potent inhibitor for the proliferation growth of mouse melanoma B16 due to the sub-G1 arrest, chromatin condensation, and fragmentation of DNA, thus inducing B16 cell apoptosis (Zhao et al. 2000). MELs from *C. Antarctica* induced cell differentiation by promoting apoptosis via destruction of DNA and chromatin condensation, and sub-G1 arrest (Coelho et al. 2020). These findings indicate that MELs biosurfactants induce the differentiation markers expression of melanoma cells. In addition, an improved melanin production was obtained, showing that MELs induced both cell differentiation and apoptosis (Coelho et al. 2020; Dey et al. 2015; Shen et al. 2020).

Other biosurfactants with biological activity for melanoma cancer cells include pseudofactin I+I (PFII), a cyclic lipopeptide biosurfactant from *Pseudomonas fluorescens* BD5 (Janek et al. 2013). This type of surfactant was applied to explore the impact of A375 and PFII on cells of melanoma (Janek et al. 2013). Melanoma A375 cells exposed to PFII had an apoptotic death through DNA fragmentation. The authors concluded that the death of melanoma A375 cell was due to permeabilization of plasma membrane through the surfactant micelles (Janek et al. 2013). Abdelli et al. (2019) studied a surfactin produced by *Bacillus safensis* and its anti-cancer activity against B16F10 mouse melanoma cells (and T47D breast cancer cells). The results showed potential cytotoxic activity against both cell lines (Abdelli et al. 2019).

8.4.5 Colon Cancer

Colon cancer normally starts from benign lesions, and due to the accumulation of DNA damage the lesions become malignant (Dienstmann et al. 2017). Besides many efforts to develop a more effective therapeutic, this type of cancer continues to be the main life threatening malignancy (ten Hoorn et al. 2018). Thus, novel effective treatment approaches for fighting colon cancer are mandatory, in which biosurfactants may play a role. In this way, three different acylpeptides (itaurin based-biosurfactants), such as mixirins A ($C_{48}H_{75}N_{12}O_{14}$ (18 unsaturations)), B ($C_{45}H_{69}N_{12}O_{14}$), and C ($C_{47}H_{73}N_{12}O_{14}$) produced by a *Bacillus* sp. were evaluated for anti-tumor activity in colon tumor cells (HCT-116) (Zhang et al. 2004). Mixirins A, B, and C demonstrated to be cytotoxic and reduced the progression of human colon cancer cells (HCT-116), variant A being the most effective (Zhang et al. 2004).

Surfactin and fengycin isoform lipopeptides produced by marine bacterium *B. circulans* DMS-2 presented selectively to fight HT-29 and HCT-15 human cells (Sivapathasekaran et al. 2010). The effect of surfactin produced by *B. subtilis* was evaluated in the anti-tumor activity of a human colon carcinoma cell line, LoVo cells (Kim et al. 2007). Surfactin strongly inhibited the propagation of these cells through survival signaling suppression, cell cycle arrest, and apoptosis induction. In this work, the anti-proliferative effect of this biosurfactant was due to the inhibition of the protein kinase (extracellular) and phosphoinositide 3-kinase/Akt stimulation (Kim et al. 2007).

A marine biosurfactant with a new isoform was evaluated as an anti-cancer agent in human colon adenocarcinoma cell line HT-29 (Das et al. 2015). The results presented for the first time an anti-proliferative activity of biosurfactants in nanomolar concentrations by programed cell death, and at the same time, with no antioxidant activity (Das et al. 2015).

Serrawettin W2 produced by *S. surfactantfaciens* showed anti-tumor activity against human colon cancer cells $CaCo_2$ (Su et al. 2016). This biosurfactant can suppress the growth of cancer cell lines, without negatively affecting the viability of healthy cell lines (Su et al. 2016). Burgos-Díaz et al. (2013) studied a new biosurfactant produced by *Sphingobacterium detergens* against $CaCo_2$ human colon cancer cells. The results showed an anti-proliferative effect and apoptosis activity on cancer cells. However, more studies are required to fully understand the apoptosis activity of this biosurfactant (Burgos-Díaz et al. 2013). Finally, surfactin from *B. subtilis* demonstrated a reduced proliferative potential in LoVo cells by apoptosis initiation, cell cycle arrest, and survival signaling suppression (Zhang et al. 2004).

8.5 Biosurfactants as Drug Delivery System (DDS)

A DDS is designed to induce a therapeutic compound introduction into the body, while improving the safety and efficacy (Jain 2008; Gangwar et al. 2012). The two main characteristics of controlled DDSs are: (1) an optimal drug loading capacity, which improves drug bioavailability and reach the target; and (2) a controlled drug release (Jain 2008; Gangwar et al. 2012). Since chemotherapy is limited by the anti-cancer drugs' poor penetration into tumor tissues, along with their severe side effects on healthy cells, novel biosurfactants-based DDS constituted by liposomes, niosomes, and nanoparticles have been developed (Ag Seleci et al. 2016).

8.5.1 Liposomes

Liposomes are bilayered lipid vesicles useful for hydrophobic and hydrophilic drugs encapsulation, sustained drug release, degradation protection, and therapeutic efficacy increase, and have low adverse effects. The performance of a gene transfection from a cationic liposome was improved by MEL-A. MEL-liposome (MEL-L) comprised of 3β-[N-(N′,N′-dimethylaminoethane)-carbamoyl] cholesterol (DC-Chol), dioleoyl phosphatidylethanolamine (DOPE), plus MEL-A exhibited efficiency in DNA transfection into cells through enhancing association of lipoplexes among serum cells (Igarashi et al. 2006; Jain et al. 2014; Gao and Huang 1991; Vigneron et al. 1996). Apart from being described as a useful vector for transfection of DNA, clinical trials with advanced melanoma patients by injection of DNA-liposome complexes into tumor nodules occurred without complications; in metastatic melanoma patients treated with catheter injection of DNA-liposomes into tumor masses were well tolerated, displaying the safety of therapeutic direct gene transfer in humans (Farhood et al. 1994, 1995; Zhou and Huang 1994; Nabel et al. 1993, 1994).

Maitani et al. (2006) produced 300-nm-sized aggregated liposome-plasmid DNA (pDNA) complexes (lipoplexes) through the addition of biosurfactants, such as β-sitosterol β-D-glucoside and MEL-A, to cationic liposomes, which can be applied in intratumoral and intravenous injections. Sit-G-liposome exhibited potential as a vector in gene-based therapy, since it demonstrated low cytotoxicity and displayed high luciferase transfection of gene performance in the human serum hepatoblastoma HepG2 cell line (Maitani et al. 2006; Hwang et al. 2001).

Overall, liposomes drug delivery has gathered attention in cancer therapy. In fact, liposomes became the first nanoparticles to reach clinical trials, mainly in breast cancer therapy (Varshochian et al. 2014). Furthermore, triggered release liposomes have been introduced in order to guarantee successful treatments through an efficient and immediate drug release in tumor tissues due to: (1) inner stimulants (pH and enzyme); and (2) outer stimulants (local heating, ultrasound, magnetic field, light) (Varshochian et al. 2014).

8.5.2 Niosomes

Niosomes are nonionic biosurfactant bilayer vesicles formed by biosurfactants with or without combinations of cholesterol or diverse lipids, whose stability, low-cost, biodegradability, biocompatibility, nonimmunogenicity, and structural characterization flexibility reinforce their potential as drug delivery vehicles (Ray et al. 2018). Their amphiphilic nature is essential for encapsulating lipophilic or hydrophilic drugs, where the hydrophilic core is the ideal medium for incorporating hydrophilic drugs and hydrophobic drugs are predominantly confined to the lipid layer (Fig. 8.15) (Ray et al. 2018). The properties of niosomes are adjustable by changing the vesicles' composition, surface charge, size, lamellarity, tapped volume and concentration. However, niosomes' stability depends on the biosurfactant type, encapsulated drug nature, storage temperature, detergents, membrane-spanning

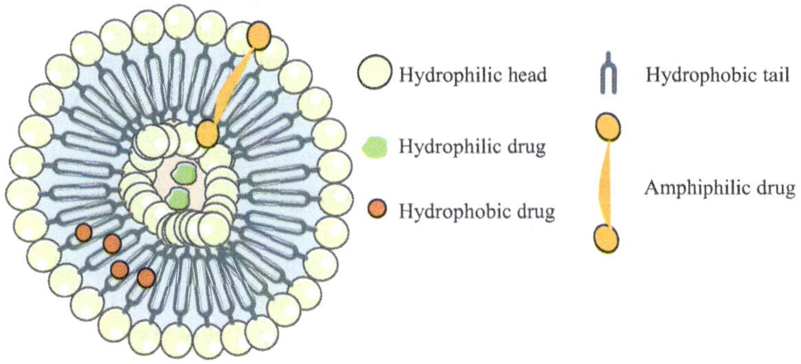

Fig. 8.15 Graphic illustration of a niosome

lipids use, interfacial polymerization of surfactant monomers in situ, and charged molecule inclusion (Ray et al. 2018; Karim et al. 2010; Naughton et al. 2019).

Wu et al. (2017) revealed that surfactin could be included into nano-formulations, such as niosomes, since it effortlessly positions itself inside their hydrophobic/hydrophilic core-shell structure because of its amphiphilic structure and surface-active property (Wu et al. 2017). Recently, Haque et al. (2017) showed sophorolipids-based niosomes for amphotericin B (AmB) delivery against *C. albicans* in a cost-effective way (Haque et al. 2017).

8.5.3 Nanoparticles

Microbial biosurfactants are growing as exciting options for quick nanoparticles synthesis (Kasture et al. 2008; Sharma et al. 2009; Reddy et al. 2009a; Rodrigues 2015). Kasture et al. (2008) described silver nanoparticles synthesis using sophorolipids biosurfactants as reducing and capping agents (Kasture et al. 2008). Reddy et al. (2009a, b) showed that surfactin can be used as a stabilizing agent for silver and gold NPs synthesis (Reddy et al. 2009a, b). Palanisamy and Raichur (2009) reported an eco-friendly alternative method, using RLs for microemulsion synthesis of spherical nickel oxide NPs (Palanisamy and Raichur 2009). Maity et al. (2011) displayed a methodology based on surfactin reverse microemulsion for nanocrystalline brushite particles (nanospheres and nanorods) synthesis (Maity et al. 2011).

Surfactin combined with other chemotherapeutic drugs can be loaded into nano-formulations and used as an adjuvant in anti-cancer treatment (Wu et al. 2017). Taking this into account, Huang et al. (2018) took advantage of the anti-cancer drug doxorubicin (DOX) to develop DOX-loaded surfactin nanoparticles (DOX@SUR), which presented higher cytotoxicity for resistant human DOX breast cancer MCF-7/ADR cells than free DOX, by exhibiting an increased cellular acceptance besides diminished cellular efflux due to inhibition of the P-glycoprotein expression (Huang et al. 2018). Furthermore, DOX@SUR presented higher in vivo tumor suppression and lower adverse effects in MCF-7/ADR-bearing nude mice (Huang et al. 2018). Consequently, DOX@SUR displayed potential as an anti-cancer drug carrier to reverse multidrug resistance in cancer chemotherapy (Huang et al. 2018). Therefore, surfactin nano-formulations have significant potential in anti-cancer nanomedicine treatment. However, their full potential still remains unexplored (Wu et al. 2017).

8.6 Conclusions and Future Challenges

The ability of biosurfactants to act on cancer cells, without negatively affecting healthy cells, makes their use as anti-cancer agents an excellent alternative to current treatments, especially when compared to chemotherapy. However, their application in this field is still a challenge. Firstly, the immensity of microorganisms and their metabolites leads to the continuous discovery of new biosurfactants, with the existing ones being unexplored in what concerns their potential of application. Secondly, only a few studies were dedicated to understanding the action mechanisms of biosurfactants. Furthermore, some studies used semi-purified fractions with biosurfactants, affecting the interpretation of results and the understanding of the underlying mechanisms. Lastly, many of the existing studies are in the stage of in vitro testing with cell lines. In order to reach the in vivo stage, a significant amount of work is still required up to their final approval by the respective health regulatory agencies.

DDS using biosurfactants are an additional area with relevant therapeutic potential. Additional research on the interactions between DDS constituents and DDS interaction with cells is still required. Overall, this field is still in its infancy with a small number of works reported up to date. However, given the promising results reported, this field of research will certainly increase in the following years.

Acknowledgments This work was developed within the scope of the project CICECO-Aveiro Institute of Materials, UIDB/50011/2020 & UIDP/50011/2020, financed by national funds through the Portuguese Foundation for Science and Technology/MCTES. This work was financially supported by POCI-01-0145-FEDER-031268—funded by FEDER, through COMPETE2020—Programa Operacional Competitividade e Internacionalização (POCI) and by national funds (OE), through FCT/MCTES. Ana P. M. Tavares acknowledges the FCT Investigator Programme and Exploratory Project (IF/01634/2015) with financing from the European Social Fund and the Human Potential Operational Programme. João C. F. Nunes acknowledges SPQ and FCT for the PhD fellowship (SFRH/BD/150671/2020). Flávia Magalhães acknowledges the SPQ and FCT PhD grant (SFRH/BD/150669/2020).

References

Abdelli F, Jardak M, Elloumi J, Stien D, Cherif S, Mnif S, Aifa S (2019) Antibacterial, anti-adherent and cytotoxic activities of surfactin(s) from a lipolytic strain Bacillus safensis F4. Biodegradation 30:287–300. https://doi.org/10.1007/s10532-018-09865-4

Abdel-Mawgoud AM, Aboulwafa MM, Hassouna NAH (2008) Characterization of surfactin produced by bacillus subtilis isolate BS5. Appl Biochem Biotechnol 150:289–303. https://doi.org/10.1007/s12010-008-8153-z

Abdel-Mawgoud AM, Lepine F, Deziel E (2010) Rhamnolipids: diversity of structures, microbial origins and roles. Appl Microbiol Biotechnol 86:1323–1336. https://doi.org/10.1007/s00253-010-2498-2

Ag Seleci D, Seleci M, Walter J-G, Stahl F, Scheper T (2016) Niosomes as nanoparticular drug carriers: fundamentals and recent applications. J Nanomater 2016:1–13. https://doi.org/10.1155/2016/7372306

Al Ageeli E (2020) Alterations of mitochondria and related metabolic pathways in leukemia: a narrative review. Saudi J Med Med Sci 8:3. https://doi.org/10.4103/sjmms.sjmms_112_18

Amani H, Müller MM, Syldatk C, Hausmann R (2013) Production of microbial rhamnolipid by Pseudomonas aeruginosa MM1011 for Ex situ enhanced oil recovery. Appl Biochem Biotechnol 170:1080–1093. https://doi.org/10.1007/s12010-013-0249-4

Arutchelvi JI, Bhaduri S, Uppara PV, Doble M (2008) Mannosylerythritol lipids: a review. J Ind Microbiol Biotechnol 35:1559–1570. https://doi.org/10.1007/s10295-008-0460-4

Bajpai Tripathy D, Mishra A (2016) Sustainable biosurfactants. In: Encyclopedia of inorganic and bioinorganic chemistry. Wiley, Chichester, pp 1–17. https://doi.org/10.1002/9781119951438.eibc2433

Banat IM, Thavasi R (2019) Microbial biosurfactants and their environmental and industrial applications. CRC, Boca Raton, FL

Banat IM, Makkar RS, Cameotra SS (2000) Potential commercial applications of microbial surfactants. Appl Microbiol Biotechnol 53:495–508. https://doi.org/10.1007/s002530051648

Banat IM, Satpute SK, Cameotra SS, Patil R, Nyayanit NV (2014) Cost effective technologies and renewable substrates for biosurfactants' production. Front Microbiol 5:1–18. https://doi.org/10.3389/fmicb.2014.00697

Baumgart F, Kluge B, Ullrich C, Vater J, Ziessow D (1991) Identification of amino acid substitutions in the lipopeptide surfactin using 2D NMR spectroscopy. Biochem Biophys Res Commun 177:998–1005. https://doi.org/10.1016/0006-291x(91)90637-m

Bednarski W, Adamczak M, Tomasik J, Płaszczyk M (2004) Application of oil refinery waste in the biosynthesis of glycolipids by yeast. Bioresour Technol 95:15–18. https://doi.org/10.1016/j.biortech.2004.01.009

Bergström S, Theorell H, Davide H (1946a) On a metabolic product of Ps. pyocyanea. Pyolipic acid, active against M. tuberculosis. Arch Chem Mineral Geol 23A:1–12

Bergström S, Theorell H, Davide H (1946b) Pyolipic acid. A metabolic product of Pseudomonas pyocyanea active against Mycobacterium tuberculosis. Arch Biochem Biophys 10:165–166

Bhattacharjee SS, Haskins RH, Gorin PAJ (1970) Location of acyl groups on two acylated glycolipids from strains of Ustilago (smut fungi). Carbohydr Res 13:235–246. https://doi.org/10.1016/S0008-6215(00)80830-7

Bonmatin JM, Genest M, Labbe H, Ptak M (1994) Solution three-dimensional structure of surfactin: a cyclic lipopeptide studied by 1H-NMR, distance geometry, and molecular dynamics. Biopolymers 34:975–986. https://doi.org/10.1002/bip.360340716

Burgos-Díaz C, Martín-Venegas R, Martínez V, Storniolo CE, Teruel JA, Aranda FJ, Ortiz A, Manresa Á, Ferrer R, Marqués AM (2013) In vitro study of the cytotoxicity and antiproliferative effects of surfactants produced by Sphingobacterium detergens. Int J Pharm 453:433–440. https://doi.org/10.1016/j.ijpharm.2013.06.029

Bursch W, Oberhammer F, Schulte-Hermann R (1992) Cell death by apoptosis and its protective role against disease. Trends Pharmacol Sci 13:245–251. https://doi.org/10.1016/0165-6147(92)90077-J

Cameotra SS, Makkar RS (2004) Recent applications of biosurfactants as biological and immunological molecules. Curr Opin Microbiol 7:262–266. https://doi.org/10.1016/j.mib.2004.04.006

Cao X, Wang AH, Jiao RZ, Wang CL, Mao DZ, Yan L, Zeng B (2009) Surfactin induces apoptosis and G2/M arrest in human breast cancer MCF-7 cells through cell cycle factor regulation. Cell Biochem Biophys 55:163–171. https://doi.org/10.1007/s12013-009-9065-4

Cao XH, Wang AH, Wang CL, Mao DZ, Lu MF, Cui YQ, Jiao RZ (2010) Surfactin induces apoptosis in human breast cancer MCF-7 cells through a ROS/JNK-mediated mitochondrial/caspase pathway. Chem Biol Interact 183:357–362

Casas JA, García De Lara S, García-Ochoa F (1997) Optimization of a synthetic medium for Candida bombicola growth using factorial design of experiments. Enzyme Microb Technol 21:221–229. https://doi.org/10.1016/S0141-0229(97)00038-0

Chen J, Song X, Zhang H, Qu YB, Miao JY (2006) Sophorolipid produced from the new yeast strain Wickerhamiella domercqiae induces apoptosis in H7402 human liver cancer cells. Appl Microbiol Biotechnol 72:52–59. https://doi.org/10.1007/s00253-005-0243-z

Chiewpattanakul P, Phonnok S, Durand A, Marie E, Thanomsub BW (2010) Bioproduction and anticancer activity of biosurfactant produced by the dematiaceous fungus Exophiala dermatitidis SK80. J Microbiol Biotechnol 20:1664–1671. https://doi.org/10.4014/jmb.1007.07052

Chu D, Lu J (2008) Novel therapies in breast cancer: what is new from ASCO 2008. J Hematol Oncol 1:16

Clements T, Ndlovu T, Khan S, Khan W (2019) Biosurfactants produced by Serratia species: classification, biosynthesis, production and application. Appl Microbiol Biotechnol 103:589–602. https://doi.org/10.1007/s00253-018-9520-5

Coelho ALS, Feuser PE, Carciofi BAM, de Andrade CJ, de Oliveira D (2020) Mannosylerythritol lipids: antimicrobial and biomedical properties. Appl Microbiol Biotechnol 76:801–810. https://doi.org/10.1007/s00253-020-10354-z

Cooper DG, Paddock DA (1984) Production of a biosurfactant from Torulopsis bombicola. Appl Environ Microbiol 47:173–176. https://doi.org/10.1128/aem.47.1.173-176.1984

Das P, Sarkar S, Mandal M, Sen R (2015) Green surfactant of marine origin exerting a cytotoxic effect on cancer cell lines. RSC Adv 5:53086–53094. https://doi.org/10.1039/c5ra07168b

Davis C, Boyd C, Mateo de Acosta Andino DA, Kumbla PA, Sanchez RJ, Kurapati S, King TW, de la Torre JI (2020) Dermal autografts in breast reconstruction. Ann Plast Surg 84(5):618–622. https://doi.org/10.1097/sap.0000000000002128

Dey G, Bharti R, Dhanarajan G, Das S, Dey KK, Kumar BNP, Sen R, Mandal M (2015) Marine lipopeptide Iturin A inhibits Akt mediated GSK3β and FoxO3a signaling and triggers apoptosis in breast cancer. Sci Rep 5:1–14. https://doi.org/10.1038/srep10316

Dienstmann R, Vermeulen L, Guinney J, Kopetz S, Tejpar S, Tabernero J (2017) Consensus molecular subtypes and the evolution of precision medicine in colorectal cancer. Nat Rev Cancer 17:79–92

Duarte C, Gudiña EJ, Lima CF, Rodrigues LR (2014) Effects of biosurfactants on the viability and proliferation of human breast cancer cells. AMB Expr 4:40. https://doi.org/10.1186/s13568-014-0040-0

Eckelmann D, Spiteller M, Kusari S (2018) Spatial-temporal profiling of prodiginines and serratamolides produced by endophytic Serratia marcescens harbored in Maytenus serrata. Sci Rep 8:5283. https://doi.org/10.1038/s41598-018-23538-5

Edwards JR, Hayashi JA (1965) Structure of a rhamnolipid from Pseudomonas aeruginosa. Arch Biochem Biophys 111:415–421. https://doi.org/10.1016/0003-9861(65)90204-3

Eraqi WA, Yassin AS, Ali AE, Amin MA (2016) Utilization of crude glycerol as a substrate for the production of rhamnolipid by Pseudomonas aeruginosa. Biotechnol Res Int 2016:1–9. https://doi.org/10.1155/2016/3464509

Farhood H, Gao X, Son K, Lazo JS, Huang L, Barsoum J, Bottega R, Epand RM (1994) Cationic liposomes for direct gene transfer in therapy of cancer and other diseases. Ann N Y Acad Sci 716:23–35. https://doi.org/10.1111/j.1749-6632.1994.tb21701.x

Farhood H, Serbina N, Huang L (1995) The role of dioleoyl phosphatidylethanolamine in cationic liposome mediated gene transfer. Biochim Biophys Acta 1235:289–295. https://doi.org/10.1016/0005-2736(95)80016-9

Farokhi F, Wielgosz-Collin G, Robic A, Debitus C, Malleter M, Roussakis C, Kornprobst JM, Barnathan G (2012) Antiproliferative activity against human non-small cell lung cancer of two O-alkyl-diglycosylglycerols from the marine sponges Myrmekioderma dendyi and Trikentrion laeve. Eur J Med Chem 49:406–410. https://doi.org/10.1016/j.ejmech.2012.01.029

Farokhi F, Grellier P, Clément M, Roussakis C, Loiseau PM, Genin-Seward E, Kornprobst JM, Barnathan G, Wielgosz-Collin G (2013) Antimalarial activity of axidjiferosides, new β-galactosylceramides from the African sponge Axinyssa djiferi. Mar Drugs 11:1304–1315. https://doi.org/10.3390/md11041304

Feinberg B, Kish J, Dokubo I, Wojtynek J, Lord K (2019) Reports of the demise of chemotherapy have been greatly exaggerated. Am J Manag Care 25:270–272

Felse PA, Shah V, Chan J, Rao KJ, Gross RA (2007) Sophorolipid biosynthesis by Candida bombicola from industrial fatty acid residues. Enzyme Microb Technol 40:316–323. https://doi.org/10.1016/j.enzmictec.2006.04.013

Fu SL, Wallner SR, Bowne WB, Hagler MD, Zenilman ME, Gross R, Bluth MH (2008) Sophorolipids and their derivatives are lethal against human pancreatic cancer cells. J Surg Res 148:77–82. https://doi.org/10.1016/j.jss.2008.03.005

Fukuoka T, Morita T, Konishi M, Imura T, Kitamoto D (2008) A basidiomycetous yeast, Pseudozyma tsukubaensis, efficiently produces a novel glycolipid biosurfactant. The identification of a new diastereomer of mannosylerythritol lipid-B. Carbohydr Res 343:555–560. https://doi.org/10.1016/j.carres.2007.11.023

Ganem A, Quintanar D, Buri P (2000) Monoolein: a review of the pharmaceutical applications. Drug Dev Ind Pharm 26:809–820. https://doi.org/10.1081/DDC-100101304

Gangwar M, Singh R, Goel RK, Nath G (2012) Recent advances in various emerging vescicular systems: an overview. Asian Pac J Trop Biomed 2:S1176–S1188. https://doi.org/10.1016/S2221-1691(12)60381-5

Gao X, Huang L (1991) A novel cationic liposome reagent for efficient transfection of mammalian cells. Biochem Biophys Res Commun 179:280–285. https://doi.org/10.1016/0006-291X(91)91366-K

Ghasemi A, Moosavi-Nasab M, Setoodeh P, Mesbahi G, Yousefi G (2019) Biosurfactant production by lactic acid bacterium Pediococcus dextrinicus SHU1593 grown on different carbon sources: strain screening followed by product characterization. Sci Rep 9:1–12. https://doi.org/10.1038/s41598-019-41589-0

Giltrap AM, Cergol KM, Pang A, Britton WJ, Payne RJ (2013) Total synthesis of fellutamide B and deoxy-fellutamides B, C, and D. Mar Drugs 11:2382–2397. https://doi.org/10.3390/md11072382

Götze S, Herbst-Irmer R, Klapper M, Görls H, Schneider KRA, Barnett R, Burks T, Neu U, Stallforth P (2017) Structure, biosynthesis, and biological activity of the cyclic lipopeptide anikasin. ACS Chem Biol 12:2498–2502. https://doi.org/10.1021/acschembio.7b00589

Gudiña EJ, Rangarajan V, Sen R, Rodrigues LR (2013) Potential therapeutic applications of biosurfactants. Trends Pharmacol Sci 34:667–675. https://doi.org/10.1016/J.TIPS.2013.10.002

Gudiña EJ, Teixeira JA, Rodrigues LR (2016) Biosurfactants produced by marine microorganisms with therapeutic applications. Mar Drugs 14:38. https://doi.org/10.3390/md14020038

Ha S, Kim HM, Chun HH, Hwang IM, Lee JH, Kim JC, Kim IS, Park HW (2018) Effect of oxygen supply on surfactin production and sporulation in submerged culture of Bacillus subtilis Y9. Appl Sci 8:1660. https://doi.org/10.3390/app8091660

Hannun YA, Bell RM (1989) Functions of sphingolipids and sphingolipid breakdown products in cellular regulation. Science (80-) 243:500–507. https://doi.org/10.1126/science.2643164

Haque F, Sajid M, Cameotra SS, Battacharyya MS (2017) Anti-biofilm activity of a sophorolipid-amphotericin B niosomal formulation against Candida albicans. Biofouling 33:768–779. https://doi.org/10.1080/08927014.2017.1363191

Hmidet N, Ben Ayed H, Jacques P, Nasri M (2017) Enhancement of Surfactin and Fengycin production by Bacillus mojavensis A21: application for diesel biodegradation. Biomed Res Int 2017:5893123. https://doi.org/10.1155/2017/5893123

Hommel R, Stiiwer O, Stuber W, Haferburg D, Kleber HP (1987) Production of water-soluble surface-active exolipids by Torulopsis apicola. Appl Microbiol Biotechnol 26:199–205. https://doi.org/10.1007/BF00286308

Huang W, Lang Y, Hakeem A, Lei Y, Gan L, Yang X (2018) Surfactin-based nanoparticles loaded with doxorubicin to overcome multidrug resistance in cancers. Int J Nanomedicine 13:1723–1736. https://doi.org/10.2147/IJN.S157368

Huq S, Maghfoor I, Perry M (2009) Non-small cell lung cancer (NSCLC): practice essentials, background, pathophysiology. https://emedicine.medscape.com/article/279960-overview. Accessed 10 Feb 2020

Hwang SH, Hayashi K, Takayama K, Maitani Y (2001) Liver-targeted gene transfer into a human hepatoblastoma cell line and in vivo by sterylglucoside-containing cationic liposomes. Gene Ther 8:1276–1280. https://doi.org/10.1038/sj.gt.3301510

Igarashi S, Hattori Y, Maitani Y (2006) Biosurfactant MEL-A enhances cellular association and gene transfection by cationic liposome. J Control Release 112:362–368. https://doi.org/10.1016/j.jconrel.2006.03.003

Igarashi Y, Shimasaki R, Miyanaga S, Oku N, Onaka H, Sakurai H, Saiki I, Kitani S, Nihira T, Wimonsiravude W, Panbangred W (2010) Rakicidin D, an inhibitor of tumor cell invasion from marine-derived Streptomyces sp. J Antibiot (Tokyo) 63:563–565. https://doi.org/10.1038/ja.2010.70

Inaba T, Tokumoto Y, Miyazaki Y, Inoue N, Maseda H, Nakajima-Kambe T, Uchiyama H, Nomura N (2013) Analysis of genes for succinoyl trehalose lipid production and increasing production in Rhodococcus sp. strain SD-74. Appl Environ Microbiol 79:7082–7090. https://doi.org/10.1128/AEM.01664-13

Isoda H, Nakahara T (1997) Mannosylerythritol lipid induces granulocytic differentiation and inhibits the tyrosine phosphorylation of human myelogenous leukemia cell line K562. Cytotechnology 25:191–195. https://doi.org/10.1023/A:1007982909932

Isoda H, Shinmoto H, Matsumura M, Nakahara T (1995) Succinoyl trehalose lipid induced differentiation of human monocytoid leukemic cell line U937 into monocyte-macrophages. Cytotechnology 19:79–88. https://doi.org/10.1007/BF00749758

Isoda H, Shinmoto H, Kitamoto D, Matsumura M, Nakahara T (1997a) Differentiation of human promyelocytic leukemia cell line HL60 by microbial extracellular glycolipids. Lipids 32:263–271. https://doi.org/10.1007/s11745-997-0033-0

Isoda H, Kitamoto D, Shinmoto H, Matsumura M, Nakahara T (1997b) Microbial extracellular glycolipid induction of differentiation and inhibition of the protein kinase C activity of human promyelocytic leukemia cell line HL60. Biosci Biotechnol Biochem 61:609–614. https://doi.org/10.1271/bbb.61.609

Jain KK (2008) Drug delivery systems—an overview. Methods Mol Biol 437:1–50. https://doi.org/10.1007/978-1-59745-210-6_1

Jain K, Sood S, Gowthamarajan K (2014) Lipid nanocarriers and molecular targets for malaria chemotherapy. Curr Drug Targets 15:292–312. https://doi.org/10.2174/13894501113146660235

Jana S, Mondal S, Kulkarni SS (2017) Chemical synthesis of biosurfactant succinoyl trehalose lipids. Org Lett 19:1784–1787. https://doi.org/10.1021/acs.orglett.7b00550

Jana S, Madhu Krishna B, Singhal J, Horne D, Awasthi S, Salgia R, Singhal SS (2020) SOX9: the master regulator of cell fate in breast cancer. Biochem Pharmacol 174:113789

Janek T, Lukaszewicz M, Krasowska A (2012) Antiadhesive activity of the biosurfactant pseudofactin II secreted by the Arctic bacterium Pseudomonas fluorescens BD5. BMC Microbiol 12:24. https://doi.org/10.1186/1471-2180-12-24

Janek T, Krasowska A, Radwańska A, Łukaszewicz M (2013) Lipopeptide biosurfactant Pseudofactin II induced apoptosis of melanoma A 375 cells by specific interaction with the plasma membrane. PLoS One 8:e57991. https://doi.org/10.1371/journal.pone.0057991

Karanth NGK, Deo PG, Veenanadig NK (1999) Microbial production of biosurfactants and their importance. Curr Sci 77:116–126

Karim K, Mandal A, Biswas N, Guha A, Chatterjee S, Behera M, Kuotsu K (2010) Niosome: a future of targeted drug delivery systems. J Adv Pharm Technol Res 1:374–380. https://doi.org/10.4103/0110-5558.76435

Karlapudi AP, Venkateswarulu TC, Srirama K, Kota RK, Mikkili I, Kodali VP (2020) Evaluation of anti-cancer, anti-microbial and anti-biofilm potential of biosurfactant extracted from an Acinetobacter M6 strain. J King Saud Univ Sci 32:223–227. https://doi.org/10.1016/j.jksus.2018.04.007

Kaskatepe B, Yildiz S (2016) Rhamnolipid biosurfactants produced by pseudomonas species. Braz Arch Biol Technol 59:1–16. https://doi.org/10.1590/1678-4324-2016160786

Kasture MB, Patel P, Prabhune AA, Ramana CV, Kulkarni AA, Prasad BLV (2008) Synthesis of silver nanoparticles by sophorolipids: effect of temperature and sophorolipid structure on the size of particles. J Chem Sci 120:515–520. https://doi.org/10.1007/s12039-008-0080-6

Kim H-S, Yoon B-D, Choung D-H, Oh H-M, Katsuragi T, Tani Y (1999) Characterization of a biosurfactant, mannosylerythritol lipid produced from Candida sp. SY16. Appl Microbiol Biotechnol 52:713–721. https://doi.org/10.1007/s002530051583

Kim SY, Kim JY, Kim SH, Bae HJ, Yi H, Yoon SH, Koo BS, Kwon M, Cho JY, Lee CE, Hong S (2007) Surfactin from Bacillus subtilis displays anti-proliferative effect via apoptosis induction, cell cycle arrest and survival signaling suppression. FEBS Lett 581:865–871. https://doi.org/10.1016/j.febslet.2007.01.059

Kitamoto D, Ikegami T, Suzuki GT, Sasaki A, Takeyama YI, Idemoto Y, Koura N, Yanagishita H (2001) Microbial conversion of n-alkanes into glycolipid biosurfactants, mannosylerythritol lipids, by Pseudozyma (Candida antarctica). Biotechnol Lett 23:1709–1714. https://doi.org/10.1023/A:1012464717259

Konishi M, Fukuoka T, Morita T, Imura T, Kitamoto D (2008) Production of new types of sophorolipids by *Candida batistae*. J Oleo Sci 57:359–369. https://doi.org/10.5650/jos.57.359

Kouhpeikar H, Butler AE, Bamian F, Barreto GE, Majeed M, Sahebkar A (2019) Curcumin as a therapeutic agent in leukemia. J Cell Physiol 234:12404–12414. https://doi.org/10.1002/jcp.28072

Kracht M, Rokos H, Özel M, Kowall M, Pauli G, Vater J (1999) Antiviral and hemolytic activities of surfactin isoforms and their methyl ester derivatives. J Antibiot (Tokyo) 52:613–619. https://doi.org/10.7164/antibiotics.52.613

Kulakovskaya E, Kulakovskaya T (2014) Chapter 1. Structure and occurrence of yeast extracellular glycolipids. In: Kulakovskaya E, Kulakovskaya T (eds) Extracellular glycolipids of yeasts. Academic, New York, pp 1–13

Kulkarni CV, Wachter W, Iglesias-Salto G, Engelskirchen S, Ahualli S (2011) Monoolein: a magic lipid? Phys Chem Chem Phys 13:3004–3021. https://doi.org/10.1039/C0CP01539C

Lee YM, Dang HT, Hong J, Lee CO, Bae KS, Kim DK, Jung JH (2010) A cytotoxic lipopeptide from the sponge-derived fungus aspergillus versicolor. Bull Korean Chem Soc 31:205–208. https://doi.org/10.5012/bkcs.2010.31.01.205

Lee YM, Dang HT, Li J, Zhang P, Hong J, Lee CO, Jung JH (2011) A cytotoxic fellutamide analogue from the sponge-derived fungus aspergillus versicolor. Bull Korean Chem Soc 32:3817–3820. https://doi.org/10.5012/bkcs.2011.32.10.3817

Liu X, Haddad NIA, Yang S, Mu B (2007) Structural characterization of eight cyclic lipopeptides produced by Bacillus subtilis HSO121. Protein Pept Lett 14:766–773. https://doi.org/10.2174/092986607781483642

Liu XD, Gu KB, Xia SS, Zhang DJ, Li YG (2018) Dolyemycins A and B, two novel cyclopeptides isolated from Streptomyces griseus subsp. griseus HYS31. J Antibiot (Tokyo) 71:838–845. https://doi.org/10.1038/s41429-018-0071-x

Maitani Y, Yano S, Hattori Y, Furuhata M, Hayashi K (2006) Liposome vector containing biosurfactant-complexed DNA as herpes simplex virus thymidine kinase gene delivery system. J Liposome Res 16:359–372. https://doi.org/10.1080/08982100600992443

Maity JP, Lin TJ, Cheng HPH, Chen CY, Reddy AS, Atla SB, Chang YF, Chen HR, Chen CC (2011) Synthesis of brushite particles in reverse microemulsions of the biosurfactant surfactin. Int J Mol Sci 12:3821–3830. https://doi.org/10.3390/ijms12063821

Marchant R, Banat IM (2012) Microbial biosurfactants: challenges and opportunities for future exploitation. Trends Biotechnol 30:558–565. https://doi.org/10.1016/j.tibtech.2012.07.003

Masuda Y, Suzuki J, Onda Y, Fujino Y, Yoshida M, Doi T (2014) Total synthesis and conformational analysis of Apratoxin C. J Org Chem 79:8000–8009. https://doi.org/10.1021/jo501130b

Meena KR, Kanwar SS (2015) Lipopeptides as the antifungal and antibacterial agents: applications in food safety and therapeutics. Biomed Res Int 2015:473050. https://doi.org/10.1155/2015/473050

Mondol MAM, Shin HJ, Islam MT (2013) Diversity of secondary metabolites from marine Bacillus species: chemistry and biological activity. Mar Drugs 11:2846–2872. https://doi.org/10.3390/md11082846

Morita T, Konishi M, Fukuoka T, Imura T, Kitamoto D (2007) Microbial conversion of glycerol into glycolipid biosurfactants, mannosylerythritol lipids, by a basidiomycete yeast, Pseudozyma antarctica JCM 10317(T). J Biosci Bioeng 104:78–81. https://doi.org/10.1263/jbb.104.78

Morita T, Fukuoka T, Konishi M, Imura T, Yamamoto S, Kitagawa M, Sogabe A, Kitamoto D (2009) Production of a novel glycolipid biosurfactant, mannosylmannitol lipid, by Pseudozyma parantarctica and its interfacial properties. Appl Microbiol Biotechnol 83:1017. https://doi.org/10.1007/s00253-009-1945-4

Mouafo TH, Mbawala A, Ndjouenkeu R (2018) Effect of different carbon sources on biosurfactants' production by three strains of Lactobacillus spp. Biomed Res Int 2018:5034783. https://doi.org/10.1155/2018/5034783

Mukherjee S, Das P, Sen R (2006) Towards commercial production of microbial surfactants. Trends Biotechnol 24:509–515

Mulligan CN, Gibbs BF (1989) Correlation of nitrogen metabolism with biosurfactant production by Pseudomonas aeruginosa. Appl Environ Microbiol 55:3016–3019. https://doi.org/10.1128/aem.55.11.3016-3019.1989

Nabel GJ, Nabel EG, Yang ZY, Fox BA, Plautz GE, Gao X, Huang L, Shu S, Gordon D, Chang AE (1993) Direct gene transfer with DNA-liposome complexes in melanoma: Expression, biologic activity, and lack of toxicity in humans. Proc Natl Acad Sci U S A 90:11307–11311. https://doi.org/10.1073/pnas.90.23.11307

Nabel EG, Yang Z, Muller D, Chang AE, Gao X, Huang L, Cho KJ, Nabel GJ (1994) Safety and toxicity of catheter gene delivery to the pulmonary vasculature in a patient with metastatic melanoma. Hum Gene Ther 5:1089–1094. https://doi.org/10.1089/hum.1994.5.9-1089

Naughton PJ, Marchant R, Naughton V, Banat IM (2019) Microbial biosurfactants: current trends and applications in agricultural and biomedical industries. J Appl Microbiol 127:12–28. https://doi.org/10.1111/jam.14243

Ndlovu T, Rautenbach M, Vosloo JA, Khan S, Khan W (2017) Characterisation and antimicrobial activity of biosurfactant extracts produced by Bacillus amyloliquefaciens and Pseudomonas aeruginosa isolated from a wastewater treatment plant. AMB Expr 7:108. https://doi.org/10.1186/s13568-017-0363-8

Nielsen TH, Sorensen D, Tobiasen C, Andersen JB, Christophersen C, Givskov M, Sorensen J (2002) Antibiotic and biosurfactant properties of cyclic lipopeptides produced by fluorescent Pseudomonas spp. from the sugar beet rhizosphere. Appl Environ Microbiol 68:3416–3423. https://doi.org/10.1128/aem.68.7.3416-3423.2002

Nunez A, Ashby R, Foglia TA, Solaiman DKY (2004) LC/MS analysis and lipase modification of the sophorolipids produced by Rhodotorula bogoriensis. Biotechnol Lett 26:1087–1093. https://doi.org/10.1023/B:BILE.0000032970.95603.6d

Nunnery JK, Mevers E, Gerwick WH (2010) Biologically active secondary metabolites from marine cyanobacteria. Curr Opin Biotechnol 21:787–793. https://doi.org/10.1016/j.copbio.2010.09.019

Oliveira FJS, Vazquez L, de Campos NP, de França FP (2009) Production of rhamnolipids by a Pseudomonas alcaligenes strain. Process Biochem 44:383–389. https://doi.org/10.1016/j.procbio.2008.11.014

Palanisamy P, Raichur AM (2009) Synthesis of spherical NiO nanoparticles through a novel biosurfactant mediated emulsion technique. Mater Sci Eng C 29:199–204. https://doi.org/10.1016/j.msec.2008.06.008

Perez Tomas R, Montaner Ramoneda B, Giralt Lledo E, Martinell Pedemonte M, Vilaseca Casas M (2005) Use of cyclic depsipeptide as a chemotherapeutic agent against cancer. pp 1–6

Pirrung MC, Zhang F, Ambadi S, Gangadhara Rao Y (2016) Total synthesis of fellutamides, lipopeptide proteasome inhibitors. More sustainable peptide bond formation. Org Biomol Chem 14:8367–8375. https://doi.org/10.1039/c6ob01233g

Poulsen TB (2011) A concise route to the macrocyclic core of the rakicidins. Chem Commun 47:12837–12839. https://doi.org/10.1039/c1cc15829e

Price NPJ, Ray KJ, Vermillion KE, Dunlap CA, Kurtzman CP (2012) Structural characterization of novel sophorolipid biosurfactants from a newly identified species of Candida yeast. Carbohydr Res 348:33–41. https://doi.org/10.1016/j.carres.2011.07.016

Raaijmakers JM, de Bruijn I, de Kock MJD (2006) Cyclic lipopeptide production by plant-associated Pseudomonas spp.: diversity, activity, biosynthesis, and regulation. Mol Plant Microbe Interact 19:699–710. https://doi.org/10.1094/MPMI-19-0699

Rahimi K, Lotfabad TB, Jabeen F, Mohammad Ganji S (2019) Cytotoxic effects of mono- and di-rhamnolipids from Pseudomonas aeruginosa MR01 on MCF-7 human breast cancer cells.

Colloids Surf B Biointerfaces 181:943–952. https://doi.org/10.1016/J.COLSURFB.2019.06.058

Ramalingam V, Varunkumar K, Ravikumar V, Rajaram R (2019) Production and structure elucidation of anticancer potential surfactin from marine actinomycete Micromonospora marina. Process Biochem 78:169–177. https://doi.org/10.1016/j.procbio.2019.01.002

Ray SK, Bano N, Shukla T, Upmanyu N, Pandey SP, Parkhe G (2018) Noisomes: as novel vesicular drug delivery system. J Drug Deliv Ther 8:335–341. https://doi.org/10.22270/jddt.v8i6.2029

Reddy AS, Chen CY, Baker SC, Chen CC, Jean JS, Fan CW, Chen HR, Wang JC (2009a) Synthesis of silver nanoparticles using surfactin: a biosurfactant as stabilizing agent. Mater Lett 63:1227–1230. https://doi.org/10.1016/j.matlet.2009.02.028

Reddy AS, Chen CY, Chen CC, Jean JS, Fan CW, Chen HR, Wang JC, Nimje VR (2009b) Synthesis of gold nanoparticles via an environmentally benign route using a biosurfactant. J Nanosci Nanotechnol 9:6693–6699. https://doi.org/10.1166/jnn.2009.1347

Reed JC (2003) Apoptosis-targeted therapies for cancer. Cancer Cell 3:17–22

Ribeiro IAC, Faustino CMC, Guerreiro PS, Frade RFM, Bronze MR, Castro MF, Ribeiro MHL (2015) Development of novel sophorolipids with improved cytotoxic activity toward MDA-MB-231 breast cancer cells. J Mol Recognit 28:155–165. https://doi.org/10.1002/jmr.2403

Rodrigues LR (2015) Microbial surfactants: fundamentals and applicability in the formulation of nano-sized drug delivery vectors. J Colloid Interface Sci 449:304–316. https://doi.org/10.1016/j.jcis.2015.01.022

Rodrigues LR, Teixeira JA (2010) Biomedical and therapeutic applications of biosurfactants. Adv Exp Med Biol 672:75–87. https://doi.org/10.1007/978-1-4419-5979-9_6

Rodrigues L, Banat IM, Teixeira J, Oliveira R (2006) Biosurfactants: potential applications in medicine. J Antimicrob Chemother 57:609–618. https://doi.org/10.1093/jac/dkl024

Ron EZ, Rosenberg E (2001) Natural roles of biosurfactants. Environ Microbiol 3:229–236

Routhu SR, Chary RN, Shaik AB, Prabhakar S, Kumar CG, Kamal A (2019) Induction of apoptosis in lung carcinoma cells by antiproliferative cyclic lipopeptides from marine algicolous isolate Bacillus atrophaeus strain AKLSR1. Process Biochem 79:142–154. https://doi.org/10.1016/j.procbio.2018.12.010

Saini HS, Barraga BE, Lebro A, Burns AM, Marron MT, Seliga CJ, Gunatilaka AAL, Maier RM (2008) Efficient purification of the biosurfactant viscosin from Pseudomonas libanensis strain M9-3 and its physicochemical and biological properties. J Nat Prod 71:1011–1015

Sak K (2012) Chemotherapy and dietary phytochemical agents. Chemother Res Pract 2012:1–11. https://doi.org/10.1155/2012/282570

Sang F, Ding Y, Wang J, Sun B, Sun J, Geng Y, Zhang Z, Ding K, Wu L-L, Liu J-W, Bai C, Yang G, Zhang Q, Li L-Y, Chen Y (2016) Structure-activity relationship study of Rakicidins: overcoming chronic myeloid leukemia resistance to Imatinib with 4-Methylester-Rakicidin A. J Med Chem 59:1184–1196. https://doi.org/10.1021/acs.jmedchem.5b01841

Santos DKF, Rufino RD, Luna JM, Santos VA, Sarubbo LA (2016) Biosurfactants: multifunctional biomolecules of the 21st century. Int J Mol Sci 17:1–31. https://doi.org/10.3390/ijms17030401

Sen S, Borah SN, Bora A, Deka S (2017) Production, characterization, and antifungal activity of a biosurfactant produced by Rhodotorula babjevae YS3. Microb Cell Fact 16:1–14. https://doi.org/10.1186/s12934-017-0711-z

Shanks RMQ, Stella NA, Lahr RM, Wang S, Veverka TI, Kowalski RP, Liu X (2012) Serratamolide is a hemolytic factor produced by Serratia marcescens. PLoS One 7:e36398. https://doi.org/10.1371/journal.pone.0036398

Sharma VK, Yngard RA, Lin Y (2009) Silver nanoparticles: Green synthesis and their antimicrobial activities. Adv Colloid Interface Sci 145:83–96. https://doi.org/10.1016/j.cis.2008.09.002

Shen C, Jiang L, Long X, Dahl KN, Meng Q (2020) Cells with higher cortical membrane tension are more sensitive to lysis by biosurfactant di-rhamnolipids. ACS Biomater Sci Eng 6:352–357. https://doi.org/10.1021/acsbiomaterials.9b01305

Shu Q, Wu J, Chen Q (2019) Synthesis, characterization of liposomes modified with biosurfactant MEL-A loading betulinic acid and its anticancer effect in HepG2 cell. Molecules 24:1–17

Singh P, Cameotra SS (2004) Potential applications of microbial surfactants in biomedical sciences. Trends Biotechnol 22:142–146. https://doi.org/10.1016/j.tibtech.2004.01.010

Sivapathasekaran C, Das P, Mukherjee S, Saravanakumar J, Mandal M, Sen R (2010) Marine bacterium derived lipopeptides: characterization and cytotoxic activity against cancer cell lines. Int J Pept Res Ther 16:215–222. https://doi.org/10.1007/s10989-010-9212-1

Su C, Xiang Z, Liu Y, Zhao X, Sun Y, Li Z, Li L, Chang F, Chen T, Wen X, Zhou Y, Zhao F (2016) Analysis of the genomic sequences and metabolites of Serratia surfactantfaciens sp. nov. YD25T that simultaneously produces prodigiosin and serrawettin W2. BMC Genomics 17:1–19. https://doi.org/10.1186/s12864-016-3171-7

Sudo T, Zhao X, Wakamatsu Y, Shibahara M, Nomura N, Nakahara T, Suzuki A, Kobayashi Y, Jin C, Murata T, Yokoyama KK (2000) Induction of the differentiation of human HL-60 promyelocytic leukemia cell line by succinoyl trehalose lipids. Cytotechnology 33:259–264

ten Hoorn S, Trinh A, de Jong J, Koens L, Vermeulen L (2018) Classification of colorectal cancer in molecular subtypes by immunohistochemistry. In: Methods in molecular biology. Humana Press, Totowa, NJ, pp 179–191

Thornburg CC, Cowley ES, Sikorska J, Shaala LA, Ishmael JE, Youssef DTA, McPhail KL (2013) Apratoxin H and apratoxin A sulfoxide from the red sea cyanobacterium Moorea producens. J Nat Prod 76:1781–1788. https://doi.org/10.1021/np4004992

Tokumoto Y, Nomura N, Uchiyama H, Imura T, Morita T, Fukuoka T, Kitamoto D (2009) Structural characterization and surface-active properties of a succinoyl trehalose lipid produced by Rhodococcus sp. SD-74. J Oleo Sci 58:97–102. https://doi.org/10.5650/jos.58.97

Trischman JA, Jensen PR, Fenical W (1994) Halobacillin: a cytotoxic cyclic acylpeptide of the iturin class produced by a marine Bacillus. Tetrahedron Lett 35:5571–5574. https://doi.org/10.1016/S0040-4039(00)77249-2

Tsan P, Volpon L, Besson F, Lancelin J-M (2007) Structure and dynamics of surfactin studied by NMR in micellar media. J Am Chem Soc 129:1968–1977. https://doi.org/10.1021/ja066117q

Uchida Y, Tsuchiya R, Chino M, Hirano J, Tabuchi T (1989) Extracellular accumulation of mono- and di-succinoyl trehalose lipids by a strain of rhodococcus erythropolis grown on n-alkanes. Agric Biol Chem 53:757–763. https://doi.org/10.1080/00021369.1989.10869385

Valliappan K, Zhang F, Li Z-Y (2015) Handbook of anticancer drugs from marine origin. Springer International, Cham, pp 253–267

Van Hamme JD, Singh A, Ward OP (2006) Physiological aspects: part 1 in a series of papers devoted to surfactants in microbiology and biotechnology. Biotechnol Adv 24:604–620. https://doi.org/10.1016/J.BIOTECHADV.2006.08.001

Varshochian R, Hosseinzadeh H, Gandomi N, Tavassolian F, Atyabi F, Dinarvand R (2014) Utilizing liposomes and lipid nanoparticles to overcome challenges in breast cancer treatment. Clin Lipidol 9:571–585. https://doi.org/10.2217/clp.14.48

Vigneron JP, Ouorhiri N, Fauquet M, Vergely L, Bradley JC, Basseville M, Lehn P, Lehn JM (1996) Guanidinium-cholesterol cationic lipids: efficient vectors for the transfection of eukaryotic cells. Proc Natl Acad Sci U S A 93:9682–9686. https://doi.org/10.1073/pnas.93.18.9682

Waghmode S, Suryavanshi M, Dama L, Kansara S, Ghattargi V, Das P, Banpurkar A, Satpute SK (2019) Genomic insights of halophilic Planococcus maritimus SAMP MCC 3013 and detail investigation of its biosurfactant production. Front Microbiol 10:1–14. https://doi.org/10.3389/fmicb.2019.00235

Waghmode S, Swami S, Sarkar D, Suryavanshi M, Roachlani S, Choudhari P, Satpute S (2020) Exploring the pharmacological potentials of biosurfactant derived from Planococcus maritimus SAMP MCC 3013. Curr Microbiol 77(3):452–459. https://doi.org/10.1007/s00284-019-01850-1

Wakamatsu Y, Zhao X, Jin C, Day N, Shibahara M, Nomura N, Nakahara T, Murata T, Yokoyama KK (2001) Mannosylerythritol lipid induces characteristics of neuronal differentiation in PC12 cells through an ERK-related signal cascade. Eur J Biochem 268:374–383. https://doi.org/10.1046/j.1432-1033.2001.01887.x

Wang CL, Ng TB, Yuan F, Liu ZK, Liu F (2007) Induction of apoptosis in human leukemia K562 cells by cyclic lipopeptide from Bacillus subtilis natto T-2. Peptides 28:1344–1350. https://doi.org/10.1016/J.PEPTIDES.2007.06.014

Wang CL, Ng TB, Cao XH, Jiang Y, Liu ZK, Wen TY, Liu F (2009) CLP induces apoptosis in human leukemia K562 cells through Ca2+ regulating extracellular-related protein kinase ERK activation. Cancer Lett 276:221–227. https://doi.org/10.1016/J.CANLET.2008.11.007

Wei YH, Chou CL, Chang JS (2005) Rhamnolipid production by indigenous Pseudomonas aeruginosa J4 originating from petrochemical wastewater. Biochem Eng J 27:146–154. https://doi.org/10.1016/j.bej.2005.08.028

WHO (2018) Latest global cancer data: cancer burden rises to 18.1 million new cases and 9.6 million cancer deaths in 2018. IARC, Lyon

Wrasidlo W, Mielgo A, Torres VA, Barbero S, Stoletov K, Suyama TL, Klemke RL, Gerwick WH, Carson DA, Stupack DG (2008) The marine lipopeptide somocystinamide A triggers apoptosis via caspase 8. Proc Natl Acad Sci U S A 105:2313–2318. https://doi.org/10.1073/pnas.0712198105

Wu Y-S, Ngai S-C, Goh B-H, Chan K-G, Lee L-H, Chuah L-H (2017) Anticancer activities of surfactin and potential application of nanotechnology assisted surfactin delivery. Front Pharmacol 8:761. https://doi.org/10.3389/fphar.2017.00761

Xie JJ, Zhou F, Li ENM, Jiang H, Du ZP, Lin R, Fang DS, Xu LY (2011) FW523-3, a novel lipopeptide compound, induces apoptosis in cancer cells. Mol Med Rep 4:759–763. https://doi.org/10.3892/mmr.2011.496

Yin H, Guo C, Wang Y, Liu D, Lv Y, Lv F, Lu Z (2013) Fengycin inhibits the growth of the human lung cancer cell line 95D through reactive oxygen species production and mitochondria-dependent apoptosis. Anticancer Drugs 24:587–598. https://doi.org/10.1097/CAD.0b013e3283611395

Yu M, Liu Z, Zeng G, Zhong H, Liu Y, Jiang Y, Li M, He X, He Y (2015) Characteristics of mannosylerythritol lipids and their environmental potential. Carbohydr Res 407:63–72. https://doi.org/10.1016/j.carres.2014.12.012

Zhang HL, Hui M, Hua Y, Hu P, Yao XS (2004) Three new cytotoxic cyclic acylpeptides from marine Bacillus sp. Chem Pharm Bull 52:1029–1030

Zhang J, Li C, Sun L, Yu G, Guan H (2015) Total synthesis of Myrmekioside A, a mono-O-alkyl-diglycosylglycerol from marine sponge Myrmekioderma sp. Eur J Org Chem 2015:4246–4253. https://doi.org/10.1002/ejoc.201500471

Zhao X, Wakamatsu Y, Shibahara M, Nomura N, Geltinger C, Nakahara T, Murata T, Yokoyama KK (1999) Mannosylerythritol lipid is a potent inducer of apoptosis and differentiation of mouse melanoma cells in culture. Cancer Res 59:482–486

Zhao X, Geltinger C, Kishikawa S, Ohshima K, Murata T, Nomura N, Nakahara T, Yokoyama KK (2000) Treatment of mouse melanoma cells with phorbol 12-myristate 13-acetate counteracts mannosylerythritol lipid-induced growth arrest and apoptosis. Cytotechnology 33:123–130

Zhao H, Yan L, Xu X, Jiang C, Shi J, Zhang Y, Liu L, Lei S, Shao D, Huang Q (2018) Potential of Bacillus subtilis lipopeptides in anti-cancer I: induction of apoptosis and paraptosis and inhibition of autophagy in K562 cells. AMB Express 8:78. https://doi.org/10.1186/s13568-018-0606-3

Zhou X, Huang L (1994) DNA transfection mediated by cationic liposomes containing lipopolylysine: characterization and mechanism of action. Biochim Biophys Acta 1189:195–203. https://doi.org/10.1016/0005-2736(94)90066-3

Biosurfactants for Oil Pollution Remediation

9

Huria Rizvi, Jitendra Singh Verma, and Ashish

Abstract

Petroleum industries are considered as major energy resources, but as simultaneously producing large amounts of hydrocarbon wastes that are discharged into soil and water bodies. Environmental pollution due to exponential development of the petrochemical industries was a major concern in the twentieth century. Oil and oil products contamination, which belong to the carcinogenic and neurotoxic organic pollutants family, pose a severe threat to general health of public, choke aquatic life to death, and accumulate in soil and disturb the ecosystem. Numerous different technologies have been used for the removal of hydrocarbon/oil pollutants from polluted sites, such as physical, chemical, and biological methods. Conventional physical and chemical methods can only immobilize at site or transfer's contaminants from one medium to another and can even result in production of toxic by-products. Hence, petroleum oil and petroleum hydrocarbons cannot be entirely eradicated with physical and chemical methods. Thus, focus is being given to biological methods generally. Biosurfactants are considered as a promising alternative for the removal of oil pollutants due to their amphiphilic nature: they have the capability to reduce interfacial tension, disperse oil particles, high surface activity, lower toxicity, biodegradability and environmental friendliness, and are active under extreme conditions of salinity, pH and temperature. This chapter briefly discusses how microorganisms produce biosurfactant when they feed on insoluble substrates such as oil/petroleum waste. It also reveals the biosurfactant mode of action to remove petroleum waste and its derivatives (heavy metals, PAHs, etc.) from oil spills, cleaning pipelines, and containers. Biosurfactants emerge as potential biomolecules in

H. Rizvi · Ashish (✉)
Department of Bioengineering, Integral University, Lucknow, Uttar Pradesh, India

J. S. Verma
Chemical Engineering-Group, Engineering Sciences and Technology Division, CSIR-North East Institute of Science and Technology, Jorhat, Assam, India

Inamuddin et al. (eds.), *Microbial Biosurfactants*, Environmental and Microbial Biotechnology, https://doi.org/10.1007/978-981-15-6607-3_9

petroleum industry waste bioremediation and need to be scaled up for the upcoming years.

Keywords

Biosurfactant · Hydrocarbon · Bioremediation · Oil waste · Oil spill

9.1 Introduction

The rapid deterioration of environmental condition due to fast industrialization, urbanization and increasing population pressure is a major concern all over the world. Petrochemical industries produce large amounts of hydrocarbon wastes that are discharged in soil and water bodies. Environmental pollution from the increased expansion of the petrochemical industries along with associated chemical industries was a major concern in the twentieth century (Okada 2002). Petroleum products are considered as the main energy source for various industries, chemical factories and routine life. For various pollutants soil and water bodies are the most common sites for disposal that could disturb the nutrient value and biodiversity of the ecosystem. Soil and water contamination with these hydrocarbon pollutants occurs by different routes, mainly through pipelines and storage tanks leakage, accidental spills, and inadequate waste disposal (Hill 2009). It was reported that the amount of crude oil seepage was 600,000 metric tons every year with 200,000 metric tons uncertainty range per year worldwide (Das and Chandran 2011). Panagos et al. (2013) reported around 342 thousand identified contaminated sites with an estimation of 2.5 million potential contaminated sites found in European countries. Petroleum hydrocarbons and heavy metals are the main pollutants responsible for approximately 60% soil contamination. Hydrocarbons pollutants include alkanes, alkenes, cycloalkanes, aromatic compounds, PAH, asphaltenes, heterocyclic nitrogen, etc. Due to these compounds' high toxicity value and their ability to accumulate and persist in nature, they are considered to be toxic to environmental and human health. Oil pollutants strongly enhance soil hydrophobicity, which induces water repellency and reduction in moisture retention, and thus, the natural flow pattern of water and air destroyed. Petroleum contaminants cause substantial changes in soil properties, such as structural, physicochemical, and biological properties (Das and Chandran 2011). The accumulation of oil/petroleum pollutants and their derivative pollutants have detrimental effects on the ecosystem in a global level because of their toxic nature and very low biodegradability, causing disturbances in the food chain and the ecosystem (Haritash and Kaushik 2009).

Numerous technologies were developed and used for removal of hydrocarbon pollutants from contaminated sites, such as physicochemical (soil washing, soil vapor extraction and centrifugation, etc.) or biological methods (in situ and ex situ *remediation*). Conventional physical and chemical methods have some limitations in the removal of spilled oil, such as removal by simply transferring pollutants from one medium or site to another that have a risk of producing toxic by-products.

Hence, physiochemical methods cannot provide a complete solution for crude oil cleanup. Thus, biological alternatives for bioremediation have received more attention lately. The physical methods used for pollutants removal involve further processing and are not economical due to relatively high application cost as compared to other remediation methods (Morgan and Atlas 1989). Pollutants removal by biodegradation is encouraged in the current era due to increasing public awareness towards more environment friendly methods. Bioremediation through microorganisms and plants offer advantages over conventional physicochemical methods as they are practical, economical, and do not leave behind any toxic by-products. Bio-remedial methods do not require any other treatment and also help in restoring the natural flora and fauna of contaminated sites (Mani and Kumar 2014).

Biosurfactants are considered as one of the promising biomolecules for the remediation of petroleum pollutants from the ecosystem. Biosurfactants are amphiphilic compounds which have the ability to decrease interfacial tension and disperse oil particles into small droplets, breaking them into non-toxic compounds (Ubalua 2011). Among different types of biosurfactants, rhamnolipid, sophorolipid, and surfactin have been widely used for the bioremediation of hydrocarbon/oil from contaminated sites. Microbial species such as *Pseudomonas*, *Bacillus*, and *Candida* have been generally used for biosurfactant production and subsequently for oil degradation, while many other microbial species were also reported for the same. Biosurfactants have several advantages compared to chemically derived surfactants, such as biodegradability, nontoxicity, biocompatibility, stability under stress conditions (pH, temperature and salinity), useful for bioremediation particularly petroleum-contaminated soils and water bodies (De et al. 2015).

This chapter provides an inclusive knowledge of various sources and hazards of petroleum oil pollutants to the environment and its clean up method through biosurfactants, which play a significant key role in oil remediation. This is because they have potency as dispersion and remediation agents, are nontoxic in nature, and have high biodegradability (Singh et al. 2009). Biosurfactants have very wide applications in different industrial processes and can hold large market value in the future. They were used in storage tanks oil residue removal, enhanced oil recovery, oil spills cleanup, and soil and water remediation (Silva et al. 2014). Biosurfactants having diverse structures with useful properties are of great industrial potential. Biosurfactants can improve the availability of low-solubility compounds such as petroleum oil pollutants, and facilitate mobilization of hydrophobic compounds within microbial cells as a substrate (Maier 2003).

This chapter also reveals the potential role and applications of biosurfactants, focusing on oil pollution and bioremediation processes, and provides brief information about the biosurfactants' mode of action on oil spills.

9.2 Oil Pollution and Its Remediation

9.2.1 Oil Pollution

Petroleum oil and petroleum oil derivative products as well as waste oil sludge from petroleum refineries are major pollutants of the environment. The most important and common path for petroleum hydrocarbon contaminants to enter into soil and marine systems are through seepage from oil deposits and oil explorations where marine environments serve as the biggest reservoirs and ultimate receivers of pollutants (Ossai et al. 2019).

Averages of 2500 accidental petroleum substances leakages are recorded every year in Poland (Hewelke et al. 2018). Oil spills due to earthquakes, storms, human error, or mechanical failure build-up to huge total losses. Over the period 1970–2018, 5.86 million metric tonnes loss of petroleum hydrocarbon into the marine environment by oil tanker spill was reported (http://www.itopf.org/fileadmin/data/Documents/Company_Lit/Oil_Spill_Stats_2018.pdf).

Häder et al. (2020) estimated that 600,000 metric tonnes oil per year are released into marine environments due to natural seepage, which account for 47% of total (Häder et al. 2020). Hydrocarbons present in petroleum oil comprise of a variety of organic substances, mainly hydrocarbons along with some mixture of oxygen-, nitrogen-, and sulfur-containing organic compounds and some inorganic components, i.e., metals (Varjani and Upasani 2017). Various hydrocarbon compounds are present in petroleum oil that can be categorized as aromatic compounds, including polycyclic aromatic hydrocarbons (PAHs) such as naphthalene; monoaromatic, e.g., ethylbenzene, toluene, benzene, xylene; paraffins or aliphatic saturated compounds, e.g., cycloalkanes, n-alkanes, unsaturated alkenes, alkynes; asphaltenes including esters, fatty acids, ketones, phenols, porphyrins; cardaxoles, pyridines, sulfonates, amides, tars and waxes (Ossai et al. 2019). Petroleum oil non-hydrocarbon content consists of sulfur compounds such as thiols, sulfides, disulfides, dibenzothiophene, benzothiophene and naphthobenzothiophene; esters, ethers, carboxylic acids, furans, and ketones; oxygen compounds include alcohols, and nitrogen compounds include benzo(a)carbazole, carbazole, pyridine, pyrrole, indole, benzo(f)quinolone, nitriles, quinoline, and indoline (Ossai et al. 2019).

These hydrocarbons pose serious threat to health and affect all life in the environment directly or indirectly, through population dynamics alteration and by disrupting ecological interaction within ecosystem. Hence, attention has been focused on the development of alternative technologies and methods for the elimination of petroleum pollutants (Effendi et al. 2018).

9.2.2 Oil Remediation in Polluted Environments

It is very well known that the presence of petroleum oil pollutants in the environment poses a massive threat to ecological balance. Remediation techniques play a crucial role in the restoration of environment by cleaning containment, removal, and destruction of petroleum oil hydrocarbons. The selection of remediation methods deployed depends mainly on the properties of contaminants and conditions of contaminated sites along with microbial community present or required augmentation considered.

Remediation methods used for hydrocarbon/oil contaminated sites (Chukwunonso et al. 2020):

1. Containment: Containment of the pollutants can be achieved by excavation and disposal either by on-site or off-site landfill; stabilization and immobilization by chemical treatment; by physical methods, e.g., site cover, vertical barriers, booms, skimmers, liners, solidification, vitrification, and encapsulation.
2. Separation method:
   (a) Oil/water separation can be achieved by chemical dosing, gravity separation, reverse osmosis, ultra-filtration, micro-filtration, air-floatation, membrane bioreactor, electrocoagulation, electro-floatation, freeze/thaw, and adsorption.
   (b) Oil/soil separation can be achieved by washing, flushing, steam stripping, vacuum extraction, solvent extraction, particle separation, ultrasound assisted-separation, electrokinetic process, microwave heating, radio frequency heating, thermal desorption, and air microbubbles.
3. Destruction method: This method is achieved by the following technologies:
   (a) Biological remediation technologies, such as bioremediation, which includes biostimulation, bioaugmentation, bioventilation, and phytoremediation, which includes phytostabilization, phytodegradation, and photovolatization.
   (b) Chemical remediation, such as solidification, dehalogenation, emulsification, chemical oxidation reduction, ultraviolet oxidation, dispersion, activated carbon treatment, supercrtitical fluid oxidation, sonochemical process, acoustic cavitation, phocatalytic process, nanoremediation.
   (c) Thermal remediation, e.g., pyrolysis, incineration, and microwave-assisted low-temperature remediation (Lim et al. 2016; Luo et al. 2019).

These techniques when used in situ or ex situ or both along with chemical, physicochemical, biological, electric, thermal, ultrasonic, and electromagnetic treatment techniques are helpful in the complete destruction of pollutants in the polluted environment. Therefore, the next section is focused on the various treatment methods mentioned in this section.

Physical and chemical methods have certain major disadvantages, such as unsuitable for weathered soil, high treatment cost, and the chemical oxidation treatment of soils making them unfit for vegetation, or further bioremediation in the future (Lim et al. 2016).

Bioremediation is an uncomplicated, sustainable, environment-friendly, and economical method for accelerating the decay of petroleum pollutants by microorganisms (fungi, yeast, and bacteria). Microorganisms degrade and neutralize crude oil and hydrocarbon pollutants in the soil pore into harmless or simpler compounds (e.g., CO_2 and H_2O) through oxidation under aerobic conditions (Chukwunonso et al. 2020; Lim et al. 2016; Prasad and Aranda 2018).

Bioremediation methods are efficient in degrading oil contaminants and various organic contaminants completely without leaving behind any detrimental effects on the environment. Besides, biological treatment methods are universally accepted by the public as they are aesthetically pleasing and an economical alternative to other remediation methods. However, bioremediation methods require long time for treatment, it can take from months to many years to achieve acceptable removal efficiency results. It was also noticed that high concentrations of oil pollutants decrease the microbial activity, which resulted in very low or insufficient degradation efficiencies (Chukwunonso et al. 2020).

9.3 Biosurfactants

Biosurfactants are amphiphilic compounds produced by microbes which act as emulsifying agents or surface-active agents with an array of biochemical structure such as fatty acids, glycolipids, lipopeptides, phospholipids, particulate, and polymeric structures (Effendi et al. 2018; Huszcza and Burczyk 2003). Surfactants are surface-active agents which help in reducing the surface tension and interfacial tension at interfaces such as water-oil and air-water (Akbari et al. 2018). Biosurfactants mainly produced by microorganism using low-cost substrates in large quantities. They are an attractive alternative compared to synthetic chemical-derived surfactants because of their bioavailability, biodegradability, non-toxicity, and high foaming capability. These properties make them safe and attractive compared to synthetic chemical-derived surfactants, especially in cosmetics, food items, and pharmaceuticals (Akbari et al. 2018).

Biosurfactants have achieved much attention from researchers worldwide because of their potential for bioremediation of petroleum pollutants and heavy metal removing capability from water and soil (Dell'Anno et al. 2018; Jimoh and Lin 2019). Biosurfactants enhance the dispersal of hydrophobic pollutants in the aqueous phase and raise its bioavailability to microbes, with subsequent eradication of such contaminants through biodegradation (Silva et al. 2014).

Glycolipids, one of the most common biosurfactant type are made up of carbohydrates attached to long chain of aliphatic fatty acids or hydroxy aliphatic acid. They comprise of trehalolipid, rhamnolipid, sophorolipid, trehalose dimycolates, and trehalolipid (Effendi et al. 2018). Biosurfactants have potential applications in the bioremediation of hydrocarbons, heavy-metal-contaminated environments such as surfactin, syringafactin, arthrofactin, lichenysin, emulsan, liposan, biodispersan, streptofactin, saponin (Dell'Anno et al. 2018; Effendi et al. 2018). Numerous studies have been done and reported for the removal of hydrocarbons and heavy metals by different biosurfactants (Table 9.1).

Table 9.1 Biosurfactants used for the remediation of hydrocarbons and heavy metals

Anthracene	*Rhamnolipids*	*Sphingomonas* sp.	Cui et al. (2008)
	Rhamnolipids	*Pseudomonas* sp.	
Phenanthrene	*Rhamnolipids*	*Sphingomonas* sp.	Gottfried et al. (2010)
	Rhamnolipids	*Paenibacillus* sp.	Pei et al. (2010)
	Rhamnolipids	*Pseudomonas aeruginosa ATCC 9027*	Shin et al. (2005)
	Rhamnolipids	*Pseudomonas putida ATCC 17,484*	Dean et al. (2001)
Pyrene	*Rhamnolipids*	*Pseudomonas aeruginosa 57SJ*	Bordas et al. (2005)
	Rhamnolipids	*Pseudomonas aeruginosa SP4*	Jorfi et al. (2013)
Crude oil	*Rhamnolipids*	*Pseudomonas aeruginosa*	Nikolopoulou et al. (2013)
	Glycolipids	*Pseudozyma* sp. *NII 08165*	Sajna et al. (2015)
	Glycolipids	*Candida bombicola*	Kang et al. (2010)
Heavy metals	*Di-rhamnolipids*	*Pseudomonas aeruginosa BS2*	Juwarkar et al. (2008)
	Lipopeptide	*Bacillus subtilis*	Müligan et al. (2001)
	Sophorolipid	*Torulopsis bombicola*	Müligan et al. (2001)
	Saponin	*Plant-derived biosurfactant*	Chen et al. (2008)
	Lichenysin	*B. licheniformis*	Zouboulis et al. (2003)

9.3.1 Synthesis of Biosurfactants

Different moieties of biosurfactant and their linkage were synthesized by following possibilities (Satpute et al. 2010):

1. Hydrophilic and hydrophobic moieties synthesis by two independent de novo pathways.
2. Substrate-induced synthesis of hydrophobic moiety and hydrophilic moiety synthesized by de novo pathway.
3. Substrate-dependent synthesis of hydrophilic moiety and hydrophobic moiety synthesized by de novo pathway.
4. Hydrophilic and hydrophobic moieties both synthesized by substrate dependent synthesis.

9.3.2 Biosurfactant Role in Oil Degradation

Biosurfactants can elevate the hydrocarbon degradation by three ways: mobilization, solubilization, and emulsification. Hydrocarbon removal mechanism of biosurfactant is affected by its concentration and molecular mass. Biosurfactants with low molecular mass remove hydrocarbons by mobilization when their concentration is below critical micelle concentration (CMC) and by solubilization when their concentration is above CMC. High-molecular-mass biosurfactants remove hydrocarbons by emulsification (Pacwa-Płociniczak et al. 2011).

During mobilization of hydrophobic compounds (oil/hydrocarbons), biosurfactants decrease the surface and interfacial tension. In reduced interfacial force, biosurfactants increase the contact angle and this will result in reduction of capillary force that hold oil and soil together. Solubilization of hydrocarbon takes place when biosurfactant concentration is above critical micelle concentration. During solubilization, biosurfactant molecules link together and form micelle, thereby increasing the solubility of oil. In biosurfactants, hydrophobic ends are connected together, while hydrophilic ends are exposed towards aqueous phase, resulting in the formation of micelles. Thus, a suitable environment for hydrophobic molecules is created by the micelle in its interior side. The process of micelle formation by incorporation of these molecules is called solubilization.

The process of emulsion formation in which very fine droplets of oil get suspended in water is called emulsification. It was reported that biosurfactants with high molecular weight show efficient emulsification which are used as a stimulant additive for removal of oil or oil-derived contaminants from the environment (Pacwa-Płociniczak et al. 2011; Urum and Pekdemir 2004).

9.4 Application of Biosurfactants Used for Oil Remediation

Biosurfactants and biosurfactant-producing microorganisms have great application in bioremediation as they are economically feasible, less toxic to environment and living beings, and biodegradable. Biosurfactants are used in various industries, such as agrochemicals, petrochemicals, mining and metallurgy (mainly bioleaching), cosmetics, fertilizers, foods and beverages, pharmaceuticals, etc. (Fig. 9.1). Also, they play a significant role in enhanced oil recovery and bioremediation of heavy crude oil pollutants because of their high interfacial and surface-tension-decreasing ability (Volkering et al. 1997). Biosurfactants are widely used in remediation of petroleum pollutants in soil and ocean spill, biodegradation of toxic contaminants such as PAHs (polycyclic aromatic hydrocarbons) and MEOR (microbial enhanced oil recovery), in cleaning tanks for oil storage, for enhancing oil flow via pipelines, in bioremediation of heavy metals, etc. (Batista et al. 2006; Behera and Prasad 2020).

9.4.1 Oil-Polluted Soil Bioremediation

The various processes of petroleum industries such as oil reserves exploration, storage, transportation, and crude oil processing lead to acute soil contamination due to improper leakage from storage tanks, improper disposal, oil spills due to mechanical or human error, etc. Oil-contaminated soil poses a high risk to public health and environmental imbalance, which needs to be addressed by remediation. Bioremediation of oil components and organic compounds has raised serious concern due to the presence of various toxic compounds such as PAHs, benzene, and its

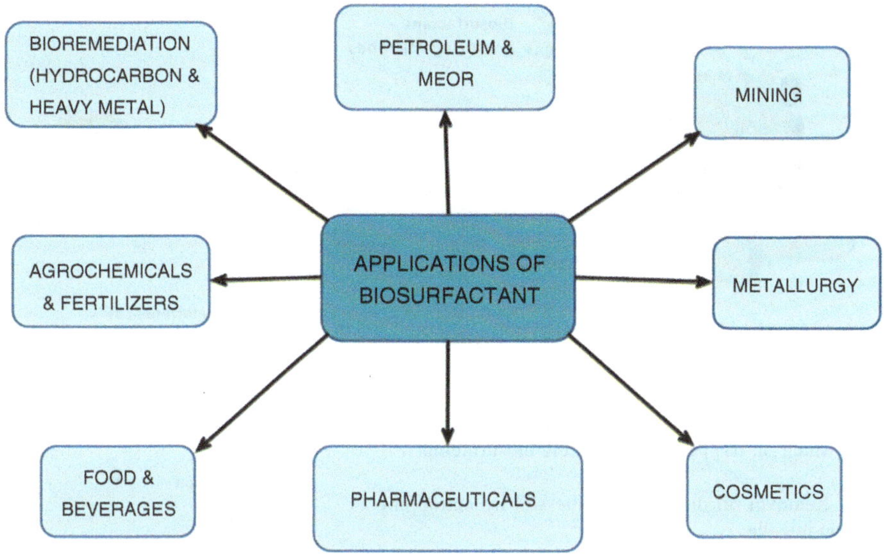

Fig. 9.1 Applications of biosurfactants

derivatives (Wang et al. 2019). High molecular weight of these compounds makes them insoluble or of very low solubility in water, thus preventing the natural biodegradation process (Erdogan and Karaca 2011). Biosurfactants emulsification property can enhance their solubility, reduce surface tension and increase the displacement of hydrophobic molecules from soil particles (Banat et al. 2000). In this way insoluble substrate (oil) gets solubilized and becomes easy to consume by microorganisms as a carbon substrate source as shown in Fig. 9.2, which is then decomposed and converted to nontoxic end products, such as CO_2, CH_4, and H_2O.

9.4.2 Bioremediation of Marine Oil Spills and Petroleum Contamination

Hydrocarbons have been part of the marine ecosystem for thousands of years. The Oil Tanker Spill Statistics 2019 has recently updated the total volume of oil lost into the marine environment in 2019 was approximately 1000 tonnes and the same quantity was recorded in 2012 (https://www.itopf.org/knowledge-resources/data-statistics/statistics/). The major causes of oil spills are drilling rigs and oil wells, spill of crude oil during transportation, off-shore platforms, and oil refineries products and their waste. These oil spills in the ocean have a severe negative effect on marine life. Oil spills have always wreaked havoc in the organisms present in affected marine environments. Massive causalities of marine life involving fish, coral reefs, otters, seals, and birds have often been witnessed (Patel et al. 2019). Crude oil forms a viscous surface slick, blocking oxygen exchange and sunlight

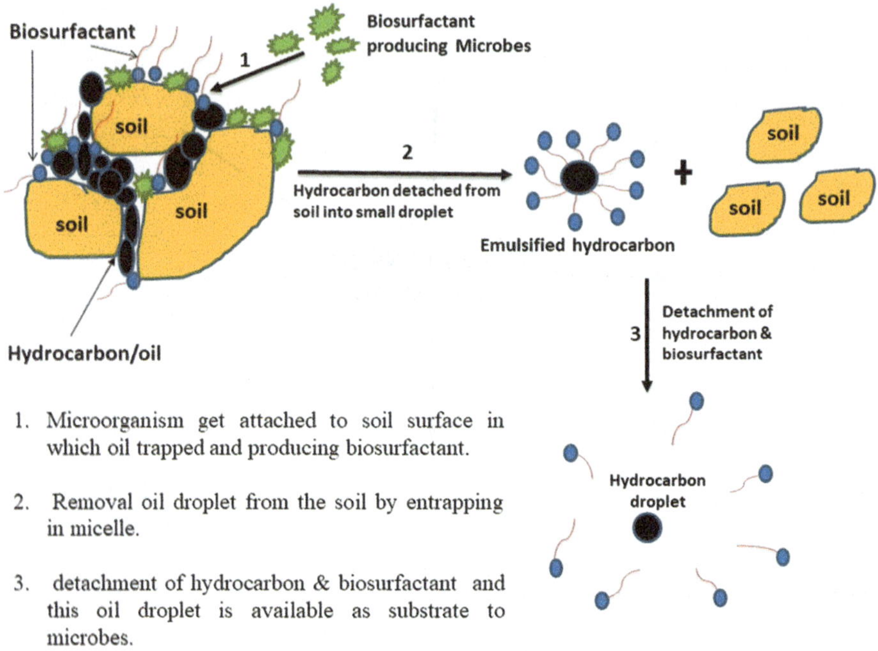

Fig. 9.2 Illustration of biosurfactant action on oil/hydrocarbon. (1) Microorganisms get attached to soil surface in which oil is trapped, producing biosurfactants. (2) Removal of oil droplet from soil by entrapping in micelles. (3) Detachment of hydrocarbon and biosurfactant: this droplet is available as substrate to microbes

penetration into water (Freitas et al. 2016). In the time of high wind and tide in ocean, oil spillage invades the coastal area, often wreaking havoc on marine life (Wang et al. 2014). The toxic effect of oil slick on aquatic life goes way further than anticipated perils. Therefore, mitigation interventions are crucial to get rid of thick oil in order to maintain ecological aquatic balance and environmental health (Brussaard et al. 2016).

The spilled oil which cannot be degraded physically/chemically can only be completely broken down through biodegradation processes. Thus, biodegradation is a key option for the eradication of oil and its derivative pollutants from the environment which remains on the surface after recovering and minimize the environmental impact of a spill (Prince et al. 2003). Biosurfactants increase the dispersal of hydrophobic pollutants (oil/hydrocarbon) in the aqueous phase and this causes an increase in their bioavailability as substrates for microorganisms, with eventual removal of such contamination through bioremediation without disturbing the aquatic ecosystem.

9.4.3 Cleaning of Oil Tanks and Pipelines

Biosurfactants have been successfully used in the removal of oil sludge deposited in storage tanks and to improve heavy crude oil transportation process though pipelines (De Almeida et al. 2016). Different types of biosurfactants (as shown in Table 9.2) degrade such oil pollutants and hydrocarbons by increasing microbial accessibility, increasing their solubility in water, and boosting oil displacement (Vijayakumar and Saravanan 2015). Biosurfactants lower the interfacial tension between oil and surface of pipelines and oil storage tanks, which raises oil availability for microbial uptake and mobilize it to detach from the surface of tank and pipe line this leads to their cleaning (Pacwa-Płociniczak et al. 2011).

9.4.4 Bioremediation of Heavy Metals and Toxic Pollutants

Industrialization and urbanization lead to excessive accretion of heavy metals present in petroleum hydrocarbon pollutants in the soil which causes serious threat to the environment (Pandey and Madhuri 2014). In the petroleum industry, the

Table 9.2 Different type of biosurfactant produced during bioremediation process

Biosurfactant group	Microorganisms	Applications	References
Rhamnolipids	Indigenous soil microflora	Degrading petroleum hydrocarbon	Benincasa (2007)
Glycolipid	*Nocardiopsis* sp. *Arthrobacter* sp. *Corynebacterium* sp. *R. wratislaviensis* BN38	Enhancement of the biodegradation of hydrocarbons in soil and marine environment	Pacwa-Płociniczak et al. (2011) and Franzetti et al. (2010)
Lipopeptides	*Bacillus licheniformis* *Bacillus subtilis* *N. alba strain* MSA10	Enhancement of oil recovery; removal of heavy metals from contaminated soil, sediment, and water	Chen et al. (2009), Bennur et al. (2015) and Jenneman et al. (1983)
Surfactin	Indigenous soil microflora	Degrading diesel oil	Whang et al. (2008)
Polymeric biosurfactants	*Saccharomyces cerevisiae* *Candida lipolytica*	Stabilization of hydrocarbons in water emulsions	Cirigliano and Carman (1985) and Cameron et al. (1988)
Sophorolipid	*Indigenous soil microflora*	Degradation of crude oil, naphthalene, hexadecane, pristane	Kang et al. (2010)
Fatty acids	*Acinetobacter* sp. *Rhodococcus erythropolis*	Increasing the tolerance of bacteria to heavy metals	Appanna et al. (1995)

generation of fuel from source rocks results in heavy metal pollution. The heavy metal contamination is prominent at terminals, tanker accidents, spills, offshore oil exploration, industrial land sources, and recreational and agricultural development. The removal of heavy metals such as Cd, Cu, Mn, Fe, Ni, Pb, and Zn is considered as one of the important areas in remediation research (Sose et al. 2018). Biosurfactants accumulate on the surface and form hemi-micelles at soil and air interface. These hemi-micelles remove heavy metals by reducing the interfacial tension and electrostatic attraction, which results in the incorporation of heavy metals into the micelle and ultimately their removal from the surface (Ochoa Loza 1998).

Persistent toxic organic pollutants found in oil waste, such as PAHs (pyrene, anthracene), are carcinogenic and exposure to such pollutants has been linked to cardiovascular disease and poor fetal development. They are hydrophobic compounds and their water solubility reduces with the incremental number of rings in their molecular structure, which aggravates the low bioavailability of these compounds, making biodegradation of PAHs difficult. Solubility in water of PAHs can be enhanced by addition of biosurfactants as it increases the surface area of hydrophobic water-insoluble compounds contributing in bioremediation of toxic pollutants (Yin et al. 2009).

9.5 Conclusion

Due to the rapid growth of human population, the demand for petroleum and petroleum-based products also increases day by day as it is considered a major energy resource. However, crude oil/petroleum is also considered as a major source of pollutants. It causes serious harm to the ocean ecosystem, soil, and human health. Due to the hydrophobic nature of oil, its degradation is quite difficult so it persists in water and soil that enhance its toxicity in ecosystem. Many traditional physicochemical methods were implemented for removal of hydrocarbon pollutants but they were not able to completely remove them. Thus, more focus is being given to biological remediation methods and the use of biosurfactants for hydrocarbon removal from environment is one of the most promising methods considered nowadays. The amphiphilic nature of biosurfactants with lower toxicity, biodegradability, ecological acceptability, high surface activity, environment friendliness, and active under extreme condition of temperature, pH, and salinity makes them most suitable candidates for removal of hydrocarbon pollutants. Many microorganisms, such as bacteria, yeast, and fungi, feed on insoluble substrates, for which they secrete biosurfactants to facilitate the insoluble substrates across cell membranes. This chapter reviewed the previous and latest research results, including the remediation of oil-contaminated sites through physicochemical methods and by the use of biosurfactants. Comparative study of these methods according to their efficiency and suitability for removing the oil/hydrocarbon from the environment was also revealed in this chapter. The mode of action of biosurfactants and biosurfactant-producing microbes for the removal of oil/hydrocarbon and other heavy metals was briefly discussed.

Finally, crude oil and petroleum products represent the major sources of oil/hydrocarbon contamination in soil and aquatic environments. Application of biosurfactants in the destruction of hydrophobic pollutants makes them favorable and promising biomolecules. Today, biosurfactants are produced mainly by microbial fermentation at the laboratory scale and therefore industrial scale-up is needed for higher production yields for subsequent applications at field levels.

This chapter provides necessary information on the application of biosurfactants as an assuring alternative to synthetic chemicals in the petroleum industry and the bioremediation of oil spills recovery, removal of heavy metals, and utilization in oil pollution control and oil storage tank cleanup.

References

Akbari S, Abdurahman NH, Yunus RM, Fayaz F, Alara OR (2018) Biosurfactants—a new frontier for social and environmental safety: a mini review. Biotechnol Res Innov 2(1):81–90

Appanna VD, Finn H, Pierre MS (1995) Exocellular phosphatidylethanolamine production and multiple-metal tolerance in Pseudomonas fluorescens. FEMS Microbiol Lett 131(1):53–56

Banat IM, Makkar RS, Cameotra SS (2000) Potential commercial applications of microbial surfactants. Appl Microbiol Biotechnol 53(5):495–508

Batista SB, Mounteer AH, Amorim FR, Totola MR (2006) Isolation and characterization of biosurfactant/bioemulsifier-producing bacteria from petroleum contaminated sites. Bioresour Technol 97(6):868–875

Behera BK, Prasad R (2020) Environmental technology and sustainability. Elsevier (ISBN: 9780128191033). https://www.elsevier.com/books/environmental-technology-and-sustainability/behera/978-0-12-819103-3

Benincasa M (2007) Rhamnolipid produced from agroindustrial wastes enhances hydrocarbon biodegradation in contaminated soil. Curr Microbiol 54(6):445–449

Bennur T, Kumar AR, Zinjarde S, Javdekar V (2015) Nocardiopsis species: incidence, ecological roles and adaptations. Microbiol Res 174:33–47

Bordas F, Lafrance P, Villemur R (2005) Conditions for effective removal of pyrene from an artificially contaminated soil using Pseudomonas aeruginosa 57SJ rhamnolipids. Environ Pollut 138(1):69–76

Brussaard CP, Peperzak L, Beggah S, Wick LY, Wuerz B, Weber J, Arey JS, Van Der Burg B, Jonas A, Huisman J, Van Der Meer JR (2016) Immediate ecotoxicological effects of short-lived oil spills on marine biota. Nat Commun 7(1):1–11

Cameron DR, Cooper DG, Neufeld RJ (1988) The mannoprotein of Saccharomyces cerevisiae is an effective bioemulsifier. Appl Environ Microbiol 54(6):1420–1425

Chen WJ, Hsiao LC, Chen KKY (2008) Metal desorption from copper (II)/nickel (II)-spiked kaolin as a soil component using plant-derived saponin biosurfactant. Process Biochem 43(5):488–498

Chen YG, Wang YX, Zhang YQ, Tang SK, Liu ZX, Xiao HD, Xu LH, Cui XL, Li WJ (2009) Nocardiopsis litoralis sp. nov., a halophilic marine actinomycete isolated from a sea anemone. Int J Syst Evol Microbiol 59(11):2708–2713

Chukwunonso I, Ahmed A, Hassan A, Shahul F (2020) Environmental Technology & Innovation Remediation of soil and water contaminated with petroleum hydrocarbon: a review. Environ Technol Innov 17:100526

Cirigliano MC, Carman GM (1985) Purification and characterization of liposan, a bioemulsifier from Candida lipolytica. Appl Environ Microbiol 50(4):846–850

Cui Z, Lai Q, Dong C, Shao Z (2008) Biodiversity of polycyclic aromatic hydrocarbon-degrading bacteria from deep sea sediments of the middle Atlantic ridge. Environ Microbiol 10:2138

Das N, Chandran P (2011) Microbial degradation of petroleum hydrocarbon contaminants: an overview. Biotechnol Res Int 2011:941810
De Almeida DG, Soares Da Silva RDCF, Luna JM, Rufino RD, Santos VA, Banat IM, Sarubbo LA (2016) Biosurfactants: promising molecules for petroleum biotechnology advances. Front Microbiol 7:1718
De S, Malik S, Ghosh A, Saha R, Saha B (2015) A review on natural surfactants. RSC Adv 5 (81):65757–65767
Dean SM, Jin Y, Cha DK, Wilson SV, Radosevich M (2001) Phenanthrene degradation in soils co-inoculated with phenanthrene-degrading and biosurfactant-producing bacteria. J Environ Qual 30(4):1126–1133
Dell'Anno F, Sansone C, Ianora A, Dell'Anno A (2018) Biosurfactant-induced remediation of contaminated marine sediments: current knowledge and future perspectives. Mar Environ Res 137:196–205
Effendi AJ, Kardena E, Helmy Q (2018) Biosurfactant-enhanced petroleum oil bioremediation. In: Microbial action on hydrocarbons. Springer, Singapore, pp 143–179
Erdogan EE, Karaca A (2011) Bioremediation of crude oil polluted soils. Asian J Biotechnol 3 (3):206–213
Franzetti A, Gandolfi I, Bestetti G, Smyth TJ, Banat IM (2010) Production and applications of trehalose lipid biosurfactants. Eur J Lipid Sci Technol 112(6):617–627
Freitas BG, Brito JM, Brasileiro PP, Rufino RD, Luna JM, Santos VA, Sarubbo LA (2016) Formulation of a commercial biosurfactant for application as a dispersant of petroleum and by-products spilled in oceans. Front Microbiol 7:1646
Gottfried A, Singhal N, Elliot R, Swift S (2010) The role of salicylate and biosurfactant in inducing phenanthrene degradation in batch soil slurries. Appl Microbiol Biotechnol 86(5):1563–1571
Häder DP, Banaszak AT, Villafañe VE, Narvarte MA, González RA, Helbling EW (2020) Anthropogenic pollution of aquatic ecosystems: emerging problems with global implications. Sci Total Environ 713:136586
Haritash AK, Kaushik CP (2009) Biodegradation aspects of polycyclic aromatic hydrocarbons (PAHs): a review. J Hazard Mater 169(1–3):1–15
Hewelke E, Szatyłowicz J, Hewelke P, Gnatowski T, Aghalarov R (2018) The impact of diesel oil pollution on the hydrophobicity and CO2 efflux of forest soils. Water Air Soil Pollut 229(2):51
Hill J (2009) Environmental costs and benefits of transportation biofuel production from food-and lignocellulose-based energy crops: a review. In: Sustainable agriculture. Springer, Dordrecht, pp 125–139. http://www.itopf.org/fileadmin/data/Documents/Company_Lit/Oil_Spill_Stats_2018.pdf; https://www.itopf.org/knowledge-resources/data-statistics/statistics/
Huszcza E, Burczyk B (2003) Biosurfactant production by Bacillus coagulans. J Surfactant Deterg 6(1):61–64
Jenneman GE, McInerney MJ, Knapp RM, Clark JB, Feero JM, Revus DE, Menzie DE (1983) Halotolerant, biosurfactant-producing Bacillus species potentially useful for enhanced oil recovery. Dev Ind Microbiol 24:485
Jimoh AA, Lin J (2019) Biosurfactant: a new frontier for greener technology and environmental sustainability. Ecotoxicol Environ Saf 184:109607
Jorfi S, Rezaee A, Moheb-ali G, Jaafarzadeh NA (2013) Pyrene removal from contaminated soils by modified Fenton oxidation using iron nano particles. J Environ Health Sci Eng 11:17. https://doi.org/10.1186/2052-336X-11-17
Juwarkar AA, Dubey KV, Nair A, Singh SK (2008) Bioremediation of multi-metal contaminated soil using biosurfactant—a novel approach. Indian J Microbiol 48:142–146
Kang SW, Kim YB, Shin JD, Kim EK (2010) Enhanced biodegradation of hydrocarbons in soil by microbial biosurfactant, sophorolipid. Appl Biochem Biotechnol 160(3):780–790
Lim MW, Von Lau E, Poh PE (2016) A comprehensive guide of remediation technologies for oil contaminated soil—present works and future directions. Mar Pollut Bull 109(1):14–45

Luo H, Wang H, Kong L, Li S, Sun Y (2019) Insights into oil recovery, soil rehabilitation and low temperature behaviors of microwave-assisted petroleum-contaminated soil remediation. J Hazard Mater 377:341–348

Maier RM (2003) Biosurfactants: evolution and diversity in bacteria. Adv Appl Microbiol 52:101–122

Mani D, Kumar C (2014) Biotechnological advances in bioremediation of heavy metals contaminated ecosystems: an overview with special reference to phytoremediation. Int J Environ Sci Technol 11(3):843–872

Morgan P, Atlas RM (1989) Hydrocarbon degradation in soils and methods for soil biotreatment. Crit Rev Biotechnol 8(4):305–333

Mülligan CN, Yong RN, Gibbs BF (2001) An evaluation of technologies for the removal of heavy metals from sediments. J Hazard Mater 85:145–163

Nikolopoulou M, Pasadakis N, Norf H, Kalogerakis N (2013) Enhanced ex situ bioremediation of crude oil contaminated beach sand by supplementation with nutrients and rhamnolipids. Mar Pollut Bull 77(1–2):37–44

Ochoa Loza FJ (1998) Physico-chemical factors affecting rhamnolipid (biosurfactant) application for removal of metal contaminants from soil. The University of Arizona, Tucson, AZ

Okada M (2002) Chemical syntheses of biodegradable polymers. Prog Polym Sci 27(1):87–133

Ossai IC, Ahmed A, Hassan A, Hamid FS (2019) Remediation of soil and water contaminated with petroleum hydrocarbon: a review. Environ Technol Innov 17

Pacwa-Płociniczak M, Płaza GA, Piotrowska-Seget Z, Cameotra SS (2011) Environmental applications of biosurfactants: recent advances. Int J Mol Sci 12(1):633–654

Panagos P, Van Liedekerke M, Yigini Y, Montanarella L (2013) Contaminated sites in Europe: review of the current situation based on data collected through a European network. J Environ Public Health 2013:158764

Pandey G, Madhuri S (2014) Heavy metals causing toxicity in animals and fishes. Res J Anim Vet Fishery Sci 2(2):17–23

Patel S, Homaei A, Patil S, Daverey A (2019) Microbial biosurfactants for oil spill remediation: pitfalls and potentials. Appl Microbiol Biotechnol 103(1):27–37

Pei D, Xu J, Zhuang Q, Tse HF, Esteban MA (2010) Induced pluripotent stem cell technology in regenerative medicine and biology. In: Bioreactor systems for tissue engineering II. Springer, Berlin, pp 127–141

Prasad R, Aranda E (2018) Approaches in bioremediation: the new era of environmental microbiology and nanobiotechnology. Springer International Publishing (ISBN: 978-3-030-02369-0). https://www.springer.com/gp/book/9783030023683

Prince RC, Lessard RR, Clark JR (2003) Bioremediation of marine oil spills. Oil Gas Sci Technol 58(4):463–468

Sajna KV, Höfer R, Sukumaran RK, Gottumukkala LD, Pandey A (2015) White biotechnology in biosurfactants. In: Industrial biorefineries & white biotechnology. Elsevier, Amsterdam, pp 499–521

Satpute SK, Banpurkar AG, Dhakephalkar PK, Banat IM, Chopade BA (2010) Methods for investigating biosurfactants and bioemulsifiers: a review. Crit Rev Biotechnol 30(2):127–144

Shin KH, Ahn Y, Kim KW (2005) Toxic effect of biosurfactant addition on the biodegradation of phenanthrene. Environ Toxicol Chem 24(11):2768–2774

Silva RDCF, Almeida DG, Rufino RD, Luna JM, Santos VA, Sarubbo LA (2014) Applications of biosurfactants in the petroleum industry and the remediation of oil spills. Int J Mol Sci 15 (7):12523–12542

Singh A, Kuhad RC, Ward OP (2009) Biological remediation of soil: an overview of global market and available technologies. In: Advances in applied bioremediation. Springer, Berlin, pp 1–19

Sose AT, Kulkarni SJ, Sose MT (2018) Oil industry–analysis, effect and removal of heavy metals. Int J Eng Sci Res Technol 6:254–257

Ubalua AO (2011) Bioremediation strategies for oil polluted marine ecosystems. Aust J Agric Eng 2(6):160

Urum K, Pekdemir T (2004) Evaluation of biosurfactants for crude oil contaminated soil washing. Chemosphere 57(9):1139–1150

Varjani SJ, Upasani VN (2017) A new look on factors affecting microbial degradation of petroleum hydrocarbon pollutants. Int Biodeter Biodegr 120:71–83

Vijayakumar S, Saravanan V (2015) Biosurfactants-types, sources and applications. Res J Microbiol 10(5):181

Volkering F, Breure AM, Rulkens WH (1997) Microbiological aspects of surfactant use for biological soil remediation. Biodegradation 8(6):401–417

Wang H, Xu J, Zhao W, Zhang J (2014) Effects and risk evaluation of oil spillage in the sea areas of Changxing Island. Int J Environ Res Public Health 11(8):8491–8507

Wang M, Zhang B, Li G, Wu T, Sun D (2019) Efficient remediation of crude oil-contaminated soil using a solvent/surfactant system. RSC Adv 9(5):2402–2411

Whang LM, Liu PWG, Ma CC, Cheng SS (2008) Application of biosurfactants, rhamnolipid, and surfactin, for enhanced biodegradation of diesel-contaminated water and soil. J Hazard Mater 151(1):155–163

Yin H, Qiang J, Jia Y, Ye J, Peng H, Qin H, Zhang N, He B (2009) Characteristics of biosurfactant produced by Pseudomonas aeruginosa S6 isolated from oil-containing wastewater. Process Biochem 44(3):302–308

Zouboulis AI, Matis KA, Lazaridis NK, Golyshin PN (2003) The use of biosurfactants in flotation: application for the removal of metal ions. Miner Eng 16(11):1231–1236

Potential Applications of Anti-Adhesive Biosurfactants

10

Maria da Gloria Conceição da Silva, Italo José Batista Durval, Maria Eduarda Pereira da Silva, and Leonie Asfora Sarubbo

Abstract

There is a growing demand for materials covered with compounds that prevent the adhesion of microorganisms that form biofilms. Surface contamination is a concern of the biomedical and food industry, due to the risks to the health of patients and consumers. Thus, the interruption of microbial adhesion in its first moments is an excellent approach for the performance of anti-adhesive compounds. The microbial biosurfactants have the potential for the application on surfaces of economic interest as agents that inhibit microbial fixation. They comprise a variety of amphiphilic molecules that can be obtained directly, synthesized by plants and microbes, or indirectly, through chemical or genetic changes. Biosurfactant production from renewable substrates is possible, and there is a tendency for the substitution of synthetic surfactants of biological origin in industrialized countries. This chapter discusses the main classes of microbial biosurfactants with anti-adhesive action, the process of microbial adhesion for the formation of biofilms, and current studies involving the application of biosurfactants as biofilm disturbing agents on different surfaces.

Keywords

Biosurfactant · Medical instruments · Food industry · Contact surfaces · Antibiofilm · Antimicrobial

M. d. G. C. da Silva (✉) · I. J. B. Durval · L. A. Sarubbo
Northeast Biotechnology Network (RENORBIO), Federal Rural College of Pernambuco, Recife, Pernambuco, Brazil

Centre of Science and Technology (CCT), Catholic College of Pernambuco, Recife, Pernambuco, Brazil

M. E. P. da Silva
Centre of Science and Technology (CCT), Catholic College of Pernambuco, Recife, Pernambuco, Brazil

Inamuddin et al. (eds.), *Microbial Biosurfactants*, Environmental and Microbial Biotechnology, https://doi.org/10.1007/978-981-15-6607-3_10

10.1 Introduction

Advances in the area of biotechnology with the use of natural products that protect surfaces increase the economic interest in the generation of inputs against biofouling of the types that occur in medical devices or in places that come into contact with food (Gopikrishnan et al. 2015; Junter et al. 2016; Giri et al. 2019).

The attachment of microbial cells to surfaces covered by particles and colloids is the first stage in the development of the biofilm structure. As adhesion is still weak at this stage, this would be an excellent time for the application of anti-adhesive constituents. Microbial biosurfactants can change the surfaces they come in contact with. When adsorbing onto polystyrene surfaces, silicon and glass microbial surfactant changes the inherent hydrophobicity of such surfaces. In this way, the biosurfactant influences the effects of fixation and ease of removal of microorganisms depending on the type of the fouled surface (Janek et al. 2012).

Biosurfactants are versatile, stable, and biocompatible molecules, obtained from various sources such as bacteria, fungi, and yeasts (Gutnick and Bach 2017) and secondary compounds extracted from plants that exhibit surfactant characteristics (Cheok et al. 2014; Zhu et al. 2019).

Biosurfactants have the same properties as synthetic surfactants. Among the properties, we can highlight high biodegradability, low toxicity, and not inducing allergic reactions. They can be used in extreme environmental situations due to the stability of their properties when exposed to unusual pH occurrence, salinity, and temperature and specific bioactivity, which give them a great potential for practical applications in several areas (Freitas et al. 2016; Zhu et al. 2019; Liu et al. 2020).

In addition to these properties, they have antibiotic, antimicrobial, anti-biofilm, and anti-adhesive activities (Rivardo et al. 2011; Padmapriya and Suganthi 2013; Banat et al. 2014; Ndlovu et al. 2017). In this way, biosurfactants are used in the fields of industries, namely cosmetics and food, and in the biomedical and pharmaceutical areas, and they also expand their use in the oil industries to improve the recovery of this product (Jimoh and Lin 2019).

There is a tendency to substitute synthetic surfactants for those of biological origin in industrialized countries, stimulated by the sustainable advantages of biosurfactants, as they can be produced using renewable substrates, derived from industrial residues, which cheapens the cost of this bioactivity (Satpute et al. 2017; Araújo et al. 2019).

Glycolipids and lipopeptides are well-established classes of biosurfactants. They exhibit broad-spectrum antimicrobial activity, anti-adhesive, and biofilm control. They are right now applied in various areas (food, beauty products, and pharmaceutical industries) as emulsifying, antimicrobial, and surfactant agents (Inès and Dhouha 2015; Mnif and Ghribi 2016).

Biosurfactants that have anti-adhesive activity can be produced by several microorganisms: *Pseudomonas aeruginosa* produces rhamnolipids, some species of *Candida* sp. produce sophorolipids, *Bacillus* sp. produces surfactin among other isoforms. Biosurfactants with anti-adhesive activity can also be released by lactic acid probiotic bacteria (LAB) (Yan et al. 2019).

This chapter deals with the main classes of biosurfactants with anti-adhesive action, microbial fixation for biofilm formation, and studies involving the use of microbial biosurfactants as disintegrating agents of this formation on different surfaces.

10.2 Biosurfactants That Display Anti-Adhesive Activity

The composition and type of microorganism are used to classify biosurfactants. Other forms of classification are low molecular weight biosurfactants (lipopeptides, glycolipids, and phospholipids) and high molecular weight biosurfactants (polysaccharides, lipopolysaccharides, proteins, and lipoproteins). Low molecular weight biosurfactants lower the surface and interfacial tension of different substances. The ones with high molecular weight are used as emulsifiers and stabilizers for different products (Sharma and Sharma 2018; Jahan et al. 2020).

The parameters of free energy and surface tension of the coated materials and the surfactant itself influence the development of the surfactant film on any solid surface. The surface orientation of the nonpolar and polar fractions of the film formed by the biosurfactants on some solids surfaces is crucial for the balance of the hydrophobic and hydrophilic properties of the covered solid. These behaviors of the biosurfactant configurations are of practical importance for the protection of areas that are often used in food handling, medical devices, and surgical implants (Zdziennicka and Jańczuk 2018).

Not all biosurfactants are of interest as anti-adhesive surfaces. The most representative classes that showed responses in reducing adhesion in different materials and against microorganisms of interest are lipopeptides, with surfactin standing out, and glycolipids being mostly represented by rhamnolipids (Cao et al. 2009; Nickzad and Déziel 2014; Abdelli et al. 2019; Ceresa et al. 2019).

Lipopeptide biosurfactants exhibit antibacterial, antifungal, antiviral, and anti-adhesive activities. They are divided into three main groups (surfactin, iturin, and fengycin). Each group presents various homologs and isoforms showing distinct constitutions of amino acids and fatty acid chains (Inès and Dhouha 2015). Due to its attractive surfactant properties and antimicrobial and antibiofilm activities, surfactin is the most powerful biosurfactant, with many isoforms that can be determined genetically or by structural chromatographic analysis techniques (Ibrar and Zhang 2020; Ohadi et al. 2020).

Another important class of interest is the glycolipids. Rhamnolipids produced by *Pseudomonas* sp. are the most important representative in this group. Rhamnolipids are easily produced as a blend of homologous molecules, specifically mono-rhamnolipids and di-rhamnolipids by *P. aeruginosa* (de Freitas et al. 2019). These biosurfactants manifest surface activities and emulsifying and biological activities. Due to this versatility, they were highlighted as versatile additives in food preparation (Nitschke and Silva 2018). Glycolipids have a polysaccharide in their main groups. So, when this group is impacted by electrolytes or undergoes pH changes, its micellar structure changes (Jahan et al. 2020). This can interfere with the process of anti-adhesion of surfaces.

Sophorolipids are other important types of glycolipid, composed of a sugar dimer formed by the glycosidic bond to a hydroxylated fatty acid that are produced and released mainly by yeasts, such as *Candida bombicola* (Shah et al. 2007). The natural diversity of sophorolipids is triggered by variations in the acetylation standard of the sophorosis unit, by the incidence of inside esterification and by the attributes of the hydroxylated fatty acid. A sample of sophorolipids may contain over 20 congeners; however, few of these forms will be dominant. Sophorolipids are structurally classified into acidic and lactonic forms (Haque et al. 2017). Acidic forms are used for cleaning purposes, while lactonic forms are primarily responsible for bioactivity (Van Bogaert et al. 2007; Dhar et al. 2011). This biosurfactant has low cytotoxicity and has been approved for use in food and the pharmaceutical industry by the US FDA (Joshi-Navare and Prabhune 2013). They have antimicrobial and anticarcinogenic properties, in addition to antifungal activity against planktonic cells of pathogenic species (Haque et al. 2016).

Finally, we have the lactobacillus microorganisms, sometimes referred to as probiotics, which are outstanding producers of anti-adhesive biosurfactants. Also, they present antimicrobial, antibiofilm, and antioxidant activities in the same molecule. Therefore, they can be applied in different industrial sectors (Meylheuc et al. 2006; Sambanthamoorthy et al. 2014; Merghni et al. 2017; Yan et al. 2019).

Table 10.1 emphasizes the anti-adhesive activity mentioned in some recent studies in the literature in the area.

10.3 Biofilms and the Adhesion Process: Mechanisms and Effects

Biofilms are complex formations of microorganisms adhered to the surface of biogenic or inert materials. They are associated with each other through extracellular polymeric substances forming an aggregation of microbial cells. The extracellular substance produced by the biofilm, besides contributing to the access to nutrients, allows the existence of these microorganisms in adverse conditions, such as competition, lack of resources, and resistance to antimicrobial treatments. Thus, biofilms are responsible for making it difficult to treat chronic diseases with antibiotics (Roy et al. 2018; Prasad et al. 2020).

The formation of biofilm on any surface involves at least three different phases. In the first phase, microbial cells are bound to a surface previously covered by particles of glycoprotein origin, in the second phase, in this slime, more microbial cells colonize forming microcolonies, and, finally, the complete development of the biofilm through the formation of channels and the formation of firm structures. With the maturation of the biofilm, the disintegration will occur by mechanical and chemical processes and will influence the renewal of the biofilm by the dispersion of the colony (Payne and Boles 2016).

Table 10.1 Examples of action of the various biosurfactants and their target microorganisms

Biosurfactant or producing microorganism	Application field	Target	References
Pediococcus acidilactici and *Lactobacillus plantarum*	Biomedical	*Staphylococcus aureus*	Yan et al. (2019)
Pseudomonas aeruginosa	Biomedical	*Staphylococcus aureus* and *Staphylococcus epidermidis*	Ceresa et al. (2019)
Surfactin	Biomedical	*Staphylococcus epidermidis*	Abdelli et al. (2019)
Lipopeptides	Biomedical and food industry	*Staphylococcus aureus, Salmonella typhimurium, and Bacillus cereus*	Giri et al. (2019)
Lipopeptides	Agribusiness (disease control in plants)	*Agrobacterium tumefaciens*	Ben et al. (2018)
Lipopeptides	Biomedical and food industry	*Staphylococcus aureus, Bacillus cereus, Micrococcus luteus, Klebsiella pneumoniae, Escherichia coli, Salmonella typhimurium, Salmonella entarica, Enterobacterium* sp., *Aspergillus Niger, Aspergillus flavus, Fusarium oxysporum, Pythium ultimum, Fusarium solani, and Rhizoctonia bataticola*	Jemil et al. (2017)
Lactobacillus plantarum subsp. plantarum	Food	*Escherichia coli, Staphylococcus aureus,* and *Salmonella enterica*	Bakhshi et al. (2017)
Rhamnolipids and surfactin	Food	*Listeria monocytogenes* and *Pseudomonas fluorescens*	Araujo et al. (2016)
Xylolipid	Biomedical	*L. monocytogenes, Escherichia coli* and *Bacillus cereus*	Sharma et al. (2015)
Glycolipid (glucose + palmitic acid)	Biomedical	*Candida albicans, Pseudomonas aeruginosa,* and the marine biofouling bacterium *Bacillus pumilus*	Dusane et al. (2011)
Surfactin/iturin A	Food	*Bacillus cereus*	Shakerifard et al. (2009)

The installation of bacteria is particularly intermediated by particle deposition, hydrodynamic forces, and Brownian motion. Adherence to the substrate is regulated by Lifshitz–Van der Waals, acid–base, hydrophobic, and electrostatic interaction forces (Van Oss 1995). Biofilms need to produce biopolymers and polymeric extracellular substances (EPS) rich in carbohydrates and proteins that function as a protective wrapper in which microorganisms are embedded. This ensures for biofilms their fixation and maintenance on surfaces in different environments (Donlan 2002).

Humidity, temperature, environmental pH value, climatic conditions, and chemical composition of the nutritive substrate are the factors that affect the growth of biofilm. Besides, biofilms contain 80–90% water, and their depth can differ between 50 and 100 μm, depending upon the inhabited area (Kaali et al. 2012).

Materials with a hydrophobic surface favor bacterial fixation and biofilm formation, except those with a superhydrophobic surface such as Teflon that has surfaced with contact angles with water >150° (Zeraik and Nitschke 2012; Li et al. 2016; Yilgör et al. 2018).

The hydrophobicity of the microbial cellular membrane and the presence of extracellular filamentous annexes can affect the ratio and degree of bacterial binding. The progress of the hydrophobic type interaction between the exposed area of the material and the microbial cell surface tends to be greater with the increase in the nonpolar constitution of those involved (Donlan 2002; Krasowska and Sigler 2014). The hydrophobic regions of the bacterial cells are partially involved in the connection with a neighboring cell (Van Oss 1995; Krasowska and Sigler 2014). Studies carried out by many laboratories concluded that the susceptibility of materials to microbial adhesion is greater on wood and latex surfaces. A reduction occurs from silicone, PVC, Teflon, polyurethane, stainless steel, and titanium materials (Stoica et al. 2016).

Biofilms can cause microbiologically influenced corrosion (MIC). Some microorganisms cause MIC through extracellular electron transfer for energy. They secrete corrosive metabolites that lead to MIC (Jia et al. 2019).

As biosurfactants reduce the surface tension between liquids and the surface, they can wet surfaces and thus make them hydrophilic (Fig. 10.1), making microbial fixation difficult. Furthermore, they allow greater penetration of different fluids, including solvents and antimicrobial agents in biofilms, which can contribute to the removal of this and other fouling.

10.4 Applications of Biosurfactants as Anti-Adhesive Agents

The pre-contact of surfaces with surfactants can lead to the adsorption of these elements on the surfaces, which can affect the development of biofilm in two ways: (1) modification of the biofilm formation capacity, as surfactants can act against cellular metabolism, favoring or impairing the adhesion forces that maintain the mechanical stability of the biofilm, and/or (2) development of biofilms with less cohesive characteristics that can lead to the detachment of biomass (Rasulev et al. 2017).

Thus, bacterial adhesion to surfaces and the consequent development of biofilm are natural phenomena in different environments, such as marine, freshwater, hospital, food, and other industrial systems (Ricker and Nuxoll 2016; Galié et al. 2018; de Carvalho 2018), and biosurfactants prove to be an effective tactic to mitigate the establishment of biofilms and other fouling organisms.

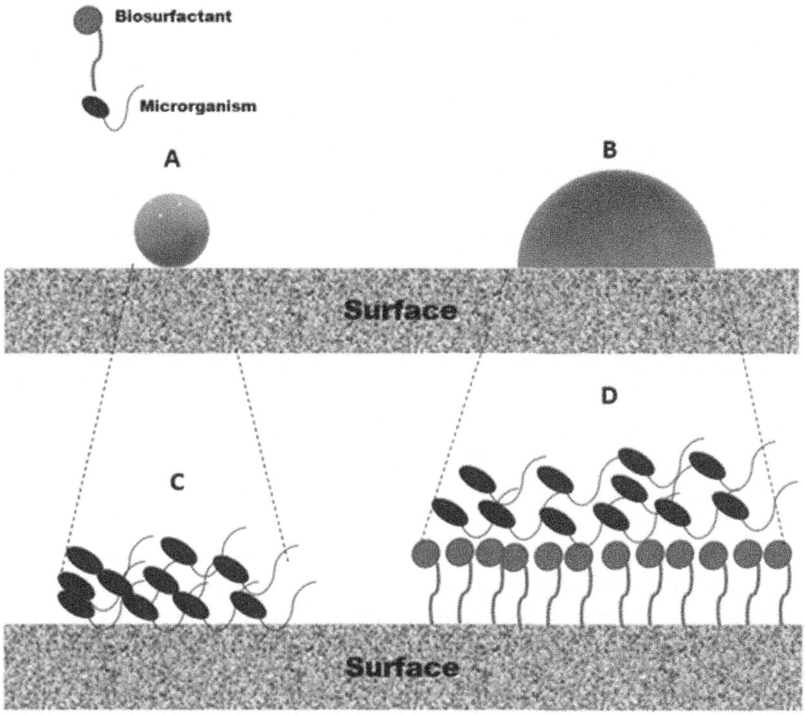

Fig. 10.1 Influence of biosurfactant on the surface hydrophilicity and microbial adhesion. (**a**) Solution without biosurfactant, hydrophobic surface; (**b**) Solution containing biosurfactant, surface becomes hydrophilic; (**c**) Emphasis on microbial adhesion influenced by surface hydrophobicity; and (**d**) Highlight on the inhibition of microbial adhesion caused by the adsorption of the biosurfactant on the surface

10.4.1 Anti-Adhesive Applications in the Biomedical Field

With the development of studies in the field of bacterial biofilms, the potential threats to health caused by infections caused by these biofilms have caused great public concern (Yan et al. 2019).

Synthetic surfactants are already used in the medical field, especially in cleaning infected lesions, in preparing the injured skin surface to receive surgical grafts (Percival et al. 2017). The presence of EPS in biofilms shows a reduced sensitivity to the host defense systems, antibiotics, among others, which contributes to bacterial persistence in chronic infections.

A lipopeptide from *Bacillus subtilis* AC7 combined with a farnesol molecule was able to neutralize biofilms of *Candida albicans* in silicone elastomer under simulated physiological conditions (Ceresa et al. 2018).

Using sophorolipid from *Candida bombicola* ATCC 22214, Ceresa et al. (2020) observed a significant reduction in the capacity of *Staphylococcus aureus* and

C. albicans to form biofilms and adhere to surfaces in 90–95% of silicone used in medical equipment. This research indicates the potential of biosurfactants as coating agents in biomedical materials to prevent infections by Gram-positive bacteria and fungi.

Satpute et al. (2019), using the glycolipoprotein biosurfactant produced by *Lactobacillus acidophilus*, observed antibiofilm and anti-adhesive activities against biofilm-producing microorganisms in medical implants based on PDMS (polydimethylsiloxane), considering a potential anti-adhesive agent on various surfaces of biomedical devices.

10.4.2 Anti-Adhesive Applications in the Food Industry Surfaces

The control of bacterial biofilms is one of the ways found by the food industry and related areas to reduce the undesirable effects of microbial contamination. The occurrence of biofilm can lead to food spoilage and disease transmission, which poses a risk to consumer health (Giri et al. 2019).

Several food manufacturing procedures present precarious sanitation environments, where microorganisms can successfully grow. These environments can include rubber surfaces, packaging machines, piping, valves, floor and walls, polystyrene materials, and stainless-steel materials (Faille and Carpentier 2009).

Several foodborne pathogens from different species of microorganisms such as *Bacillus cereus, Escherichia coli, Shigella* sp., *Staphylococcus aureus* (Sharma and Anand 2002; Sharma et al. 2015), *Listeria monocytogenes, Salmonella typhi, Pseudomonas fragi,* and *Leuconostoc citreum* (Dzieciol et al. 2016) among others are of great apprehension in food processing and preparation spaces.

Biosurfactants can either serve as a colonization factor for a specific microorganism or are also able to prevent or delay the establishment of other microorganisms. Scientific research with microbiological surfactants has already indicated anti-adhesive activities of these molecules against food-borne microbial pathogens. Therefore, microbiological surfactants, acting as antimicrobials, affected the growth of free and fixed forms of microbial cells of these organisms (Nitschke and Silva 2018).

When adsorbed on the surfaces of different materials that come into contact with food, biosurfactants were able to inhibit adhesion and biofilm formation. On polystyrene and AISI 304 stainless steel surfaces, the bacterium surfactin *Bacillus subtilis* ATCC 21332 and rhamnolipid from *Pseudomonas aeruginosa* PA1 (Petrobras) were tested against Gram-positive and -negative microorganisms. These biosurfactants significantly reduced the formation of biofilm pathogens from Gram-positive food sources (*Listeria monocytogenes* ATCC 19112 and ATCC 7644) and Gram-negative microorganisms (*P. fluorescens* ATCC 13525) (Araujo et al. 2016).

The biosurfactants of *Bacillus subtilis* VSG4 and *Bacillus licheniformis* VS16 are demonstrated to be notable blockers of microbial adherence and biofilm generation of microorganisms associated with food contamination. Thus, the authors propose that both the biosurfactants have the potential to be exploited as an antioxidant, antimicrobial, and anti-adhesive and thus mitigate the development of microbial biofilms in the biomedical and food industries (Giri et al. 2019).

The DCS1 lipopeptide synthesized by *Bacillus methylotrophicus* DCS1 showed antimicrobial activity against several tested microorganisms. Besides, they interrupted the preformed biofilm and also presented anti-adhesive activity in the formation of biofilm. Thus, the authors suggested the viable use of the DCS1 lipopeptide as a substance that inhibits oxidation, acting as antimicrobial and anti-adhesive in reducing microbial adhesion and biofilm formation and its applicability also in biomedical devices and the food sector (Jemil et al. 2017).

Biosurfactants isolated from *Lactobacillus paracasei* showed antimicrobial properties and anti-adhesive and antimicrobial properties against various food pathogens at different levels of inhibition. Therefore, they recommend the biosurfactant tested against various food pathogens as an alternative antimicrobial agent (Gudiña et al. 2010).

Biosurfactants also show differences in anti-adhesive and antibiofilm action depending on the type of surface material treated. In a test carried out by Araujo et al. (2016), rhamnolipids reduced the fixation on the polystyrene surface up to 79% and on stainless steel up to 83%. Surfactin reduced 54% and 73%, respectively, in the same materials. pH is an important factor to be considered in the development of strategies based on rhamnolipids for the control of food pathogens. Rhamnolipid showed antimicrobial action against Gram-positive pathogens (*Bacillus cereus*, *Listeria monocytogenes*, and *Staphylococcus aureus*). This activity was related to the increase in the acidity of the environment caused by the different pH levels. The susceptibility of these pathogens was associated with a reduction in the hydrophobicity of the microbial surface layer and consequent deterioration of the cytoplasmic membrane (de Freitas et al. 2019).

10.5 Future Trends and Conclusions

Due to a large number of applications, biosurfactants are exceptionally useful molecules. The surfactant, antimicrobial, and emulsifying properties of these molecules have been implemented in several industries, such as pharmaceutical, cosmetic, food, and biotechnological. The application of anti-adhesive activity is of great importance mainly in the pharmaceutical and food industries. Recent advances related to the identification of microorganisms that produce biosurfactants, purification, and characterization of their compounds, as well as cultivation with different residual raw materials and scale-up studies, have enabled the production of biosurfactants with different functionalities. However, the high production cost does not allow for large-scale synthesis, limiting the availability of these molecules.

The biosurfactants mentioned here can be used in the development of new strategies to delay the colonization of the surface, i.e., be used as antifouling agents.

Smart antibacterial coatings may contain fixed microorganisms that release biosurfactants with anti-adhesive action. These same coatings may also be antimicrobial by encapsulating these agents. Both are promising strategies since they can be doubly effective in presenting anti-adhesive and antimicrobial functions in the same product. Hence, investments in research for the development and industrial production of natural anti-adhesive products based on biosurfactants are necessary for the elucidation of chemical structures and their application in different sectors.

References

Abdelli F, Jardak M, Elloumi J et al (2019) Antibacterial, anti-adherent and cytotoxic activities of surfactin(s) from a lipolytic strain Bacillus safensis F4. Biodegradation 30:287–300. https://doi.org/10.1007/s10532-018-09865-4

Araujo LV, Guimarães CR, da Silva Marquita RL et al (2016) Rhamnolipid and surfactin: anti-adhesion/antibiofilm and antimicrobial effects. Food Control 63:171–178. https://doi.org/10.1016/j.foodcont.2015.11.036

Araújo HWC, Andrade RFS, Montero-Rodríguez D et al (2019) Sustainable biosurfactant produced by Serratia marcescens UCP 1549 and its suitability for agricultural and marine bioremediation applications. Microb Cell Fact 18:1–13. https://doi.org/10.1186/s12934-018-1046-0

Bakhshi N, Soleimanian-Zad S, Sheikh-Zeinoddin M (2017) Dynamic surface tension measurement for the screening of biosurfactants produced by lactobacillus plantarum subsp. plantarum PTCC 1896. Enzyme Microb Technol 101:1–8. https://doi.org/10.1016/j.enzmictec.2017.02.010

Banat IM, De Rienzo MAD, Quinn GA (2014) Microbial biofilms: biosurfactants as antibiofilm agents. Appl Microbiol Biotechnol 98:9915–9929. https://doi.org/10.1007/s00253-014-6169-6

Ben AD, Tounsi S, Gharsallah H et al (2018) Lipopeptides from Bacillus amyloliquefaciens strain 32a as promising biocontrol compounds against the plant pathogen Agrobacterium tumefaciens. Environ Sci Pollut Res 25:36518–36529. https://doi.org/10.1007/s11356-018-3570-1

Cao X-H, Liao Z-Y, Wang C-L et al (2009) Evaluation of a lipopeptide biosurfactant from Bacillus Natto Tk-1 as a potential source of anti-adhesive, antimicrobial and antitumor activities. Braz J Microbiol 40:373–379. https://doi.org/10.1590/S1517-83822009000200030

Ceresa C, Tessarolo F, Maniglio D et al (2018) Inhibition of Candida albicans biofilm by lipopeptide AC7 coated medical-grade silicone in combination with farnesol. AIMS Bioeng 5:192–208. https://doi.org/10.3934/bioeng.2018.3.192

Ceresa C, Tessarolo F, Maniglio D et al (2019) Medical-grade silicone coated with rhamnolipid R89 is effective against Staphylococcus spp. biofilms. Molecules 24:3843. https://doi.org/10.3390/molecules24213843

Ceresa C, Fracchia L, Williams M et al (2020) The effect of sophorolipids against microbial biofilms on medical-grade silicone. J Biotechnol 309:34–43. https://doi.org/10.1016/j.jbiotec.2019.12.019

Cheok CY, Salman HAK, Sulaiman R (2014) Extraction and quantification of saponins: a review. Food Res Int 59:16–40. https://doi.org/10.1016/j.foodres.2014.01.057

de Carvalho CCCR (2018) Marine biofilms: a successful microbial strategy with economic implications. Front Mar Sci 5:126. https://doi.org/10.3389/fmars.2018.00126

de Freitas FJ, Vieira EA, Nitschke M (2019) The antibacterial activity of rhamnolipid biosurfactant is pH dependent. Food Res Int 116:737–744. https://doi.org/10.1016/j.foodres.2018.09.005

Dhar S, Reddy EM, Prabhune A et al (2011) Cytotoxicity of sophorolipid-gellan gum-gold nanoparticle conjugates and their doxorubicin loaded derivatives towards human glioma and human glioma stem cell lines. Nanoscale 3:575–580. https://doi.org/10.1039/c0nr00598c

Donlan RM (2002) Biofilms: microbial life on surfaces. Emerg Infect Dis 8:881–890. https://doi.org/10.3201/eid0809.020063

Dusane DH, Pawar VS, Nancharaiah YV et al (2011) Anti-biofilm potential of a glycolipid surfactant produced by a tropical marine strain of *Serratia marcescens*. Biofouling 27:645–654. https://doi.org/10.1080/08927014.2011.594883

Dzieciol M, Schornsteiner E, Muhterem-Uyar M et al (2016) Bacterial diversity of floor drain biofilms and drain waters in a Listeria monocytogenes contaminated food processing environment. Int J Food Microbiol 223:33–40. https://doi.org/10.1016/j.ijfoodmicro.2016.02.004

Faille C, Carpentier B (2009) Food contact surfaces, surface soiling and biofilm formation. Woodhead Publishing, Cambridge

Freitas BG, Brito JGM, Brasileiro PPF et al (2016) Formulation of a commercial biosurfactant for application as a dispersant of petroleum and by-products spilled in oceans. Front Microbiol 7:1–9. https://doi.org/10.3389/fmicb.2016.01646

Galié S, García-Gutiérrez C, Miguélez EM et al (2018) Biofilms in the food industry: health aspects and control methods. Front Microbiol 9:1–18. https://doi.org/10.3389/fmicb.2018.00898

Giri SS, Ryu EC, Sukumaran V, Park SC (2019) Antioxidant, antibacterial, and anti-adhesive activities of biosurfactants isolated from Bacillus strains. Microb Pathog 132:66–72. https://doi.org/10.1016/j.micpath.2019.04.035

Gopikrishnan V, Radhakrishnan M, Pazhanimurugan R et al (2015) Natural products: potential and less explored source for antifouling compounds. J Chem Pharm Res 7:1144–1153

Gudiña EJ, Rocha V, Teixeira JA, Rodrigues LR (2010) Antimicrobial and antiadhesive properties of a biosurfactant isolated from *Lactobacillus paracasei* ssp. *paracasei* A20. Lett Appl Microbiol 50:419–424. https://doi.org/10.1111/j.1472-765X.2010.02818.x

Gutnick DL, Bach H (2017) Biosurfactants. In: Reference module in life sciences. Elsevier, Amsterdam

Haque F, Alfatah M, Ganesan K, Bhattacharyya MS (2016) Inhibitory effect of Sophorolipid on Candida albicans biofilm formation and hyphal growth. Sci Rep 6:1–11. https://doi.org/10.1038/srep23575

Haque F, Sajid M, Cameotra SS, Battacharyya MS (2017) Anti-biofilm activity of a sophorolipid-amphotericin B niosomal formulation against Candida albicans. Biofouling 33:768–779. https://doi.org/10.1080/08927014.2017.1363191

Ibrar M, Zhang H (2020) Construction of a hydrocarbon-degrading consortium and characterization of two new lipopeptides biosurfactants. Sci Total Environ 714:136400. https://doi.org/10.1016/j.scitotenv.2019.136400

Inès M, Dhouha G (2015) Lipopeptide surfactants: production, recovery and pore forming capacity. Peptides 71:100–112. https://doi.org/10.1016/j.peptides.2015.07.006

Jahan R, Bodratti AM, Tsianou M, Alexandridis P (2020) Biosurfactants, natural alternatives to synthetic surfactants: physicochemical properties and applications. Adv Colloid Interface Sci 275:102061. https://doi.org/10.1016/j.cis.2019.102061

Janek T, Łukaszewicz M, Krasowska A (2012) Antiadhesive activity of the biosurfactant pseudofactin II secreted by the Arctic bacterium Pseudomonas fluorescens BD5. BMC Microbiol 12:24. https://doi.org/10.1186/1471-2180-12-24

Jemil N, Ben Ayed H, Manresa A et al (2017) Antioxidant properties, antimicrobial and anti-adhesive activities of DCS1 lipopeptides from Bacillus methylotrophicus DCS1. BMC Microbiol 17:1–11. https://doi.org/10.1186/s12866-017-1050-2

Jia R, Unsal T, Xu D et al (2019) Microbiologically influenced corrosion and current mitigation strategies: a state of the art review. Int Biodeter Biodegr 137:42–58. https://doi.org/10.1016/J.IBIOD.2018.11.007

Jimoh AA, Lin J (2019) Biosurfactant: a new frontier for greener technology and environmental sustainability. Ecotoxicol Environ Saf 184:109607. https://doi.org/10.1016/j.ecoenv.2019.109607

Joshi-Navare K, Prabhune A (2013) A biosurfactant-sophorolipid acts in synergy with antibiotics to enhance their efficiency. Biomed Res Int 2013:512495. https://doi.org/10.1155/2013/512495

Junter GA, Thébault P, Lebrun L (2016) Polysaccharide-based antibiofilm surfaces. Acta Biomater 30:13–25. https://doi.org/10.1016/j.actbio.2015.11.010

Kaali P, Strmberg E, Karlsso S (2012) Prevention of biofilm associated infections and degradation of polymeric materials used in biomedical applications. In: Biomedical engineering, trends in materials science. IntechOpen, London. https://doi.org/10.5772/12983

Krasowska A, Sigler K (2014) How microorganisms use hydrophobicity and what does this mean for human needs? Front Cell Infect Microbiol 4:112. https://doi.org/10.3389/fcimb.2014.00112
Li S, Huang J, Chen Z, Lai Y (2016) Models, fabrication technologies and. J Mater Chem A Mater Energy Sustain 5:31–55. https://doi.org/10.1039/C6TA07984A
Liu K, Sun Y, Cao M et al (2020) Rational design, properties, and applications of biosurfactants: a short review of recent advances. Curr Opin Colloid Interface Sci 45:57–67. https://doi.org/10.1016/j.cocis.2019.12.005
Merghni A, Dallel I, Noumi E et al (2017) Antioxidant and antiproliferative potential of biosurfactants isolated from lactobacillus casei and their anti-biofilm effect in oral Staphylococcus aureus strains. Microb Pathog 104:84–89. https://doi.org/10.1016/j.micpath.2017.01.017
Meylheuc T, Methivier C, Renault M et al (2006) Adsorption on stainless steel surfaces of biosurfactants produced by gram-negative and gram-positive bacteria: consequence on the bioadhesive behavior of Listeria monocytogenes. Colloids Surf B Biointerfaces 52:128–137. https://doi.org/10.1016/j.colsurfb.2006.04.016
Mnif I, Ghribi D (2016) Glycolipid biosurfactants: main properties and potential applications in agriculture and food industry. J Sci Food Agric 96:4310–4320. https://doi.org/10.1002/jsfa.7759
Ndlovu T, Rautenbach M, Vosloo JA et al (2017) Characterisation and antimicrobial activity of biosurfactant extracts produced by Bacillus amyloliquefaciens and Pseudomonas aeruginosa isolated from a wastewater treatment plant. AMB Express 7:108. https://doi.org/10.1186/s13568-017-0363-8
Nickzad A, Déziel E (2014) The involvement of rhamnolipids in microbial cell adhesion and biofilm development—an approach for control? Lett Appl Microbiol 58:447–453. https://doi.org/10.1111/lam.12211
Nitschke M, Silva SS (2018) Recent food applications of microbial surfactants. Crit Rev Food Sci Nutr 58:631–638. https://doi.org/10.1080/10408398.2016.1208635
Ohadi M, Forootanfar H, Dehghannoudeh G et al (2020) Antimicrobial, anti-biofilm, and anti-proliferative activities of lipopeptide biosurfactant produced by Acinetobacter junii B6. Microb Pathog 138:103806. https://doi.org/10.1016/j.micpath.2019.103806
Padmapriya B, Suganthi S (2013) Antimicrobial and anti adhesive activity of purified biosurfactants produced by candida species. Middle East J Sci Res 14:1359–1369. https://doi.org/10.5829/idosi.mejsr.2013.14.10.73221
Payne DE, Boles BR (2016) Emerging interactions between matrix components during biofilm development. Curr Genet 62:137–141. https://doi.org/10.1007/s00294-015-0527-5
Percival SL, Mayer D, Malone M et al (2017) Surfactants and their role in wound cleansing and biofilm management. J Wound Care 26:680–690. https://doi.org/10.12968/jowc.2017.26.11.680
Prasad R, Siddhardha B, Dyavaiah M (2020) Nanostructures for antimicrobial and antibiofilm applications. Springer International Publishing (ISBN 978-3-030-40337-9). https://www.springer.com/gp/book/9783030403362
Rasulev B, Jabeen F, Stafslien S et al (2017) Polymer coating materials and their fouling release activity: a cheminformatics approach to predict properties. ACS Appl Mater Interfaces 9:1781–1792. https://doi.org/10.1021/acsami.6b12766
Ricker EB, Nuxoll E (2016) Thermal mitigation of *Pseudomonas aeruginosa* biofilms. Biofouling 31:665–675. https://doi.org/10.1080/08927014.2015.1083985.Thermal
Rivardo F, Martinotti MG, Turner RJ, Ceri H (2011) Synergistic effect of lipopeptide biosurfactant with antibiotics against Escherichia coli CFT073 biofilm. Int J Antimicrob Agents 37:324–331. https://doi.org/10.1016/j.ijantimicag.2010.12.011
Roy R, Tiwari M, Donelli G, Tiwari V (2018) Strategies for combating bacterial biofilms: a focus on anti-biofilm agents and their mechanisms of action. Virulence 9:522–554. https://doi.org/10.1080/21505594.2017.1313372

Sambanthamoorthy K, Feng X, Patel R et al (2014) Antimicrobial and antibiofilm potential of biosurfactants isolated from lactobacilli against multi-drug-resistant pathogens. BMC Microbiol 14:197. https://doi.org/10.1186/1471-2180-14-197

Satpute SK, Płaza GA, Banpurkar AG (2017) Biosurfactants' production from renewable natural resources: example of Innovativeand smart Technology in Circular Bioeconomy. Manag Syst Prod Eng 25:46. https://doi.org/10.1515/mspe-2017-0007

Satpute SK, Mone NS, Das P et al (2019) Inhibition of pathogenic bacterial biofilms on PDMS based implants by L. acidophilus derived biosurfactant. BMC Microbiol 19:39. https://doi.org/10.1186/s12866-019-1412-z

Shah V, Badia D, Ratsep P (2007) Sophorolipids having enhanced antibacterial activity [1]. Antimicrob Agents Chemother 51:397–400. https://doi.org/10.1128/AAC.01118-06

Shakerifard P, Gancel F, Jacques P, Faille C (2009) Effect of different Bacillus subtilis lipopeptides on surface hydrophobicity and adhesion of Bacillus cereus 98/4 spores to stainless steel and Teflon. Biofouling 25:533–541. https://doi.org/10.1080/08927010902977943

Sharma M, Anand SK (2002) Characterization of constitutive microflora of biofilms in dairy processing lines. Food Microbiol 19:627–636. https://doi.org/10.1006/yfmic.472

Sharma V, Sharma D (2018) Microbial biosurfactants: future active food ingredients. In: Microbial bioprospecting for sustainable development, pp 265–276. https://doi.org/10.1007/978-981-13-0053-0_14

Sharma D, Saharan BS, Chauhan N et al (2015) Isolation and functional characterization of novel biosurfactant produced by Enterococcus faecium. SpringerPlus 4:1–14. https://doi.org/10.1186/2193-1801-4-4

Stoica P, Chifiriuc MC, Rapa M, Lazăr V (2016) Overview of biofilm-related problems in medical devices. In: Biofilms implant med devices infect control. Woodhead Publishing, Cambridge, pp 3–23. https://doi.org/10.1016/B978-0-08-100382-4.00001-0

Van Bogaert INA, Saerens K, De Muynck C et al (2007) Microbial production and application of sophorolipids. Appl Microbiol Biotechnol 76:23–34. https://doi.org/10.1007/s00253-007-0988-7

Van Oss CJ (1995) Hydrophobicity of biosurfaces - Origin, quantitative determination and interaction energies. Colloids Surf B Biointerfaces 5:91–110. https://doi.org/10.1016/0927-765(95)01217-7

Yan X, Gu S, Cui X et al (2019) Antimicrobial, anti-adhesive and anti-biofilm potential of biosurfactants isolated from Pediococcus acidilactici and lactobacillus plantarum against Staphylococcus aureus CMCC26003. Microb Pathog 127:12–20. https://doi.org/10.1016/j.micpath.2018.11.039

Yilgör E, Söz CK, Yilgör I (2018) Wetting behavior of superhydrophobic poly(methyl methacrylate). Prog Org Coat 125:530. https://doi.org/10.1016/j.porgcoat.2018.07.018

Zdziennicka A, Jańczuk B (2018) Wetting and adhesion properties of rhamnolipid and surfactin. Int J Adhes Adhes 84:275–282. https://doi.org/10.1016/j.ijadhadh.2018.04.005

Zeraik AE, Nitschke M (2012) Influence of growth media and temperature on bacterial adhesion to polystyrene surfaces. Braz Arch Biol Technol 55:569–576. https://doi.org/10.1590/S1516-89132012000400012

Zhu Z, Wen Y, Yi J et al (2019) Comparison of natural and synthetic surfactants at forming and stabilizing nanoemulsions: tea saponin, Quillaja saponin, and tween 80. J Colloid Interface Sci 536:80–87. https://doi.org/10.1016/j.jcis.2018.10.024

Applications of Biosurfactant for Microbial Bioenergy/Value-Added Bio-Metabolite Recovery from Waste Activated Sludge

11

Aijuan Zhou, Yanqing Duan, Yaoli Wei, Alimzhanova Shyryn, Mengxuan Deng, Zenan Zhang, and Xiuping Yue

Abstract

Waste activated sludge (WAS) is in dire need of prudent disposal due to its abundant organics, as well as the latent refractory contaminants and heavy metals, for instance. Applications of surfactants, especially biosurfactant, provided comprehensive opportunities for WAS treatment via alleviating the stiff protection of extracellular polymeric substances (EPS) matrix and microbial cell wall, facilitating the value-added bio-metabolites and energy recovery, decontaminating the refractory organics, dehydrating WAS flocs, and removing heavy metals. In this chapter, applications of surfactants for the short-chain fatty acid (SCFA) extraction was covered with specific attention on the promotion of both hydrolysis and acidification via increasing available organics and hydrolyzing enzyme, as well as inhibiting the methanogenesis step. Also, the effect of surfactants on bio-energy recovery, including methane and hydrogen, was discussed. Benefited from the surfactant pretreatment in anaerobic digestion (AD) process, the performance in sludge dewaterability was comprehended. Due to the hydrophobic nature of some refractory organics, the surfactant micelles were employed to decontaminate polycyclic aromatic hydrocarbons (PAHs), dyes and, polychlorinated biphenyl (PCB) in WAS. Furthermore, recent efforts in heavy metal desorption from sludge flocs were addressed. Finally, state-of-the-art processes to promote organics biotransformation from WAS were presented,

Aijuan Zhou and Yanqing Duan contributed equally to this work.

A. Zhou (✉) · Y. Duan · Y. Wei · A. Shyryn · M. Deng · Z. Zhang · X. Yue
College of Environmental Science and Engineering, Taiyuan University of Technology, Taiyuan, China

Shanxi Engineer Research Institute of Sludge Disposition and Resources, Taiyuan, China
e-mail: zhouaijuan@tyut.edu.cn

Inamuddin et al. (eds.), *Microbial Biosurfactants*, Environmental and Microbial Biotechnology, https://doi.org/10.1007/978-981-15-6607-3_11

including co-pretreatment, interfacing AD with bioelectrochemical systems and optimizing process conditions.

Keywords

Waste activated sludge (WAS) · Anaerobic digestion (AD) · Value-added bio-metabolite recovery · Energy production · Dewatering · Heavy metals

11.1 Introduction

Recently, with the increasing sewage disposal rate in China, large amount of wastewater treatment plants (WWTPs) have been developed. As a result, massive waste activated sludge (WAS) generated as the inevitable by-product, which caused increased operational cost and severe environmental risks (Zhen et al. 2017). As reported previously, cost in sludge disposal ranked second contributor of the total operation cost, right after the aeration power during the traditional wastewater treatment process (Chai et al. 2015). Besides, the environmental impact related with inadequate treatment/disposal of WAS has attracted more social attentions with the increased awareness in environmental protection (Pradel et al. 2016; Cao and Pawłowski 2013). Taking the current sludge disposal in China as example, most Chinese WWTPs used the basic configuration of "thickening-coagulation-mechanical dewatering"; however, the dewatered sludge still contained 80% moisture, far beyond the safety threshold of 60% (Yang et al. 2015). Thus, promising and alternative dewatering technologies were needed. Even for the disposed sludge via landfilling, which accounted for more than 50% application rate in China (Zhang et al. 2016), the related environmental footprint could not be negligible. As recorded by Yi and colleagues, landfill would make a contribution in net energy consumption of 4.9 GJ/t dry matter (DM) and global warming potential (GMP) of 1302 kg CO_2 eq./t DM (Yi et al. 2013). Notably, some contaminants, e.g., heavy metals, antibiotic, and pesticide, were introduced in the generated sludge due to inadequate regulation of sewage into the pipe, further complicating the sludge disposal (Feng et al. 2019; Jang et al. 2018). Therefore, an efficient, cost-effective, and promising handling route is strongly desired for WAS treatment. Taken the composition of WAS into consideration, it is better if the wrapped organic resource could be reused; meanwhile, the heavy metals and refractory contaminants could properly handle before entering the environment.

Anaerobic digestion (AD) has shown great potential in energy and resource recovery from WAS. However, its efficiency was constrained by the limited hydrolysis step, and only 30–50% of the organic resources in raw WAS could be degraded with a long time of 20–30 days (Ruffino et al. 2014). It was the stiff structure, consisted of extracellular polymeric substances (EPS) and microbial cells, that protected both the intracellular and extracellular organic matters from being released

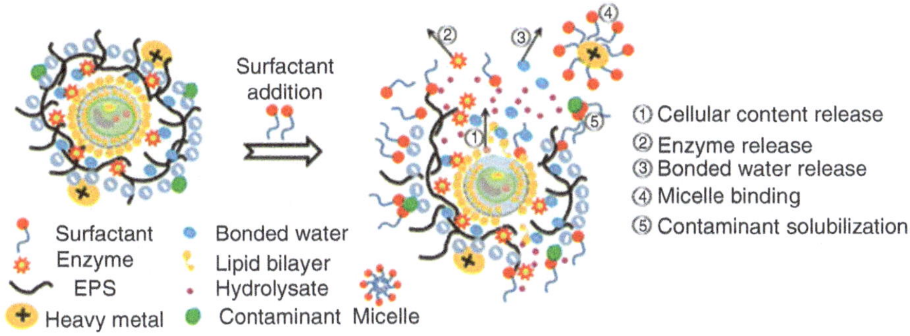

Fig. 11.1 Schematic of interaction between surfactants and WAS components

and available for enzyme attack, thereby reducing the effective decomposition of WAS (Guangyin and Youcai 2017). As elaborated in Fig. 11.1, a double-reinforced barrier is formed in the sludge flake. As a result, not only the EPS itself but also the other organics being wrapped in it cannot be effectively extracted due to the low bio-availability. Several refractory contaminants were included in the loosely bound EPS (LB-EPS) layer, like antibiotic and pesticide. Similarly, bound water was locked, and strong repulsion forces were developed between flocs, which made dehydration more difficult (Guan et al. 2017). This was attributed to the negatively charged surface of the microbial cells, EPS, and sludge flake, resulted from the ionization of anionic groups (i.e., carboxylic and phosphate) (Liu and Fang 2003). Thus, WAS pretreatment, focusing on EPS solubilization, was very crucial to disintegrate the sludge matrix and release both organic and inorganic compositions, thereby facilitating the subsequent biotransformation of WAS. In practice, extensive researches concentrating on WAS pretreatment have been undertaken, including thermal, mechanical, chemical, and biological methods (Zhen et al. 2017).

Surfactants have been recently investigated as a novel pretreatment to dissolve the EPS matrix and facilitate energy and resource recovery from WAS (He et al. 2019a). With both hydrophilic and lipophilic groups, it can reduce the surface/interfacial tension and change the sludge structure, organics solubility, microbial activity, and metal speciation by interacting with organics and metals. There are various categories of surfactants, including cationic, anionic, amphoteric, non-ionic, and complex types. When applied as WAS pretreatment, surfactants could break up the EPS matrix and reduce the surface tension between water and sludge particulate (Fig. 11.1). Subsequently, polysaccharide and proteins were released (Zhou et al. 2015), as well as the bound water and exoenzyme were blocked in the sludge flake (Huang et al. 2015). Concomitantly, both acidification and methanogenesis could be improved with more available organic matter and enzyme activity. Notably, the binding force caused by surfactant micelles was more powerful than that of the flocs, devoting to the desorb of refractory pollutants from WAS flocs (He et al. 2019a). Heavy metal irons could be either replaced by cationic surfactant via ion exchange or bound in the form of metal–surfactant complexes via ionic bonds (He et al. 2019a).

Linear alkylbenzene sulfonates (LAS), sodium dodecyl benzene sulfonate (SDBS), sodium dodecyl sulfate (SDS), glucolipid (GL), rhamnolipid (RL), lipopeptide (LP), surfactin (SF), saponin (SP), and phospholipid (PL) have been widely used as WAS pretreatments (Ji et al. 2010; Mayer et al. 1999).

Given the comprehensive function of surfactants in the transformation of WAS components, a review focused on the recent developments of surfactants on value-added bio-metabolites and bio-energy recovery, dewatering, organic contaminant decontamination, and heavy metal removal was covered. Besides, state-of-the-art processes to promote organics biotransformation from WAS were also addressed, including combined pretreatment, combination with microbial electrolysis and feedstock conditioning.

11.2 Applications of Surfactants for Value-Added Bio-Metabolites Recovery from WAS

Anaerobic digestion of WAS is mainly composed of three steps, hydrolysis, acidification, and methanogenesis. In contrast to the time-consuming methane generation process (20–30 days) and low utilization efficiency of organics (30–50% of dry weight) from WAS, the production of higher value-added soluble metabolites, such as SCFAs (0.33–2.09 €/kg) and ethanol (0.3–1.5 €/kg) vs. 0.09–0.20 €/kg methane, could be regarded as a promising recovery route due to the relatively short production cycle (3–8 days) (Moscoviz et al. 2018). Specifically, SCFAs could be used as an ideal carbon source for many bioprocesses. More importantly, it was believed that SCFAs played a crucial role in biological nutrient removal (BNR) process as electron donors. Acetate (HAc) held the highest denitrification rate compared with other carbon sources, while propionate (HPr) was critical in phosphorus removal, with the capacity of improving the activity of phosphorus-accumulating bacteria (PAO) but hampering the activity of glycogen accumulating organisms (GAOs) (He et al. 2019a). However, for example, most of the HAc produced nowadays is oil-based, with only 10% being bio-based in 2015. Thus, SCFA extraction via anaerobic fermentation of WAS has been considered as one of the most cost-effective alternatives for both sludge reduction and resource recovery.

Therefore, researches on boosting SCFA production from WAS has attracted many attentions, especially for the studies on various pretreatments undertaken to break the bottleneck of particulate hydrolysis. Surfactants has been investigated as an alternative pretreatment method for SCFA recovery from WAS bio-refinery, mainly because it is easier to operate, requires fewer treatment facilities, and has less corrosion to equipment. When surfactants are added into WAS treatment system, it could be absorbed on the solid–liquid interface and reduce the surface tension, thereby improve the dissociation of sludge flocs due to its high solubility and surface activity (Fig. 11.2). In terms of composition transformation of WAS, for instance, some tightly bound EPS (TB-EPS) and LB-EPS would transform into slime layer EPS assisted by the surfactants solubilization, leading to the release of organics wrapped in sludge flake into the aqueous phase (Guan et al. 2017). The

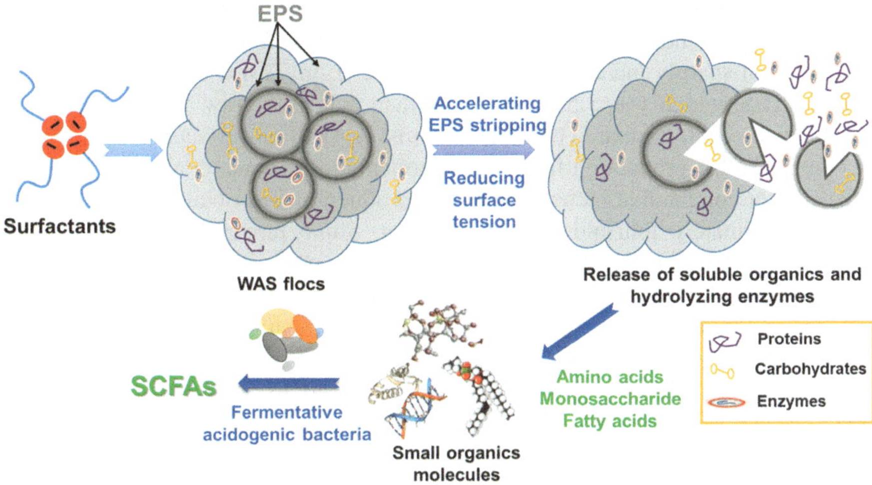

Fig. 11.2 The mechanism of surfactants on anaerobic fermentation of WAS

surfactant can not only accelerate degradation of the organics wrapped in the WAS flocs to facilitate its hydrolysis but also make liberation of hydrolyzing enzymes bounded by EPS matrix. It was both the increase of enzyme activity and the availability for organic matters that enhance the efficiency of acidification by fermentative acidogenic bacteria.

Jiang and colleagues investigated the effect of SDS on the SCFAs yield from WAS during the AD process (Jiang et al. 2007a). It recorded the maximum concentration of SCFAs of 2243.04 mg chemical oxygen demand (COD)/L at 6 days with SDS dosage of 100 mg/g vs. 191.10 mg COD/L in the untreated group. A maximum value of SCFAs with 2599.1 mg COD/L at 6 days was reached (with only 339.1 mg COD/L in the control), when 20 mg/g SDBS was added into the sludge (Jiang et al. 2007b). However, when higher concentrations of SDS or SDBS were employed, longer lag time was observed to reach the maximum yield of SCFAs, which was attributed to the toxicity of the chemical surfactants on the acidogenic bacteria (Feitkenhauer 2003). Meanwhile, the inhibition of methanogenesis step also appeared, which would devote to the SCFA accumulation coupling with the efficient hydrolysis and acidification steps (Jiang et al. 2007b; Su et al. 2007).

Except the toxicity to acidogenic bacteria, chemical surfactants had the unavoidable adverse impact on the environment due to its low biodegradability. By contrast, due to their biodegradability, biocompatibility, and low toxicity, biosurfactants have broad prospects in environmental applications. Lipopeptides and glycolipids were two kinds of well-known biosurfactants (Kosaric et al. 1987). Our previous study believed that RL, as a kind of microbial-derived biosurfactants, was effective to boost the solubilization and acidification of WAS. When the RL dosage was 0.04 g/g TSS, the maximum SCFA concentration (5844 ± 97 mg COD/L) was 1.16-, 3.63-, and 5.24-fold higher than that obtained from the SDS-, SDBS-, and untreated WAS

(Zhou et al. 2013b). In addition, SP, a kind of plant-derived biosurfactant, was also applied in the promotion of SCFA accumulation during WAS fermentation. SCFA concentration sharply increased to 4047 mg COD/L in 72 h with the SP dosage of 0.20 g/g TSS, which was 3.51-fold higher than that in the raw WAS (Zhou et al. 2015). Huang et al. compared the effect of three types of biosurfactants (SP, SF, and RL) on the SCFA production from WAS fermentation (Huang et al. 2015). Results revealed that comparable increase in SCFA production was obtained in all groups, with approximately fourfold over the control. Some researchers realized subtle differences for the mechanism of promoting SCFA production from WAS among different kinds of biosurfactants. Lipopeptides promoted the SCFA recovery mainly via increasing the amounts of available organics, while glycolipids not only increased the organics for SCFAs production but also protect the accumulated SCFAs from consuming by methanogenesis, due to its inhibition on microbial activity.

11.3 Applications of Surfactants for Energy Recovery from WAS

Compared with the SCFA production, relatively few researches were focused on methane promotion when employing surfactants on WAS treatment. Kavitha et al. observed the enhanced methane production with SDS addition. A maximum 50 mL/g volatile solid (VS) yield was achieved at the SDS dosage of 0.02 g/g SS, which was mainly due to the enough available substrates (Kavitha et al. 2016). Another study reported that low dosage of SDS with 0.02 g/g SS can increase the biochemical methane potential (BMP) of WAS by 49% (0.315 L/g VS) vs. 0.212 L/g VS in the raw WAS (Kavitha et al. 2014). Ushani employed 0.009 g/g SS dioctyl sodium sulfosuccinate (DOSS) as the pretreatment and harvested 0.225 g methane per gram WAS (as COD) (Ushani et al. 2017). External addition of LAS homologs (5–10 g/g SS) could also slightly improve the methane yield (Garcia et al. 2006).

Some researchers found that the introduction of surfactant could inhibit the methane production to some extent. For example, an increase of SDS dosage from 20 to 300 mg/g would definitely inhibit methane production from a hindrance ratio of 3%–100% (He et al. 2019a). SDBS with dosage higher than 0.1 g/g SS also prevented the methane production procedure with high toxicity on the acetoclastic methanogenesis (Shcherbakova et al. 1999). Similarly, inhibition of benzalkonium chlorides (BACs) was also recorded (He et al. 2019b). One possible reason for the inhibition of methanogens was attributed to the lowered pH derived by the accumulation of SCFAs, which was out from the optimal neutral pH range for methanogenesis. The activity of key enzymes were also affected. For example, after RL addition, the activity of F420 and acetate kinase reduced by 40% and 26%, respectively (Huang et al. 2015). Meanwhile, the inhibition was significantly dependent on the characteristics of surfactants. For example, SP showed weaker inhibition than RL, because it was more biocompatibility (Huang et al. 2016b). As revealed by Garcia and colleagues, while studying the effect of different LAS homologs on biogas inhibition, the higher alkyl chain length was correlates closely with lower toxicity (Garcia et al. 2006).

Hydrogen has tremendous potential as a promising alternative clean energy; its combustion product has zero pollution (only water), and it possesses a higher energy density of 120 MJ/kg than other gaseous fuels (i.e., methane (50 MJ/kg) and ethanol (26.8 MJ/kg) (Kadier et al. 2016). During traditional AD process, hydrogen can be produced accompanying with acidification step. However, with rapid consumption of hydrogen by methanogens, little hydrogen could be accumulated, with a recovery rate of only 23–25%. Yang et al. observed less than 3 mL hydrogen per gram sludge (as VS) was obtained all through the fermentation period (Yang et al. 2020). Lower hydrogen accumulation value was also recorded with less than 1.5 mL/g VS by Liu et al. (2020). Therefore, some researchers employed efficient pretreatments to disintegrate the particulate organics in WAS and avoid hydrogen consumption by methanogens. Liu et al. presented the peak value of hydrogen yields with 6.4 ± 0.3 mL/g VS (156 h) and 4.5 ± 0.2 mL/g VS (144 h) from dark fermentation of sole freezing-pretreated (−5 °C, 4 h) and nitrite-pretreated (400 mg/L) WAS, respectively. Co-pretreatment further raised the H_2 yields to 19.40 mL/g VS (Liu et al. 2020). A synergetic effect driven by freeze and nitrite pretreatments was believed to boost the WAS disintegration, with more available organics releasing. Besides, the related hydrogen-consuming microorganisms, methanogens, homoacetogens, and sulfate-reducing bacteria were severely suppressed. A comparable yield with 19.2 mL/g VS was also recorded by Yang et al. via dark fermentation of pH 9.5 + K_2FeO_4 pretreated WAS (Yang et al. 2020).

Notably, high hydrogen production was expected by sole surfactant pretreatment with abundance substrates from upstream hydrolysis while less consumption in the downstream methanogenesis. As previously reported, with the addition of 0.20 SP g/g TSS, the WAS hydrolysis was dramatically improved within 48 h, with 4.77-fold and 5.87-fold increase in soluble proteins and polysaccharides concentrations compared with the control test (Zhou et al. 2015). As revealed by Huang et al., several biosurfactants (RL, SF, and SP) can enhance the activity of hydrolyzing enzymes in different degrees (Huang et al. 2015). As a result, SCFA production was significantly enhanced, particularly with higher concentration of HAc, which would contribute to hydrogen production. Meanwhile, the inhibition of methanogen was remarkable. As recorded, methanogenesis is suppressed with a conversion efficiency of $0.18 \pm 0.03\%$, $1.89 \pm 0.15\%$, and $6.63 \pm 0.77\%$ for RL, SDS, and SDBS, respectively (Zhou et al. 2017). Till now, hydrogen recovery from surfactant-pretreated WAS was relatively rare. Microbial electrolysis cells (MECs) emerged as a promising solution for hydrogen production, with dual advantages of fermentation organisms and electroactive bacteria (EAB). Single-chamber MECs have been a prevailing application for hydrogen recovery from WAS. Wang et al. recovered 8.5 mg H_2/g VSS from SDS-treated sludge fermentation liquid (SFL) with an energy efficiency of $138\% \pm 8\%$ in single-chamber MECs (Wang et al. 2014a). As a biosurfactant, RL outperformed chemical surfactants in hydrogen production. Our previous study produced the largest hydrogen production of 12.90, 9.36, and

3.18 mg H_2/g VSS in a MECs fed with RL-pretreated, SDS-pretreated, and SDBS-pretreated WAS (Zhou et al. 2017).

11.4 Applications of Surfactants for Refractory Organic Decontamination from WAS

Refractory organics contained in WAS are generally hydrophobic organic compounds, e.g., polycyclic aromatic hydrocarbons (PAHs), polychlorinated biphenyl (PCB), dyes, benzopyrene, antibiotics, and oil (Guan et al. 2017). Inadequate removal of these organics could trigger the contamination in soil, air, and groundwater via landfilling and land-application, for instance. Traditional WAS disposal methods were insufficient to decontaminate these pollutants. EPS matrix processed a strong ability to absorb complex pollutants onto sludge surface, with various active functional groups and hydrophobic regions (Sheng et al. 2008). The latter one may be the main driven for organic compounds adsorbed by EPS (Rogers 1996). Besides, EPS usually carry a negative charge, which can also bind positively charged refractory organics through electrostatic interactions (He et al. 2019a). External addition of surfactants has been regarded as an efficient alternative to significantly improve the dissolution of organics due to the characteristics of hydrophilic and hydrophobic parts (Fig. 11.3). Three aspects were attributed, i.e., the emulsification of refractory organics, micellar solubilization, and accelerated transport, which can promote the contact between microbial cells and refractory organics, thereby strengthening the decontamination of these organics (He et al. 2019a). Emulsification occurred directly between the cells and refractory organics, due to the change of the hydrophobicity in microbial cells assisted by external surfactant addition. Besides, micelle could be formed by surfactants, encasing the refractory organics in the micelle core and improving its solubility. Furthermore, due to the reduced

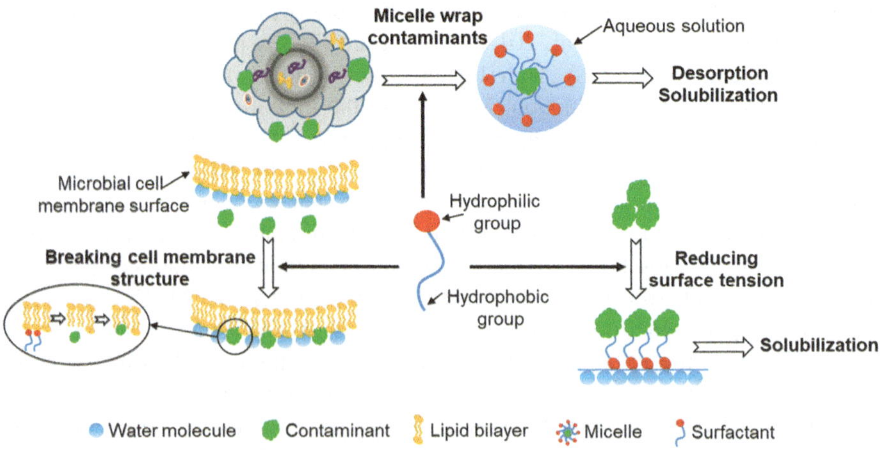

Fig. 11.3 The mechanisms of surfactants on the decontamination of refractory organics in WAS

surface tension, more refractory organics can be transformed into the aqueous phase with raised availability, facilitating the biodegradation by the microorganisms (Yu et al. 2017).

11.4.1 PAHs Decontamination

PAHs are a kind of volatile hydrocarbon produced by incomplete combustion of organic polymers. They belong to persistent organic pollutants (POPs) with stable physical and chemical structures (Zheng et al. 2007). Cai et al. detected the concentration of 44 semi-volatile organic contaminants in the sludge from 11 wastewater treatment plants in nine Chinese cities. Their research showed that PAHs were the most abundant contaminants, the total concentration ranging from 1.4 to 79 mg/kg dry sludge (Cai et al. 2007). Surfactants with the combination of other treatments were efficient in solubilization of PAHs. Surfactant was tend to aggregate and self-assemble to form micelles, which could encapsulate PAHs into it via a form of host–guest complexes, increasing its availability (Guan et al. 2017). Previous research showed that the lower weight compounds (two-ring, three-ring, and four-ring PAHs) behaved better degradability than the higher ones ($\geq$5-ring PAHs) (Cerniglia 1992). Zheng et al. investigated the PAH removal from WAS by adding Tw80 (Tween 80, a nonionic surfactant) during mesophilic aerobic digestion (MAD) and chemical metal leaching (METIX-AC) processes (Zheng et al. 2007). Results showed that 90% of three-ring PAHs (0.5 g/L) were rapidly removed in both MAD and MAD + Tw80 tests. The total removal rates of PAHs in MAD and MAD + Tw80 reached $54 \pm 2.9\%$ and $60 \pm 2.0\%$, respectively. However, chemical metal leaching process coupling with Tw80 was less efficient in PAH removal than sole METIX-AC. This was attributed to the easily cracking characteristics of double bond by Fenton-like reagents in Tw80. Then Tw80 would lose its functionality completely. To avoid this clash, similar surfactants with saturated lipophilic side chains could be served as substitutes. Bernal-Martínez et al. studied the effect of three surfactants (tyloxapol, tergitol, and Brij-35) on PAH removal from anaerobic digestion of urban sludge by the combination of ozone and H_2O_2 (Bernal-Martínez et al. 2005). It revealed that PAH removal was enhanced by the sole ozone pretreatment with the removal efficiency of 61%. With synergistic effect of ozone pretreatment and the three surfactants addition (1 g/L), PAHs could be efficiently removed with a removal rate of 77–81%.

11.4.2 Dye Decontamination

Dyes are widely used in many industrial fields, resulting in large amounts of dye-containing wastewater. Because it is difficult to remove dyes completely by conventional biological treatment process, some dyes remain in WAS, causing serious colored issue (Davis et al. 1994; Han et al. 2009). In recent years, the application of surfactants for the removal of dyes had been attracting attentions due to the interactions between dyes and oppositely charged surfactants. Petzold et al. prepared a kind of complex with the polyelectrolyte P-SSNa and the surfactant dodecylamidoethyl-dimethylbenzyl-ammoniumchloride (Quartolan) to remove the dyes in sludge (Petzold and Schwarz 2006). The results showed that the removal rate of dyes can be increased by the polyelectrolyte–surfactant complexes, which achieved 99% for "pure" dye green. Previous studies also indicated that this complex can remove dyes due to the formation of a triple complex (Buchhammer et al. 2001; Zliobaite 1998; Klimavičiūtė 2004). It is not only the hydrophobic interaction between the complex and dyes but also the complex charge that played a great role in promoting the dye removal.

11.4.3 PCB Decontamination

PCB is a synthetic organic compound, which is widely used in the industry as heat carriers, insulating oils, and lubricants. PCB belongs to POPs, which is also harmful to human health and environment (Clarke et al. 2010). PCB is poorly water soluble, so it is easy to precipitate and then absorb on the sludge particles. The concentration of PCB in sludge can vary from 1.0 to 10.0 mg/kg dry sludge (Rosinska and Karwowska 2017; El-Hadj et al. 2007). Previous study showed that the desorption of PCB could be significantly promoted via the combination pretreatment of ethylenediamine disuccinic acid (EDDS) and SP (Cao et al. 2013). The maximum desorption of PCB was 45.7% with the addition of 3 g/L SP and 10 mM EDDS. Results further confirmed that EDDS had little effect on the desorption of PCB. However, SP had significant effects on the removal of PCB and phenanthrene, for instance (Song et al. 2008; Xia et al. 2009; Cao et al. 2013). Laha et al. concluded that the nonionic micelle-forming surfactant was efficient for increasing the solubilization of various hydrophobic organic compounds (HOCs) (Laha et al. 2009).

11.5 Applications of Surfactants for WAS Dewatering

Dewatered WAS with low moisture content and smaller volume would be easier to transport and store (Chen et al. 2001). Thus, it is necessary to execute sludge dewatering before the subsequent disposal or utilization due to the high moisture content of WAS (95%–99%), although it is hard to decrease to less than 80% by the basic configuration. Moreover, the recovering value-added bio-metabolites from fermented sludge is generally associated with sludge dewaterability. Free and bound water are the two forms of water in sludge. The former can be easily removed by simple mechanical technology, while the latter needed special treatment, like adding polyelectrolytes (Liu and Fang 2003). Bound water could be further subdivided into interstitial (driven by capillary forces), surface (bound by adsorption and adhesive forces), intracellular, and hydrated water. Only the first two forms could be partially stripped by inorganic and organic polymer flocculants (Chen et al. 2004; Neyens et al. 2004), which leads to the low dehydration efficiency. EPS

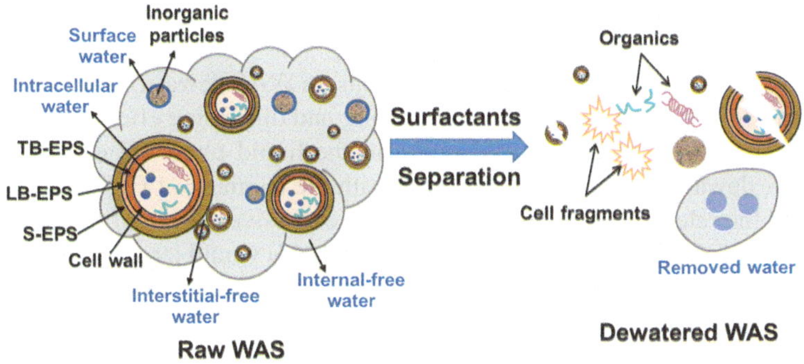

Fig. 11.4 The mechanism of surfactants on WAS dehydration

played a critical role in the sludge dewatering process. Generally, microbial cells are cross-linked within EPS, jointly forming a polymer network covered with pores and channels, which contain abundant free and bond water. As depicted in Fig. 11.4, effective EPS decomposition is the prerequisite for the water release. In fact, this can not only improve the dehydration efficiency of fermented/digested WAS but also affect the previous SCFAs and methane production during AD process. Zhu et al. found that thermal-alkaline co-pretreatment led to a higher filtration resistance and viscosity, which is 522-fold higher than that of raw WAS, leading to efficient EPS release with severe disintegration of sludge flake (Zhu et al. 2015). Anaerobic fermentation can alleviate the filtration resistance by almost 99% via consuming EPS to SCFAs. In this sense, the dewatering performance is closely correlated with the release and consumption of EPS.

Accordingly, surfactants with efficient performance of EPS release, combining with AD is expected to improve sludge dehydration. Via weakening or definitely breaking the stiff structure within sludge flake by dissolving EPS, the bound water could be released in AD system with surfactant addition (Chen et al. 2013). On the one hand, surfactant addition is benefited for enzyme release, leading to more organics transferred into the liquid phase (Chen et al. 2001). On the other hand, it can neutralize the negative charge distributed on the surface, reducing the repulsive force among sludge floc, resulting in compact structure with more water released (Jiang et al. 2007b). However, higher concentration of surfactants is adverse to WAS dewatering, which would be adsorbed on the sludge flocs to regenerate electrostatic repulsion, causing flocs dispersion (Flemming and Wingender 2010). Furthermore, micelles would be formed once the concentration of surfactants surpass critical micelle concentration (CMC), leading to the connection of polar groups between micelles and macromolecular organics of EPS (e.g., proteins and nucleic acid) (Flemming and Wingender 2010). The hydrophilic groups prefer to the liquid phase, reducing the liquid–solid interfacial tension (Muthukumar et al. 2007), further enhancing the solubility of organics existed in the liquid phase, achieving WAS disintegration and cells apoptosis (Wang et al. 2014b). Ultimately, the dewatering performance of WAS is promoted with the intracellular water released (Huang et al. 2015).

As mentioned above, the effect of surfactants on WAS dewatering is depended on EPS composition and content. The solubilizing and dispersing effects of surfactants would transfer EPS from the solid phase to the liquid phase (Chen et al. 2001). Generally, dewatering rate would increase with the augment of EPS entering liquid phase and the release of bond water. However, excessive release of EPS under high dosage of surfactants (beyond a reasonable threshold) would lead the stripped water re-absorbed in the sludge flocs through hydrogen bonding and electrostatic force, which is unconducive to sludge dehydration (Tang et al. 2017). Furthermore, the effect of surfactants on EPS species (S (slime layer)-EPS, LB (loosely bound)-EPS and TB (tightly bound)-EPS) and their components (i.e., proteins, carbohydrates, and nucleic acids) has been demonstrated to be consistent with the mentioned trend of EPS content (Sun et al. 2014; Raszka et al. 2006). For instance, dodecyl dimethyl benzyl ammonium chloride (DDBAC) can weaken the combination of sludge flocs with TB-EPS and LB-EPS, which would transform into S-EPS. Our previous study showed the detaching of LB-EPS and TB-EPS from WAS was evidently improved by SB treatment (Zhou et al. 2015). Accordingly, the solubility of proteins and polysaccharides can be significantly improved accompanying with the release of bound water.

Surfactants can also dissolve macromolecular organics like membrane proteins, existed on microbial cells, via the above-mentioned mechanism (Andersen and Otzen 2014). For example, cetyltrimethylammonium bromide (CTAB) can be adsorbed on the surface of sludge due to its small molecular volume, which can easily penetrate the internal pores of sludge flocs. CTAB can be adsorbed on cells by electrostatic attraction, hydrogen bonding and hydrophobic interaction forces, leading to the aggregate of cell walls, further inhibiting nutrient intake and cause cell decay (Huang et al. 2015). Ultimately, it is reflected as the release of intracellular substances and water. However, the proportion of the intracellular water is relatively small in the sludge flocs. Generally, the sum of surface water and internal-free water of sludge flocs accounts for around 10% of the total moisture (Wang et al. 2014b). Therefore, the contribution of surfactants to the release of internal-free water through cell disintegration is relatively small in the overall WAS dewatering process.

11.6 Applications of Surfactants for Heavy Metal Removal from WAS

As mentioned before, inadequate regulation led some undesirable pollutants into sewage and sludge, which may raise risks in both environmental and human health. Among these, heavy metals are attracting increasing attentions recently as serious environmental contaminants and very harmful to the environment (Wang et al. 2015). Therefore, the removal or passivation of the heavy metals before final disposal of sludge was an urgent demand. Application of surfactants in heavy metal removal had confirmed to be feasible and effective by several literatures.

Benefited from more powerful force between surfactants and heavy metals than that between flocs and heavy metals, heavy metals can be easily desorbed from sludge flocs with less residue on the sludge surface (Kuczajowska-Zadrożna et al. 2015; Guan et al. 2017). Ren et al. investigated the performance of heavy metal removal from pre-acidified and Fenton oxidized sludge with various surfactant addition (Ren et al. 2014). Four most common and cheap surfactants were undertaken with various removal efficiency of heavy metal, including the anionic SDBS, nonionic Tween-20 and Tween-60, and a cationic hexadecyl trimethyl ammonium chloride (HTAC). Especially, Tween-60 nonionic surfactant significantly contributed to promoting the sludge leaching process. Specifically, the removal of copper reached 85%, followed by 60% for cadmium and 30% for lead. By contrast, HTAC and SDBS showed the inhibition effects. Kiliç et al. used SP to treat tannery sludge for chromium (Cr) recovery and achieved 24% extraction (Kiliç et al. 2011).

Combination of surfactants with other technologies can also enhance the heavy metal removal, bioleaching, and electro-kinetic (EK) process, for instance. Yuan and Weng explored three enhanced electro-kinetic processes, i.e., processing fluids of tap water (TW), SDS, and citric acid (CA), for the removal of different heavy metals (Yuan and Weng 2006). In the EK-SDS process, removal priority varied with each heavy metal due to different mobility. The investigated rank is: Cu > Pb > Ni > Fe > Zn > Cr; and the removal efficiency is in the range of 37–77%, with Ni and Pb reaching to 77% and 51%, respectively. Tang and colleagues used the combined RL and SP treatment for metal removal from sludge. The maximum specific removal efficiency was 74% for Zn, followed by 68% for Ni, and 60%–64% for Mn, Cu, and Cr, while the removal to Pb reached to only 15% (Tang et al. 2019).

11.7 State-of-the-Art Processes to Promote Organics Biotransformation from WAS

Due to the limitation of low hydrolysis efficiency, nutrient out-of-balance, instability during traditional AD biotransformation for WAS, various strategies were undertaken to maximize the resource and energy recovery, which included but not limited to the following descriptions (Fig. 11.5): (1) releasing the abundant organics both in EPS matrix and microbial cells by sole/co-pretreatments; (2) balancing the nutrients via co-digestion with diverse carbon-rich feedstocks; (3) accelerating electron transfer efficiency by interfacing AD with bioelectrochemical systems; and (4) optimizing the process conditions, e.g., pH, temperature, HRT, or introducing multi-stage process.

11.7.1 Co-Pretreatment

Due to the relatively mild effect of surfactants on EPS decomposition and cell wall disintegration, combination with other pretreatments, i.e., co-pretreatment, was

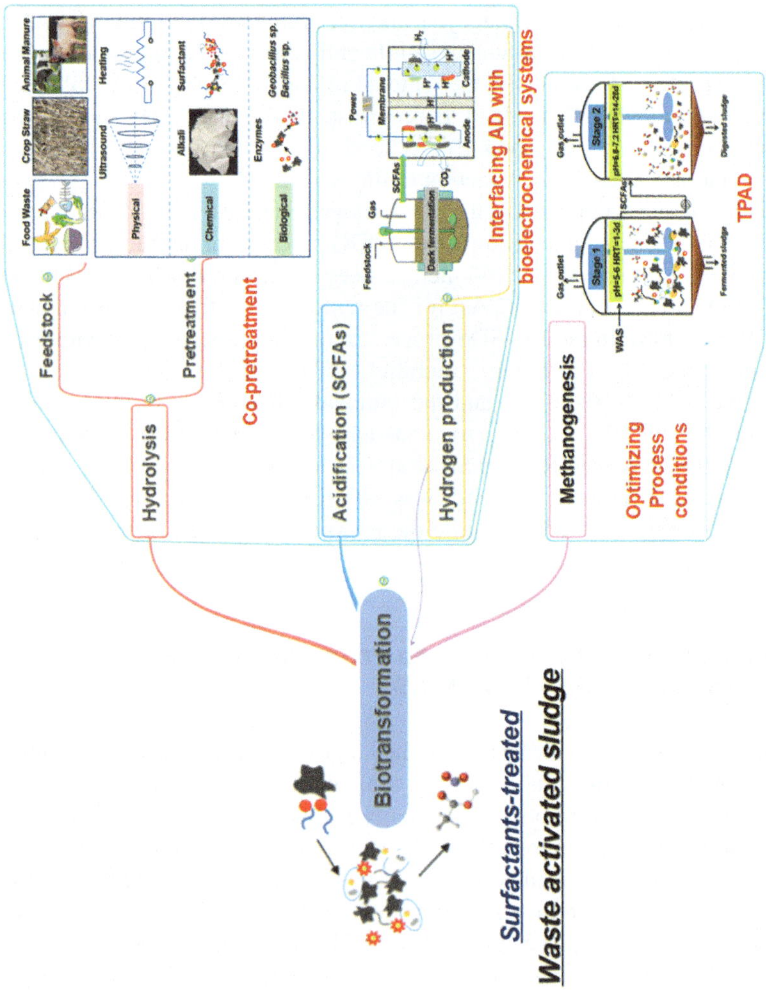

Fig. 11.5 Processes and strategies to promote organic biotransformation from WAS

alternative for efficiently organics biotransformation to resource and energy from WAS. The effects of surfactants combined with other assistant technologies on producing SCFAs from WAS fermentation were investigated. Luo et al. employed SDS and mixed enzyme pretreatments and harvested only 628.07 mg COD/L in the control test that increased 1.82- and 2.32-fold for the SDS and SDS and mixed enzyme system (Luo et al. 2011b). Chen and colleagues attained 2056 mg COD/L SCFAs from SDBS and pH 10 co-pretreated sludge, which were 1142 and 910 mg COD/L in the sole SDBS and pH 10 tests (Chen et al. 2013). The combination of biosurfactants with alkaline pretreatment could further boost the SCFAs yield, for example, 420 mg COD/g VSS was harvested from SP and pH 10 pretreated WAS fermentation, which was 1.5 and 4.7 times of the sole pH 10 and the control tests, respectively (Huang et al. 2016a).

As mentioned before (in the Sect. 11.2), the inhibition of surfactant on functional microbiomes, e.g., acetogens and methanogens, was documented during WAS digestion, calling for alternative strategies to offset this status. As a result, several pretreatments were employed assisting with surfactants. Tanaka and Ichikawa advocated the detoxification of surfactants to improve the methane production during WAS digestion via photolytic pretreatment with UV irradiation by a TiO_2 catalyst. As a result, the toxicity was reduced by 50–60%, due to a cleavage of the benzene ring in the surfactants, including anionic SDS, sodium dodecyl benzene sulfonate (ABS), and nonionic p-nonylphenyl poly(oxyethylene) ether (NPE). Particularly, more methane was produced after 100% reduction in the toxicity of the cationic tetradecyldimethylbenzyl ammonium chloride (TBC), 30–50% higher than that in the control (Tanaka and Ichikawa 2015). Similarly, S. Kavitha recorded the improved biogas production with 0.467 L/g VS by SDS pretreatment of 0.02 g/g SS mediated with extracellular enzymes secreted by thermophilic bacteria (Kavitha et al. 2014).

11.7.2 Interfacing AD with Bioelectrochemical Systems

Although AD has achieved great success in degrading organics and producing biogas from WAS, it still has limitations, such as instability, weak substrate decomposition, inadequate thermodynamic input, and low particulate organic transformation. This underscores the need for interfacing complementary technologies into AD process. Recent researches showed that aforementioned bottlenecks could be solved by electrical stimulation in the bioelectrochemical systems (BES) by adding a small voltage. Furthermore, inserting electrodes into the traditional AD system is an effective way to enhance organics biotransformation from WAS, which could also enhance organics recovery due to the CO_2 capture and conversion to methane/hydrogen. Liu et al. reported a tripled methane production rate in the AD process stimulated by the microbial electrolysis (ME), with 91.8 g CH_4/m^3-reactor/d in ME-AD reactor and 30.6 g CH_4/m^3 reactor/d in the control reactor (Liu et al. 2016). In order to avoid the influence of direct interspecies electron transfer (DIET) in the anode biofilm on methane production, Cai et al. designed an innovative MEC-AD reactor with separated anode and cathode chambers via an anion exchange membrane, which produced higher methane production rate than AD reactor (0.070 vs. 0.027 m^3 $CH_4/m^3/d$) (Cai et al. 2016). As mentioned in the Sect. 11.3, our previous studies showed that combining surfactants with the AD-MEC cascading system were more beneficial for the organics transformation to recover bio-energy from WAS (Zhou et al. 2017; Wang et al. 2014a).

Meanwhile, the operating parameters, i.e., applied voltage, hydraulic retention time (HRT), influent COD, and temperature, were also very important for the cascading treatment system. Escapa et al. found that hydrogen production rate (0.18–1.42 L/L anode/d) and COD removal (46–94%) increased followed with applied voltage (0.6–1.0 V) and HRT (8–12 h) (Escapa et al. 2013). Xu et al. optimized the external voltage of 0.8 V for maximum hydrogen production in the MECs feeding with sludge fermentation liquid (Xu et al. 2013). Beegle et al. optimized the fermentation stage in an anaerobic baffled reactor, and higher H_2 production rate (1.31 L/L d) and TCOD removal efficiency (99.0%) were obtained in a single-chamber MECs feeding with the acetic acid-rich fermentation effluent (Beegle and Borole 2018).

11.7.3 Optimizing Process Conditions

Except the pretreatments mentioned above, the stable operation of AD feeding with WAS was also significantly affected by several parameters, including nutrient balance, pH value, temperature, and end-product inhibition. Therefore, the optimization of operating parameters was required according to the specific aim of efficient organics biotransformation. Due to various proportion of proteins and polysaccharides in WAS, the nutrient was severely unbalanced with the ratio of carbon to nitrogen (C/N) lower than 7.0. As a result, much more ammonia was accumulated with the hydrolysis of proteins, inhibiting the normal growth and metabolism of functional microbe and leading to failure of the AD system. Accordingly, various substrates with rich carbon sources were applied to balance the C/N ratio of raw WAS. The conditioning feedstock included food waste (FW), crop straw, animal manure, and fats, oils, and grease (FOGs) (Kübler et al. 2000; Scaglione et al. 2008; Liu et al. 2013; Esteban-Gutierrez et al. 2018; Zhou et al. 2013a).

The optimal pH value for WAS solubilization was reported to be 10.0 (Huang et al. 2016b). Coupling with the strong decomposition of alkaline agents on sludge flake and cell walls, both EPS and intracellular material were easily released, raising the SCFA production by adequate substrates (Zhang et al. 2009). Demand on alkaline agents can be reduced when other pretreatment was a combination of ultrasonic (Sahinkaya 2015) and surfactants (Huang et al. 2016b), for instance. It was worth noting that alkaline could inhibit the activity of methanogens, whose optimal pH condition has been reported in the range of 6.5–7.2. Therefore, pH should be regulated with caution, consistent with the specific purpose of either SCFA extraction or biogas production. AD was usually performed in either thermophilic or mesophilic conditions. Higher temperature tended to lead to higher hydrolysis of particulates in WAS, facilitating the acidification, while methane production was more sensitive with environmental conditions. The elevated temperature was also beneficial for increase in WAS dewaterability due to the increase of released water and reduction of sludge viscosity.

Given the distinct difference in metabolic pathways and living conditions for acid-producing bacteria and methanogen, a system permits the selection, and enrichment of functional bacteria in separate reactor is preferable for the stable and efficient operation of AD system. Generally, the first stage for acid-producing bacteria was with pH value of 5–6 and a short HRT of 1–3 days, and the second one for methanogens with neutral pH of 6.8–7.2 and HRT of 14–28 days (Luo et al. 2011a; Muha et al. 2013). Eliminating the inhibition of VFA accumulation on the methanogen, both the recovery of SCFAs and methane under optimal conditions were anticipated with higher efficiency accompanying with higher VS reduction, more stable system, and fewer lag phase (Kumaran et al. 2016). Liu et al. gained 106.4 mL H_2/g VS and 353.5 mL CH_4/g VS by the two-stage AD treatment of food waste (FW) and WAS, respectively (Liu et al. 2013). Martín-Pascual et al. optimized the first stage for methane production from WAS, in parallel with the conventional one-stage AD system. As revealed, at least 10.8 days can be reduced for the total HRT with comparable performance in VS reduction. Besides, methane production was maximized with the optimal HRT of 2.18 days for the acid stage (Martín-Pascual et al. 2017). Similarly, temperature-phased anaerobic digestion (TPAD) was evolved with superior performance in methane production, consisting of a thermophilic phase operated at high organic loading rate (OLR) (~15 g COD/L/d) under low HRT of 2–5 days and a mesophilic one at OLR of 2–5 COD/L/d with 10–20 days (De Vrieze et al. 2016). Hydrolysis and acidification of macromolecular organics are proceeded in the thermophilic phase, while the processes of syntrophic acetogenesis and methanogenesis occurred in mesophilic phase. Better process control could be achieved in TPAD, due to the different condition optimization (i.e., pH, HRT, and OLR) of SCFA accumulation and methanogenesis (Pervin et al. 2013).

11.8 Conclusion

As an inevitable by-product in the biological wastewater treatment process, WAS includes a large amount of organic resources, meanwhile accompanying with certain amount of pollutants, e.g., heavy metals and refractory contaminants. Although AD has shown great potential in energy and resource recovery from WAS, its benefits were partially offset by the limited available organics due to the protection of the stiff EPS matrix and microbial cells. As discussed above, surfactants can efficiently dissolve the EPS and facilitate the AD efficiency with comprehensive roles including resource and energy recovery, refractory organics decontamination, dewaterability promotion, and heavy metal removal. However, more studies need to be addressed but not limited due to the following deficiencies: (1) the mechanism of resource and/or energy recovery via surfactant combined with other pretreatments; (2) the microbial response to the external addition of surfactants; (3) the interplay of value-added resource recovery and pollutants removal, if multiple goals were simultaneously addressed; (4) the specific environmental footprint for surfactants addition, when the upstream data were incorporated.

Notably, biosurfactants were reported in improving the biotransformation of WAS organics with excellent microbial compatibility and low environmental risks. Nevertheless, there was still a long way to implement the large-scale application of biosurfactants, due to its high costs than chemical surfactants. Thus, solutions addressing in razing the biosurfactant with low cost is worthy of further study. It would be more prominent during WAS digestion for in situ synthesis of some biosurfactants, RL, for instance. Moreover, despite the multiple advantages, the applications of biosurfactant was closely related with the disposal routes for WAS. With the widespread practice of value-added resources and/or bio-energy recovery and the heightened concern on the contaminants removal, the application of biosurfactants for WAS treatment could be improved.

Acknowledgments This chapter was supported by the National Natural Science Foundation of China (NSFC, No. 52070139), by the Key Research and Development (R&D) Project of Shanxi Province (No. 201903D321057 and 201903D321055) and the Ministry of Environmental Protection of the People's Republic of China (Major Science and Technology Program, No. 2019YFC0408601 and 2019YFC0408602).

References

Andersen KK, Otzen DE (2014) Folding of outer membrane protein a in the anionic biosurfactant rhamnolipid. FEBS Lett 588(10):1955–1960. https://doi.org/10.1016/j.febslet.2014.04.004

Beegle JR, Borole AP (2018) Energy production from waste: evaluation of anaerobic digestion and bioelectrochemical systems based on energy efficiency and economic factors. Renew Sustain Energy Rev 96:343–351. https://doi.org/10.1016/j.rser.2018.07.057

Bernal-Martínez A, Carrère H, Patureau D, Delgenès JP (2005) Combining anaerobic digestion and ozonation to remove PAH from urban sludge. Process Biochem 40(10):3244–3250. https://doi.org/10.1016/j.procbio.2005.03.028

Buchhammer H-M, Oelmann M, Petzold G (2001) Flocculation of disperse dyes in effluents with polyelectrolyte complexes. Melliand Int 82(5):104–105

Cai QY, Mo CH, Wu QT, Zeng QY, Katsoyiannis A (2007) Quantitative determination of organic priority pollutants in the composts of sewage sludge with rice straw by gas chromatography coupled with mass spectrometry. J Chromatogr A 1143(1-2):207–214. https://doi.org/10.1016/j.chroma.2007.01.007

Cai W, Han T, Guo Z, Varrone C, Wang A, Liu W (2016) Methane production enhancement by an independent cathode in integrated anaerobic reactor with microbial electrolysis. Bioresour Technol 208:13–18. https://doi.org/10.1016/j.biortech.2016.02.028

Cao Y, Pawłowski A (2013) Life cycle assessment of two emerging sewage sludge-to-energy systems: evaluating energy and greenhouse gas emissions implications. Bioresour Technol 127(1):81–91. https://doi.org/10.1016/j.biortech.2012.09.135

Cao M, Hu Y, Sun Q, Wang L, Chen J, Lu X (2013) Enhanced desorption of PCB and trace metal elements (Pb and cu) from contaminated soils by saponin and EDDS mixed solution. Environ Pollut 174:93–99. https://doi.org/10.1016/j.envpol.2012.11.015

Cerniglia CE (1992) Biodegradation of polycyclic aromatic hydrocarbons. J Ind Microbiol 3 (2–3):351–368. https://doi.org/10.1007/BF00129093
Chai CY, Zhang DW, Yu YL, Feng YJ, Mansing W (2015) Carbon footprint analyses of mainstream wastewater treatment technologies under different sludge treatment scenarios in China. Water 7(3):918–938. https://doi.org/10.3390/w7030918
Chen Y, Yang H, Gu G (2001) Effect of acid and surfactant treatment on activated sludge dewatering and settling. Water Res 35(11):2615–2620. https://doi.org/10.1016/s0043-1354(00)00565-0
Chen Y, Chen Y-S, Gu G (2004) Influence of pretreating activated sludge with acid and surfactant prior to conventional conditioning on filtration dewatering. Chem Eng J 99(2):137–143. https://doi.org/10.1016/j.cej.2003.08.027
Chen Y, Liu K, Su Y, Zheng X, Wang Q (2013) Continuous bioproduction of short-chain fatty acids from sludge enhanced by the combined use of surfactant and alkaline pH. Bioresour Technol 140:97–102. https://doi.org/10.1016/j.biortech.2013.04.075
Clarke BO, Porter NA, Marriott PJ, Blackbeard JR (2010) Investigating the levels and trends of organochlorine pesticides and polychlorinated biphenyl in sewage sludge. Environ Int 36 (4):323–329. https://doi.org/10.1016/j.envint.2010.01.004
Davis RJ, Gainer JL, O'Neal G, Wu W (1994) Photocatalytic decolorization of wastewater dyes. Water Environ Res 66(1):50–53. https://doi.org/10.2175/WER.66.1.8
De Vrieze J, Smet D, Klok J, Colsen J, Angenent LT, Vlaeminck SE (2016) Thermophilic sludge digestion improves energy balance and nutrient recovery potential in full-scale municipal wastewater treatment plants. Bioresour Technol 218:1237–1245. https://doi.org/10.1016/j.biortech.2016.06.119
El-Hadj TB, Dosta J, Torres R, Mata-álvarez J (2007) PCB and AOX removal in mesophilic and thermophilic sewage sludge digestion. Biochem Eng J 36(3):281–287. https://doi.org/10.1016/j.bej.2007.03.001
Escapa A, Lobato A, Garcia D, Morán A (2013) Hydrogen production and COD elimination rate in a continuous microbial electrolysis cell: the influence of hydraulic retention time and applied voltage. Environ Prog Sustain 32(2):263–268. https://doi.org/10.1002/ep.11619
Esteban-Gutierrez M, Garcia-Aguirre J, Irizar I, Aymerich E (2018) From sewage sludge and Agri-food waste to VFA: individual acid production potential and up-scaling. Waste Manag 77:203–212. https://doi.org/10.1016/j.wasman.2018.05.027
Feitkenhauer H (2003) Anaerobic digestion of desizing wastewater: influence of pretreatment and anionic surfactant on degradation and intermediate accumulation. Enzyme Microb Technol 33 (2):250–258. https://doi.org/10.1016/S0141-0229(03)00125-X
Feng L, Chen J, Wang F, Chen Y, Luo J (2019) Acidogenic fermentation facilitates anaerobic biodegradation of polycyclic aromatic hydrocarbons in waste activated sludge. ACS Sustain Chem Eng 7(5):5404–5411. https://doi.org/10.1016/j.scitotenv.2020.136500
Flemming HC, Wingender J (2010) The biofilm matrix. Nat Rev Microbiol 8(9):623–633. https://doi.org/10.1038/nrmicro2415
Garcia MT, Campos E, Dalmau M, Illan P, Sanchez-Leal J (2006) Inhibition of biogas production by alkyl benzene sulfonates (LAS) in a screening test for anaerobic biodegradability. Biodegradation 17(1):39–46. https://doi.org/10.1007/s10532-005-2798-x
Guan R, Yuan X, Wu Z, Wang H, Jiang L, Li Y, Zeng G (2017) Functionality of surfactants in waste-activated sludge treatment: a review. Sci Total Environ 609:1433–1442. https://doi.org/10.1016/j.scitotenv.2017.07.189
Guangyin Z, Youcai Z (2017) Chapter one. Sewage sludge generation and characteristics. In: Pollution control and resource recovery, pp 1–11
Han F, Srinivasan VSRKM, Rajarathnam D, Naidu R (2009) Tailored titanium dioxide photocatalysts for the degradation of organic dyes in wastewater treatment: a review. Appl Catal A-Gen 359(1–2):25–40. https://doi.org/10.1016/j.apcata.2009.02.043
He Q, Xu P, Zhang C, Zeng G, Liu Z, Wang D, Tang W, Dong H, Tan X, Duan A (2019a) Influence of surfactants on anaerobic digestion of waste activated sludge: acid and methane production

and pollution removal. Crit Rev Biotechnol 39(5):746–757. https://doi.org/10.1080/07388551.2018.1530635

He Z-W, Liu W-Z, Tang C-C, Liang B, Guo Z-C, Wang L, Ren Y-X, Wang A-J (2019b) Performance and microbial community responses of anaerobic digestion of waste activated sludge to residual benzalkonium chlorides. Energ Conver Manage 202:112211. https://doi.org/10.1016/j.enconman.2019.112211

Huang X, Shen C, Liu J, Lu L (2015) Improved volatile fatty acid production during waste activated sludge anaerobic fermentation by different bio-surfactants. Chem Eng J 264:280–290. https://doi.org/10.1016/j.cej.2014.11.078

Huang X, Mu T, Shen C, Lu L, Liu J (2016a) Alkaline fermentation of waste activated sludge stimulated by saponin: volatile fatty acid production, mechanisms and pilot-scale application. Water Sci Technol 74:2860. https://doi.org/10.2166/wst.2016.459

Huang X, Mu T, Shen C, Lu L, Liu J (2016b) Effects of bio-surfactants combined with alkaline conditions on volatile fatty acid production and microbial community in the anaerobic fermentation of waste activated sludge. Int Biodeter Biodegr 114:24–30. https://doi.org/10.1016/j.ibiod.2016.05.014

Jang HM, Lee J, Choi S, Shin J, Kan E, Kim YM (2018) Response of antibiotic and heavy metal resistance genes to two different temperature sequences in anaerobic digestion of waste activated sludge. Bioresour Technol 267:303–310. https://doi.org/10.1016/j.biortech.2018.07.051

Ji Z, Chen G, Chen Y (2010) Effects of waste activated sludge and surfactant addition on primary sludge hydrolysis and short-chain fatty acids accumulation. Bioresour Technol 101(10):3457–3462. https://doi.org/10.1016/j.biortech.2009.12.117

Jiang S, Chen Y, Zhou Q (2007a) Effect of sodium dodecyl sulfate on waste activated sludge hydrolysis and acidification. Chem Eng J 132(1–3):311–317. https://doi.org/10.1016/j.cej.2007.01.017

Jiang S, Chen Y, Zhou Q, Gu G (2007b) Biological short-chain fatty acids (SCFAs) production from waste-activated sludge affected by surfactant. Water Res 41(14):3112–3120. https://doi.org/10.1016/j.watres.2007.03.039

Kadier A, Simayi Y, Abdeshahian P, Azman NF, Chandrasekhar K, Kalil MS (2016) A comprehensive review of microbial electrolysis cells (MEC) reactor designs and configurations for sustainable hydrogen gas production. Alex Eng J 55(1):427–443. https://doi.org/10.1016/j.aej.2015.10.008

Kavitha S, Jayashree C, Kumar SA, Yeom IT, Banu JR (2014) The enhancement of anaerobic biodegradability of waste activated sludge by surfactant mediated biological pretreatment. Bioresour Technol 168:159–166. https://doi.org/10.1016/j.biortech.2014.01.118

Kavitha S, Pray SS, Yogalakshmi K, Kumar SA, Yeom IT, Banu JR (2016) Effect of chemo-mechanical disintegration on sludge anaerobic digestion for enhanced biogas production. Environ Sci Pollut Res 23(3):1–13. https://doi.org/10.1007/s11356-015-5461-z

Kiliç E, Font J, Puig R, Çolak S, Çelik D (2011) Chromium recovery from tannery sludge with saponin and oxidative remediation. J Hazard Mater 185(1):456–462. https://doi.org/10.1016/j.jhazmat.2010.09.054

Klimavičiūtė R (2004) Starch-cationic polyelectrolyte complexes and their interaction with acid dye. Fibres Text East Eur 2(46):74–78

Kosaric N, Cairns WL, Gray NCC, Gray NCC (1987) Biosurfactants and biotechnology. M. Dekker, New York

Kübler H, Hoppenheidt K, Hirsch P, Kottmair A, Nimmrichter R, Nordsieck H, Mücke W, Swerev M (2000) Full scale co-digestion of organic waste. Water Sci Technol 41(3):195–202. https://doi.org/10.2166/wst.2000.0072

Kuczajowska-Zadrożna M, Filipkowska U, Jóźwiak T (2015) Application of biosurfactants for heavy netals leaching from immobilized activated sludge/Zastosowanie biosurfaktantów do wymywania metali ciężkich z immobilizowanego osadu czynnego. Nephron Clin Pract 41(1):43–52. https://doi.org/10.1515/aep-2015-0006

Kumaran P, Hephzibah D, Sivasankari R, Saifuddin N, Shamsuddin AH (2016) A review on industrial scale anaerobic digestion systems deployment in Malaysia: opportunities and challenges. J Renew Sustain Energy 56:929–940. https://doi.org/10.1016/j.rser.2015.11.069

Laha S, Tansel B, Ussawarujikulchai A (2009) Surfactant–soil interactions during surfactant-amended remediation of contaminated soils by hydrophobic organic compounds: a review. J Environ Manage 90(1):95–100. https://doi.org/10.1016/j.jenvman.2008.08.006

Liu Y, Fang HH (2003) Influences of extracellular polymeric substances (EPS) on flocculation, settling, and dewatering of activated sludge. Crit Rev Environ Sci Technol 33(3):237–273. https://doi.org/10.1080/10643380390814479

Liu X, Li R, Ji M, Han L (2013) Hydrogen and methane production by co-digestion of waste activated sludge and food waste in the two-stage fermentation process: substrate conversion and energy yield. Bioresour Technol 146:317–323. https://doi.org/10.1016/j.biortech.2013.07.096

Liu W, Cai W, Guo Z, Wang L, Yang C, Varrone C, Wang A (2016) Microbial electrolysis contribution to anaerobic digestion of waste activated sludge, leading to accelerated methane production. Renew Energy 91:334–339. https://doi.org/10.1016/j.renene.2016.01.082

Liu X, He D, Wu Y, Xu Q, Wang D, Yang Q, Liu Y, Ni B-J, Wang Q, Li X (2020) Freezing in the presence of nitrite pretreatment enhances hydrogen production from dark fermentation of waste activated sludge. J Clean Prod 248:119305. https://doi.org/10.1016/j.jclepro.2019.119305

Luo G, Xie L, Zhou Q, Angelidaki I (2011a) Enhancement of bioenergy production from organic wastes by two-stage anaerobic hydrogen and methane production process. Bioresour Technol 102(18):8700–8706. https://doi.org/10.1016/j.biortech.2011.02.012

Luo K, Yang Q, Yu J, X-m L, Yang G-j, Xie B-x, Yang F, Zheng W, Zeng GM (2011b) Combined effect of sodium dodecyl sulfate and enzyme on waste activated sludge hydrolysis and acidification. Bioresour Technol 102(14):7103–7110. https://doi.org/10.1016/j.biortech.2011.04.023

Martín-Pascual J, Rueda-Pérez JJ, Jóver M, Hontoria E, Poyatos JM (2017) Optimization of the acid stage of a two-stage anaerobic digestion process to treat urban wastewater sludge. J Environ Eng 143(9):1–7. https://doi.org/10.1061/(ASCE)EE.1943-7870.0001237

Mayer AS, Zhong L, Pope GA (1999) Measurement of mass-transfer rates for surfactant-enhanced Solubilization of nonaqueous phase liquids. Environ Sci Technol 33(17):2965–2972. https://doi.org/10.1021/es9813515

Moscoviz R, Trably E, Bernet N, Carrère H (2018) The environmental biorefinery: state-of-the-art on the production of hydrogen and value-added biomolecules in mixed-culture fermentation. Green Chem 20(14):3159–3179. https://doi.org/10.1039/c8gc00572a

Muha I, Zielonka S, Lemmer A, Schonberg M, Linke B, Grillo A, Wittum G (2013) Do two-phase biogas plants separate anaerobic digestion phases? – a mathematical model for the distribution of anaerobic digestion phases among reactor stages. Bioresour Technol 132:414–418. https://doi.org/10.1016/j.biortech.2012.12.031

Muthukumar N, Maruthamuthu S, Palaniswamy N (2007) Role of cationic and nonionic surfactants on biocidal efficiency in diesel-water interface. Colloids Surf B 57(2):152–160. https://doi.org/10.1016/j.colsurfb.2007.01.019

Neyens E, Baeyens J, Dewil R, De Heyder B (2004) Advanced sludge treatment affects extracellular polymeric substances to improve activated sludge dewatering. J Hazard Mater 106 (2–3):83–92. https://doi.org/10.1016/j.jhazmat.2003.11.014

Pervin HM, Dennis PG, Lim HJ, Tyson GW, Batstone DJ, Bond PL (2013) Drivers of microbial community composition in mesophilic and thermophilic temperature-phased anaerobic digestion pre-treatment reactors. Water Res 47(19):7098–7108. https://doi.org/10.1016/j.watres.2013.07.053

Petzold G, Schwarz S (2006) Dye removal from solutions and sludges by using polyelectrolytes and polyelectrolyte–surfactant complexes. Sep Purif Technol 51(3):318–324. https://doi.org/10.1016/j.seppur.2006.02.016

Pradel M, Aissani L, Villot J, Baudez J-C, Laforest V (2016) From waste to added value product: towards a paradigm shift in life cycle assessment applied to wastewater sludge—a review. J Clean Prod 131:60–75. https://doi.org/10.1016/j.jclepro.2016.05.076

Raszka A, Chorvatova M, Wanner J (2006) The role and significance of extracellular polymers in activated sludge. Part I: literature review. Clean 34(5):411–424. https://doi.org/10.1002/aheh.200500640

Ren MM, Yuan XZ, Zhu Y, Huang HJ, Zeng GM, Li H, Chen M, Wang H, Chen CY, Lin NB (2014) Effect of different surfactants on removal efficiency of heavy metals in sewage sludge treated by a novel method combining bio-acidification with Fenton oxidation. J Cent South Univ 21(12):4623–4629. https://doi.org/10.1007/s11771-014-2469-3

Rogers HR (1996) Sources, behaviour and fate of organic contaminants during sewage treatment and in sewage sludges. Sci Total Environ 185(1–3):3–26. https://doi.org/10.1016/0048-9697(96)05039-5

Rosinska A, Karwowska B (2017) Dynamics of changes in coplanar and indicator PCB in sewage sludge during mesophilic methane digestion. J Hazard Mater 323:341–349. https://doi.org/10.1016/j.jhazmat.2016.04.016

Ruffino B, Campo G, Genon G, Lorenzi E, Novarino D, Scibilia G, Zanetti M (2014) Improvement of anaerobic digestion of sewage sludge in a wastewater treatment plant by means of mechanical and thermal pre-treatments: performance, energy and economical assessment. Bioresour Technol 175:298–308. https://doi.org/10.1016/j.biortech.2014.10.071

Sahinkaya S (2015) Disintegration of municipal waste activated sludge by simultaneous combination of acid and ultrasonic pretreatment. Process Saf Environ Prot 93:201–205. https://doi.org/10.1016/j.psep.2014.04.002

Scaglione D, Caffaz S, Ficara E, Malpei F, Lubello C (2008) A simple method to evaluate the short-term biogas yield in anaerobic codigestion of WAS and organic wastes. Water Sci Technol 58(8):1615–1622. https://doi.org/10.2166/wst.2008.502

Shcherbakova VA, Laurinavichius KS, Akimenko VK (1999) Toxic effect of surfactants and probable products of their biodegradation on methanogenesis in an anaerobic microbial community. Chemosphere 39(11):1861–1870. https://doi.org/10.1016/S0045-6535(99)00081-8

Sheng GP, Zhang M-L, Yu H-Q (2008) Characterization of adsorption properties of extracellular polymeric substances (EPS) extracted from sludge. Colloids Surf B 62(1):83–90. https://doi.org/10.1016/j.colsurfb.2007.09.024

Song S, Zhu L, Zhou W (2008) Simultaneous removal of phenanthrene and cadmium from contaminated soils by saponin, a plant-derived biosurfactant. Environ Pollut 156(3):1368–1370. https://doi.org/10.1016/j.envpol.2008.06.018

Su J, Chen Y, Zhou Q (2007) Effect of sodium dodecyl sulfate on waste activated sludge hydrolysis and acidification. Chem Eng J 132(1–3):311–317. https://doi.org/10.1016/j.cej.2007.01.017

Sun YJ, Zheng HL, Zhai J, Teng HK, Zhao C, Zhao CL, Liao Y (2014) Effects of surfactants on the improvement of sludge Dewaterability using cationic flocculants. PLoS One 9(10):e111036. https://doi.org/10.1371/journal.pone.0111036

Tanaka S, Ichikawa T (2015) Effects of photolytic pretreatment on biodegradation and detoxification of surfactants in anaerobic digestion. Water Sci Technol 28(7):103–110. https://doi.org/10.2166/wst.1993.0149

Tang J, He JG, Liu TT, Xin XD (2017) Extraction and environmental risk assessment of heavy metal in the municipal dewatered sludge using rhamnolipid treatment. Hum Ecol Risk Assess 23(6):1522–1538. https://doi.org/10.1080/10807039.2017.1331728

Tang J, He J, Qiu Z, Xin X (2019) Metal removal effectiveness, fractions, and binding intensity in the sludge during the multiple washing steps using the combined rhamnolipid and saponin. J Soil Sediment 19(3):1286–1296. https://doi.org/10.1007/s11368-018-2106-0

Ushani U, Kavitha S, Johnson M, Yeom IT, Banu JR (2017) Upgrading the hydrolytic potential of immobilized bacterial pretreatment to boost biogas production. Environ Sci Pollut R 24(1):813–826. https://doi.org/10.1007/s11356-016-7819-2

Wang L, Liu W, Kang L, Yang C, Zhou A, Wang A (2014a) Enhanced biohydrogen production from waste activated sludge in combined strategy of chemical pretreatment and microbial electrolysis. Int J Hydrogen Energy 39(23):11913–11919. https://doi.org/10.1016/j.ijhydene.2014.06.006

Wang LF, He DQ, Tong ZH, Li WW, Yu HQ (2014b) Characterization of dewatering process of activated sludge assisted by cationic surfactants. Biochem Eng J 91:174–178. https://doi.org/10.1016/j.bej.2014.08.008

Wang H, Yuan X, Wu Y, Zeng G, Chen X, Leng L, Wu Z, Jiang L, Li H (2015) Facile synthesis of amino-functionalized titanium metal-organic frameworks and their superior visible-light photocatalytic activity for Cr (VI) reduction. J Hazard Mater 286:187–194. https://doi.org/10.1016/j.jhazmat.2014.11.039

Xia H, Chi X, Yan Z, Cheng W (2009) Enhancing plant uptake of polychlorinated biphenyls and cadmium using tea saponin. Bioresour Technol 100(20):4649–4653. https://doi.org/10.1016/j.biortech.2009.04.069

Xu L, Liu W, Wu Y, Wang A, Li S, Ji W (2013) Optimizing external voltage for enhanced energy recovery from sludge fermentation liquid in microbial electrolysis cell. Int J Hydrogen Energy 38(35):15801–15806. https://doi.org/10.1016/j.ijhydene.2013.05.084

Yang G, Zhang G, Wang H (2015) Current state of sludge production, management, treatment and disposal in China. Water Res 78:60–73. https://doi.org/10.1016/j.watres.2015.04.002

Yang J, Liu X, Liu X, Xu Q, Wang W, Wang D, Yang G, Fu Q, Kang Z, Yang Q (2020) Enhanced dark fermentative hydrogen production from waste activated sludge by combining potassium ferrate with alkaline pretreatment. Sci Total Environ 707:136105. https://doi.org/10.1016/j.scitotenv.2019.136105

Yi L, Luo X, Huang X, Wang D, Zhang W (2013) Life cycle assessment of a municipal wastewater treatment plant: a case study in Suzhou, China. J Clean Prod 57:221–227. https://doi.org/10.1016/j.jclepro.2013.05.035

Yu Z, Zhang C, Zheng Z, Hu L, Li X, Yang Z, Ma C, Zeng G (2017) Enhancing phosphate adsorption capacity of SDS-based magnetite by surface modification of citric acid. Appl Surf Sci 403:413–425. https://doi.org/10.1016/j.apsusc.2017.01.163

Yuan C, Weng C-HJC (2006) Electrokinetic enhancement removal of heavy metals from industrial wastewater sludge. Chemosphere 65(1):88–96. https://doi.org/10.1016/j.chemosphere.2006.02.050

Zhang P, Chen Y, Zhou Q (2009) Waste activated sludge hydrolysis and short-chain fatty acids accumulation under mesophilic and thermophilic conditions: effect of pH. Water Res 43 (15):3735–3742. https://doi.org/10.1016/j.watres.2009.05.036

Zhang QH, Yang WN, Ngo HH, Guo WS, Jin PK, Dzakpasu M, Yang SJ, Wang Q, Wang XC, Ao D (2016) Current status of urban wastewater treatment plants in China. Environ Int 92-93:11–22. https://doi.org/10.1016/j.envint.2016.03.024

Zhen G, Lu X, Kato H (2017) Overview of pretreatment strategies for enhancing sewage sludge disintegration and subsequent anaerobic digestion: current advances, full-scale application and future perspectives. Renew Sustain Energy Rev 69:559–577. https://doi.org/10.1016/j.rser.2016.11.187

Zheng X-J, Blais J-F, Mercier G, Bergeron M, Drogui P (2007) PAH removal from spiked municipal wastewater sewage sludge using biological, chemical and electrochemical treatments. Chemosphere 68(6):1143–1152. https://doi.org/10.1016/j.chemosphere.2007.01.052

Zhou A, Guo Z, Yang C, Kong F, Liu W, Wang A (2013a) Volatile fatty acids productivity by anaerobic co-digesting waste activated sludge and corn straw: effect of feedstock proportion. J Biotechnol 168(2):234–239. https://doi.org/10.1016/j.jbiotec.2013.05.015

Zhou A, Yang C, Guo Z, Hou Y, Liu W, Wang A (2013b) Volatile fatty acids accumulation and rhamnolipid generation in situ from waste activated sludge fermentation stimulated by external rhamnolipid addition. Biochem Eng J 77:240–245. https://doi.org/10.1016/j.bej.2013.06.007

Zhou A, Luo H, Varrone C, Wang Y, Liu W, Wang A, Yue X (2015) Enhanced anaerobic digestibility of waste activated sludge by plant-derived biosurfactant. Process Biochem 50 (9):1413–1421. https://doi.org/10.1016/j.procbio.2015.04.023

Zhou A, Zhang J, Cai W, Sun R, Wang G, Liu W, Wang A, Yue X (2017) Comparison of chemosynthetic and biological surfactants on accelerating hydrogen production from waste activated sludge in a short-cut fermentation-bioelectrochemical system. Int J Hydrogen Energy 42(14):9044–9050. https://doi.org/10.1016/j.ijhydene.2016.02.075

Zhu Y, Liu H, Liu H, Huang S, Ma H, Tian Y (2015) Filtration characteristics of anaerobic fermented sewage sludge for fatty acids production. Sep Purif Technol 142:8–13. https://doi.org/10.1016/j.seppur.2014.11.037

Zliobaite E (1998) Binding of dyes with cationic polyelectrolytes in textile wastewater. PhD Thesis, Kaunas University of Technology, Lithuania

Application of Microbial Biosurfactants in the Pharmaceutical Industry

12

Sib Sankar Giri

Abstract

Numerous studies on the characterization and application of microbial biosurfactants have been carried out. Low in toxicity, environmental compatibility, and higher biodegradability of biosurfactants make them an attractive choice for numerous applications. Their structural novelty, diverse properties, and versatility make them an attractive group of compounds for potential use in a wide variety of industrial and biotechnological applications. Biosurfactants like glycolipids or lipopeptides are able to damage cell membranes and inhibit the proliferation of cancerous cells, which eventually lead to cell lysis via apoptosis pathways. As drug delivery molecules, biosurfactants can have promising applications in the biomedical field. Surfactin, a lipopeptide biosurfactant, exhibits interesting properties like insecticidal, anti-microbial, antitumor, and anti-mycoplasma activities. In this chapter, the current status of biosurfactant research for its potential application in pharmaceutical industry is discussed. Potential for the development of biosurfactants as novel molecules with multifarious functions and numerous applications are also discussed.

Keywords

Surfactin · Lipopeptide · Anticancer · Wound healing · Rhamnolipid

S. S. Giri (✉)
Laboratory of Aquatic Biomedicine, College of Veterinary Medicine and Research Institute for Veterinary Science, Seoul National University, Seoul, South Korea

Inamuddin et al. (eds.), *Microbial Biosurfactants*, Environmental and Microbial Biotechnology, https://doi.org/10.1007/978-981-15-6607-3_12

12.1 Introduction

Surfactants are amphiphilic compounds and their use worldwide has increased in the last decade. They contain hydrophobic and hydrophilic moieties which reduce surface tension. Due to dispersing, emulsifying, foaming, solubilizing, and wetting properties, surfactants are widely utilized in cosmetic, pharmaceutical, and food formulations (Varvaresou and Iakovou 2015). The unfavorable environmental concerns during the synthesis and use of chemical surfactants constitute an increasing concern worldwide. With increased environmental awareness and rapid biotechnological advancement, there has been a demand for environmentally friendly surfactants that can replace chemically synthesized compounds. Recently, biosurfactants are in great demand due to their advantages over chemical surfactants. Microbial surfactants or biosurfactants are surface-active molecules produced by various classes of microorganisms (Sen 2010). Microbial biosurfactants are basically secondary metabolites, which are either secreted outside the cells or remain adhered to microbial cells. Biosurfactants have several potential advantages over synthetic surfactants: (1) susceptible to biodegradation by microorganisms, (2) specificity of biosurfactants usually better than currently available commercial surfactants, (3) lower toxicity, and (4) higher foaming (Cameotra et al. 2010; Jahan et al. 2020). Surfactants are amphiphilic in nature and their molecular weight is low. They have a hydrophobic tail and hydrophilic head group. The balance between two groups (i.e., hydrophobic and hydrophilic) is the driving force behind surface activity (Martins and Martins 2018). Microbial genera, such as *Aspergillus, Acinetobacter*, *Brevibacterium*, *Bacillus*, *Candida, Saccharomyces*, *Citrobacter, Pseudomonas*, *Corynebacterium*, *Leuconostoc*, *Clostridium*, *Penicillium, Enterobacter*, *Rhodococcus*, *Lactobacillus*, and *Thiobacillus* are well-known biosurfactant producers (Jimoh and Lin 2019).

The estimated annual production of surfactants is above 15 million tons worldwide. The production is projected to arrive at 24 million tons approximately with estimated cost of 42,120 million USD per annum (Gudiña and Rodrigues 2019). Commercial productions of biosurfactants are expensive. Generally, microbial biosurfactants are a complex blend of congeners (Marchant and Banat 2012). High similarities in physical properties and molecular characteristics make it complicated to purify and take apart one element of the surfactant from another. Comprehensive knowledge about the physiochemical properties and detailed molecular structures of biosurfactant molecules are not yet elucidated fully (Jana et al. 2017). In this chapter, the author attempts to elaborate the broad overview of biosurfactants that are exploited for its usefulness in the pharmaceutical industry.

12.2 Mechanism of Interaction of Biosurfactants

Biosurfactants in a heterogeneous system tend to network with the phase boundary between two phases, known as interface (Perfumo et al. 2018). For all interfacial systems, organic molecules have a tendency to immobilize at the solid interface from the aqueous phase. The conditioning film formed at the interface is responsible for changing the surface energy and wettability of the original surface (Rodrigues et al. 2006). Interaction of biosurfactants with interfaces may influence bacterial adhesion. A certain degree of uniformity may be observed after 4 h, and the composition of the adsorbed material becomes substratum independent (Neu and Marshall 1990).

As previously mentioned, biosurfactants are amphiphilic in nature, and due to this numerous interactions are involved in the possible adsorption of charged biosurfactants to the interface. To investigate the interactions of ionic biosurfactants at interfaces, ionic condition and pH are to be considered as important parameters (Craig et al. 1993). The behavior of a surfactant at interfaces is influenced by its molecular structure. It is well known that the condition in a natural system is multifarious and requires consideration of several added parameters.

12.3 Physiochemical Properties

The adsorption of biosurfactants to interfaces has practical importance. For successful application of biosurfactants, we need to know their physicochemical properties. Here, we briefly discussed about the production and origin of microbial or biosurfactants.

12.3.1 Surface Tension

The main role of biosurfactants is to lessen surface tensions at interfaces. At interfaces (air/liquid, liquid/liquid, solid/liquid), biosurfactants are absorbed due to their twofold nature, i.e., hydrophilic and hydrophobic. Water or oil molecules at the interface are replaced by surfactant molecules, then the reduction of intermolecular forces between solvent molecules takes place, which reduces the interfacial or surface tension (Jahan et al. 2020).

Various isotherms described adsorption of biosurfactant at air/water interface. Lateral interactions of adsorbed molecules have been described by the Frumkin model (Onaizi et al. 2016). The adsorption of biosurfactants can significantly reduce surface tension at air/water interface. At the interface, rhamnolipid forms a dense monolayer. This is an example of adsorption isotherm described in the Frumkin model (Chen 2004). When an ionic surfactant adsorbs at the surface of an aqueous solution, a reduction in the surface tension is usually observed (Kolev et al. 2002). This phenomenon enhances surface electric charge, and develops an electric double layer. Biosurfactant (rhamnolipid) extracted from *P. aeruginosa* LBI was capable of reducing the surface tension of water (Benincasa et al. 2004).

The head group and hydrophobic compartment of biosurfactants are involved in the reduction of interfacial tension. One of the promising applications of biosurfactants is the increment of transportation of thinly soluble hydrocarbons to microbes (Hung and Shreve 2001).

Adsorption of biosurfactants to solid/liquid interfaces can be applied in various fields. For solid/liquid interfaces, the Langmuir isotherm is usually used (Feng et al. 2013). Rhamnolipid increases cell hydrophobicity, which in turn decreases the surface tension on the bacterial cells in water. Feng et al. (2013) reported that *Pseudomonas putida* 852 hydrophobicity was increased by rhamnolipid and tergitol. The same authors reported the decrease of hydrophobicity of *Rhodococcus erythropolis* 3586 by rhamnolipid and tergitol (Feng et al. 2013).

12.3.2 Biosurfactant and Self-Assembly

Above the CMC, micelles are formed from surfactants in aqueous solution. Formation of micelle is a process for equilibrium. Van der Waals interactions and hydrophobic and hydrogen bonding are the main reasons behind the self-assembly tendency of biosurfactants (Kitamoto et al. 2009). Biosurfactant concentration, temperature, pH, salt content, and pressure modulate the shape or size of the micelles. The repulsive forces between headgroups constrain the micelle association number. Therefore, larger micelles cannot be seen beyond CMC, even at higher surfactant concentration; however, it increases the number of micelles (Benincasa et al. 2004).

Chemical composition and chemical environment is associated with the CMC of rhamnolipids (Manko et al. 2014). The CMC of monorhamnolipids is lesser than dirhamnolipids and the difference is about 36% (Raza et al. 2010). Rhamnolipids form lamella, micelles, or vesicles at concentrations higher than CMC. However, it depends on the concentration and pH of the solution, and on the presence of electrolytes (Nitschke et al. 2011). The self-assembly behavior of rhamnolipid is driven by the solution's ionic strength and pH (Helvaci et al. 2004). Further, electrostatic effects of the metal ions may control the morphology of self-assembled structures in surfactants solutions (Raza et al. 2010).

It has been reported that in aqueous solution, surfactin can form various self-assembled nanostructures (Arutchelvi et al. 2014). The surfactin biosurfactant isolated from *B. subtilis* YB7 formed CMC in aqueous solution. However, addition of just 200 μM of divalent ions (e.g., Cd^{2+}, Ca^{2+}, Zn^{2+}, and Ni^{2+}) decreased CMC formation (Arutchelvi et al. 2014; Janek et al. 2018). Using isothermal titration calorimetry (ITC) spontaneous binding of surfactin analogs with phospholipid vesicles is determined. This is an endothermic reaction and driven by entropy process (Razafindralambo et al. 2009). At the air/water interface, surfactin molecules could form a robust film in association with perdeuterated leucine groups. The variation of pH or the number of absorbed surfactant molecules did not influence the film structure (Goussous et al. 2017).

12.3.3 Emulsification Activity

Emulsions are kinetically stabilized, non-equilibrium systems. The stability, structure, and appearance of emulsions depend on the composition, components, and conditions involved in their preparation (e.g., temperature and pressure) and processes (equipment type, mixing duration, input energy) (Kaizu and Alexandridis 2016). Mechanisms like flocculation, creaming, coalescence, coagulation, and Ostwald ripening can break emulsions over time (Heeres et al. 2014).

Bacillus subtilis LAMI005 produced a surfactin biosurfactant. The LAMI005 surfactin showed a high emulsification index ($E_{24} > 50\%$) on soybean oil and kerosene (de Oliveira et al. 2013). The emulsion behavior is controlled by the pH and salinity of the system. *B. subtilis* ATCC 21332 surfactin at pH >7.4 produced stable emulsions. Kerosene exhibited emulsification ratio of about 98%, at pH 7.4 but no emulsification was observed below pH 3 (Long et al. 2017). The interfacial adsorption properties of rhamnolipids are influenced by pH range. This in turn influences the surface tension, CMC, and elasticity coefficient. The E_{24} value for rhamnolipid was increased from 0 to 70% in a saline kerosene system when the pH augmented from 3 to 6. Increased E_{24} is possibly associated with the long chain size of the hydrophobic sources (Lovaglio et al. 2011).

Surfactants enhance the stability of emulsions. Surfactants can form self-assembled structures, which can better stabilize emulsions through electrostatic and steric barriers (Santos et al. 2016). When electrostatic effects are present, more complexity is observed in emulsion stability. If pH varies then ionization and protonation of the carboxyl groups in rhamnolipids can be seen. The ionized carboxyl group prevails with a negative charge at pH above 5.6. Increased concentration of OH^- enhances the emulsifying activity and stability of rhamnolipid extracted from *P. aeruginosa* LBI (Lovaglio et al. 2011).

12.4 Application of Biosurfactants in Pharmaceutical Industry

Basically, microbial origin and composition are considered to classify biosurfactants. Sometimes they are categorized based on their molecular weight. Lipopolysaccharide is a high-molecular-weight biosurfactant. Lipopeptide is a common example of low-molecular-weight biosurfactants (Naughton et al. 2019). Among glycolipids, sophorolipids (source: fungi), rhamnolipids (source: *Pseudomonas*), glycolipids, cellobiose lipids, trehalolipids, and mannosylerythritol lipids (MELs) are mostly studied. In terms of pharmaceutical potential, glycolipids (Marchant and Banat 2012) and the LPs are of particular interest. Most low-molecular-weight biosurfactants have been reported to be released extracellularly. They have profound applications in transplantations and device manufacturing units.

Biosurfactant decrease the surface tension between two immiscible or miscible liquids, block hydrogen-bonding, and augment hydrophobic/hydrophilic interactions (Jimoh and Lin 2019). Biosurfactants are broadly utilized in various fields, for example, agriculture, cosmetics, chemistry, horticulture, food sector, and

Table 12.1 Application of biosurfactants in medical/pharmaceutical fields

Biosurfactant type	Microorganisms	Application
Surfactin	*Bacillus subtilis*	Antimicrobial and antifungal activities
		Antitumor activity against Ehrlich's ascites carcinoma cells
		Inhibition of fibrin clot formation
Iturin	*Bacillus subtilis*	Antimicrobial activity and antifungal activity against profound mycosis
		Increase in the electrical conductance of biomolecular lipid membranes
		Nontoxic and nonpyrogenic immunological adjuvant
Lichenysin	*Bacillus licheniformis*	Antibacterial activity
		Chelating properties (membrane-disrupting effect of lipopeptides)
Rhamnolipid	*Pseudomonas aeruginosa*	Antiadhesive activity
		Antimicrobial activity against *Mycobacterium tuberculosis*
Mannosylerythritol lipid	*Candida antartica*	Antimicrobial activity
		Immunological and neurological properties
		Induction of cell differentiation in the leukemia cell line HL60
Treahalose lipid	*Rhodococcus erythropolis*	Antiviral activity against herpes simplex virus
		Antiviral activity against influenza virus
Surlactin	Lactobacillus	Antiadhesive activity against several pathogens, including enteric bacteria
Glycolipid	*Streptococcus thermophilus*	Antiadhesive activity against several bacterial and yeast strains

in pharmaceutical industry (Table 12.1). Here, we discuss about the potential applicability of a few important biosurfactants in pharmaceutical industry and their notable biological activities.

12.4.1 Biosurfactant as an Antitumor/AntiCancer Agent

Glycolipids-led apoptosis and growth blocking of mouse B16 cells (malignant melanoma) have already been proven. In the sub-G0/G1 phase, the accumulation of B16 cells was observed when exposed to increasing concentration of mannosylerythritol lipids (MELs). Chromatin condensation and fragmentation of DNA was observed (Zhao et al. 2000).

The exposure of HL60 cells to MELs markedly increased general differentiation-related characters in granulocytes (e.g., expression of Fc receptors, ability to reduce nitroblue tetrazolium, and phagocytic activities). MELs can suppress the action of PKC. Thus, both apoptotic and differentiation mechanisms can be triggered by MEL biosurfactants (Zhao et al. 2010; Isoda et al. 1997).

Wickerhamiella domercqiae produced sophorolipid biosurfactant. Exposure H7402 cell (human liver cancer cells (H7402) to this biosurfactant resulted in apoptosis by activating caspase-3, blockage of cell cycles (during G1), and by enhancing concentration of Ca^{2+} in cytoplasm (Chen et al. 2006). Exposure of pancreatic carcinoma cells to sophorolipids resulted in cytotoxic activity (Fu et al. 2008). The effect of various sophorolipids on human esophageal cancer cell lines were determined (Shao et al. 2012). Sophorolipids having higher degrees of acetylation exhibited stronger inhibitory activities. The completed inhibition of cells was observed in cells exposed to diacetylated lactonic sophorolipid (30 mg/mL). However, in the case of monoacetylated lactonic sophorolipid, double concentration was necessary to inhibit the cells completely. The strongest cytotoxic effect was recorded in sophorolipid possessing a double bond in the fatty acid portion. Antitumor activities of acidic sophorolipids are very scarce. Those researchers speculated that anticancer activities of various sophorolipids may have different anticancer mechanisms.

Biosurfactant monoolein was extracted from the fungus *Exophiala dermatitidis* SK80 (Chiewpattanakul et al. 2010). Monoolein efficiently inhibited the proliferation of leukemia (U937) and cervical cancer cell lines (HeLa) in a dose-dependent manner. It did not show cytotoxicity even at higher concentrations. In both cancer cell lines, morphological changes of DNA and cell were observed, which include DNA fragmentation, membrane blebbing, and cell shrinkage. The biosurfactant extracted from *Acetinobacter indicus* M6 inhibited the propagation of A549 (lung cancer cells) at G1 phase and confirmed its antitumor activity (Karlapudi et al. 2020). This biosurfactant showed low toxicity and strong drug-like properties.

Surface activity profiles of biosurfactants will be altered if there is any change in lipid profiles. Preetha et al. (2005) studied the phospholipid profiles of both cancerous as well as normal cervical tissues. They found that phosphatidylcholine levels were higher (around five times) in cancerous tissue than that of normal tissue. These authors suggested that presence of phospholipids may change the membrane permeability of cancer cell membranes. Biosurfactants are capable of altering the lipid content to modulate interfacial properties and fluidizing the rigid cancerous tissues. If rigidity increased then surface tension will be lower and the drug penetration through such membranes will be reduced. Use of fluidizers can reverse those rigidifying effects and may improve penetration of drug into cancerous tissues (Gudiña et al. 2015). Therefore, biosurfactants could be exploited further for their potential use in drug delivery systems.

12.4.2 Biosurfactants as Drug Delivery Agents

In the case of passive immunization, biosurfactants may be suitable for delivering drugs. Due to the side effects and limited availability of antifungal drugs, the treatment of candidiasis seems to be difficult (Naughton et al. 2019). If antifungal drugs incorporate into suitable drug delivery systems then this problem can be solved. Various types of drug carriers (e.g., particulate, polymeric, cellular, and macromolecular carriers) are being investigated and also being used in drug delivery (Prasad et al. 2017). In biomedical and agricultural industries, micelles, lipid particles, microspheres, and vesicular systems (noisome, liposomes, sphingosomes, and virosomes) are currently being used (Gangwar et al. 2012).

Microemulsions are isotropic mixtures of oil and water. They are transparent, and thermodynamically stable. To stabilize the system, an interfacial film of surfactant molecules in combination with a co-surfactant is often used. A microemulsion system mainly consists of surfactants. Depending on self-aggregation, the structure of the microemulsion varies. Hydrophobic or hydrophilic drugs can be encapsulated or solubilized into those structures in the presence of a dispersed phase within its structural vicinity (mostly spherical). Thereby, dispersed phase partitioned from the continuous phase (Israelachvili 1994). Example: for microemulsion formulation, sucrose esters (non-ionic surfactants) possessing a sucrose (hydrophilic) and lipophillic group are used often (Csizmazia et al. 2012). Recently, as greener alternatives to synthetic surfactants, use of biosurfactants as templates for nanoparticle synthesis is gaining momentum (Palanisamy 2008).

Mannosylerythritol lipid (MEL)-A is a glycolipid biosurfactant and its liposomes can increase the efficiency of gene transfections in mammalian cell culture (Ueno et al. 2007). Noisomes have also been used as drug delivery systems (Khan and Irichhaiya 2016). In a recent study, the potential of SL-AmB niosome was compared with a commercial one (Haque et al. 2017). Noisome-treated biofilm had less fungal hyphae. However, treatment of biofilm with phosome (AmB) had budding cells (AmB). In *Candida albicans*, one of the crucial virulence factors is pseudohyphae/true hyphae (Mayer et al. 2013). Cheng et al. (2009) demonstrated that SL-AmB may downregulate the expression of hyphal genes by interfering with its gene expression (Cheng et al. 2009).

12.4.3 Wound Healing and Dermatological Applications

Zouari et al. (2016) isolated lipopeptide from *B. subtilis* SPB1. They evaluated the wound healing capacity of lipopeptide rats (Zouari et al. 2016). These researchers observed that CICAFLORA™ treatment significantly increased wound healing in experimental rats.

Gupta et al. (2017) in an in vivo experiment demonstrated the potential of biosurfactant glycolipid in healing of wounds in rats. Another study showed that lipopeptides from *Acinetobacter junii* B6 enhanced histopathological remission and free-radical scavenging activities (Ohadi et al. 2017). Lydon et al. (2017) isolated glycolipid biosurfactant sophorolipids from yeast *Starmerella bombicola.* Purified sophorolipids inhibited the growth of *Pseudomonas aeruginosa* and *Enterococcus faecalis*. The sophorolipid (< 0.5 mg/mL) had no undesirable effects on endothelial or keratinocyte-derived cell lines.

Probiotic Lactobacilli strains are nontoxic, environmentally friendly, and provide beneficial effects on the host (Satpute et al. 2016). The biosurfactant isolated from *C. lipolytica* UCP 0988 (yeast) exhibited expected antiadhesive activity against

E. coli compared with those produced from other probiotic bacteria *L. helveticus* and *L. paracasei* (Sharma and Saharan 2016; Gudiña et al. 2010). Rhamnolipids and lactonic sophorolipid inhibited oral pathogens (*Actinomyces naeslundii*, *Neisseria mucosa*, *Streptococcus oralis*, *Streptococcus mutans*, and *Streptococcus sanguinis*) both in planktonic and oral biofilm states (Elshikh et al. 2017).

The antiadhesive and antimicrobial properties of *Lactobacillus pentosus*-derived glycolipid (PEB) were investigated and its activity was compared with glycolipids isolated from *Lactobacillus paracasei* (PAB) (Vecino et al. 2018). Both PEB and PAB exhibited antimicrobial activity against *E. coli*, *S. aureus*, *P. aeruginosa*, *Streptococcus agalactiae*, *C. albicans*, and *Streptococcus pyogenes*. The lipid content in PAB biosurfactant was lower than that of PEB biosurfactant. The *Lactobacillus helveticus*-derived glycolipid exhibited elevated antimicrobial potential against *S. epidermidis* and *E. coli* (Sharma and Saharan 2016). These properties suggest the potential of biosurfactant for various pharmaceutical applications.

12.4.4 Potential Antimicrobial Application

As various bacteria are showing resistance to different antibiotics, there is demand for new antimicrobial compounds. Lipopeptide biosurfactants have been used widely as antimicrobial agents. The antimicrobial lipopeptides produced by probiotic *Bacillus* inhibited the growth of pathogenic bacteria present in GI (gastrointestinal) tract (Hong et al. 2005). A cell-free biosurfactant was isolated from *Lactobacillus* spp. and it exhibited a broad spectrum of antimicrobial activity against *B. cereus*, *E. faecalis*, *Escherichia coli*, and *Salmonella* spp. (Augustin and Hippolyte 2012). The cell-free biosurfactant from *Lactobacillus rhamnosus* inhibited the growth of bacteria causing urinary tract infections (Salman and Alimer 2014). Biosurfactants isolated from *B. licheniformis* VS16 and *B. subtilis* VSG4 inhibited the growth of various Gram-positive and Gram-negative bacteria (Giri et al. 2019). Moreover, those biosurfactants have excellent antiadhesive activities against tested organisms. Similarly, biosurfactants from *L. lactis* and *B. licheniformis* exhibited antimicrobial activities against pathogenic methicillin-resistant *S. aureus* (MRSA) and *E. coli* (Saravanakumari and Mani 2010). *Corynebacterium xerosis* NS5 produced a novel lipopeptide biosurfactant coryxin (Dalili et al. 2015). Coryxin disrupted the biofilms of *Pseudomonas aeruginosa* (30%), *E. coli* (66%), *Streptococcus mutans* (80%), and *S. aureus* (82.5%).

Various probiotic bacilli produce antimicrobial lipopeptides (Hong et al. 2005). Lipopeptide biosurfactant marine *Bacillus circulans* was effective against *Alcaligens faecalis*, *Proteus vulgaris*, MRSA, and other multidrug-resistant pathogens (Das et al. 2008). Lipopeptide biosurfactants are useful in preventing fungal diseases in plants. Surfactin, fengycin C, and Iturin A isolated from *B. subtilis* strain EA-CB0015 were effective against fungus *Mycosphaerella fijiensis* (González-Jaramillo et al. 2017). Mannosylerythritol lipid (MEL) produced by *Pseudozyma aphidis* exhibited bactericidal and bacteriostatic effects on spores and vegetative cells of *Bacillus cereus*. The minimum inhibitory concentration (MIC) of MEL against *B. cereus* cells was 1.25 mg/mL and minimum bactericidal concentration (MBC) of MEL against the same bacteria was 2.50 mg/mL (Shu et al. 2019). Recently, chitosan biopolymer was used in combination with rhamnolipid biosurfactants as an antimicrobial substance (Marangon et al. 2020). This combination had low toxicity and was highly effective against *S. aureus* and *S. epidermis* biofilm nanoparticles combining the biopolymer chitosan with the biosurfactant rhamnolipid.

In a recent study, uniformly dispersed silver nanoparticles (AgNPs) were synthesized in lipopeptide biosurfactant reverse micelles (Bezza et al. 2020). These nanoparticles displayed remarkable inhibitory effect against *B. subtilis* CN2 and *P. aeruginosa* CB1 strains in a dose-dependent decline of cell viability and loss of membrane integrity (Bezza et al. 2020; Aziz et al. 2014, 2015, 2016). Waghmode et al. (2020) explored the therapeutic potential of glycolipid biosurfactant produced by *Planococcus maritimus*. The biosurfactant was capable of inhibiting the growth of *Mycobacterium tuberculosis* H37Ra at IC_{50}:64.11 ± 1.64 μg/mL and MIC at 160.8 ± 1.64 μg/mL. Biosurfactants inhibited the growth of *P. falciparum* at EC50 34.56 ± 0.26 μM. Ohadi et al. (2020) assessed the in vitro biofilm inhibition potential of lipopeptide biosurfactant isolated from *Acinetobacter junii*. These authors observed that biosurfactants at concentrations just below CMC inhibited bacterial growth. Further, the biosurfactant damaged the biofilm formed by *P. aeruginosa*, *S. aureus*, and *Proteus mirabilis* (Ohadi et al. 2020).

Sophorolipid biosurfactants were isolated from *Candida albicans* SC5314 and *Candida glabrata* CBS138 (Gaur et al. 2019). Isolated sophorolipid biosurfactant was effective against *S. aureus* MTCC9886, *B. subtilis* MTCC44, *E. coli* MTCC723, and *P. aeruginosa* MTCC 424. Flow cytometry analysis demonstrated that biosurfactants at 60 mg/L killed 65.8% of *B. subtilis* (Gaur et al. 2019). Recently, Bucci et al. (2018) reported that surfactin could enhance the effect of plant natural product terpinen-4-ol (TP) as antiadhesion and antimicrobial agent. Synthetic surfactants and surfactin promote the antimicrobial activity of TP against *S. mutans*. *S. mutans* is the responsible for tooth decay.

Lahkar et al. (2018) investigated the antifungal role of biosurfactant. Biosurfactant isolated from *P. aeruginosa* JS29 was tested for its efficacy in controlling anthracnose disease. They demonstrated the significant disease reduction in fungal spore. Moreover, in detached-fruit assay it was observed that biosurfactant could effectively inhibit the growth of fungus in different storage conditions. *Planococcus halotolerans* IITR55 and *P. rifietoensis* IITR53 produced rhamnolipid biosurfactants (Gaur et al. 2020). Both biosurfactants inhibited the growth of a variety of bacteria. Biosurfactants released extracellular DNA and protein content at a concentration of 40 mg/mL. At very low concentrations, they generated significant amount of ROS. These results showed their antimicrobial potential.

Lactococcus lactis produced a polycyclic peptide, namely, nisin. Sophorolipids and nisin exhibited inhibitory activities against *Staphylococcus aureus* with MICs of 32 and 0.5 μg/mL, respectively (Chen et al. 2020). Biosurfactant produced from *Candida bombicola* URM 3718 using low-cost substrates exhibited noteworthy emulsifying and antioxidant activities, which shows its potential for application in food systems (Da Silva et al. 2020).

12.4.5 Other Applications in the Pharmaceutical Field

Sambanthamoorthy et al. (2014) isolated biosurfactants from *Lactobacillus rhamnosus* and *Lactobacillus jensenii* cell surfaces. *Acinetobacter baumannii*, *E. coli*, and MRSA are known to form biofilms on wounds, medical implants, and industrial surfaces. Biosurfactants isolated from *L. rhamnosus* or *L. jensenii* exhibited antibacterial activity against those bacteria at concentrations between 25 to 100 mg/mL. Microscopic analysis revealed that biosurfactants damaged either cell wall or cell membrane of *A. baumannii* and *S. aureus*.

Bielinska et al. (2007) used nanoemulsions to prepared recombinant *Bacillus anthracis* protective antigen (rPA) against anthrax. They used soybean oil-water nanoemulsion used for the preparation. The formulated nanoemulsion was stable and functioned as an effective mucosal adjuvant in inducing robust, long-lasting, and precise cellular responses, without any negative effect (Bielinska et al. 2007).

Ohadi et al. (2017) investigated the antioxidant and related biological properties of a lipopeptide biosurfactant isolated from *Acinetobacter junii* B6 (Ohadi et al. 2017). Rats were wounded in the depilated thoracic region and treated with biosurfactant. A biosurfactant dose of 5 mg/mL exhibited best histopathological remission of scar wounds. A recent study revealed that biosurfactants could be used to remove tetracycline contamination in aquatic environment (Liu et al. 2020). Via biotransformation mechanism, co-culture of *B. amyloliquefaciens* and *B. clausii* efficiently bio-removed oxytetracycline (76.6%) and chlortetracycline (88.9%).

Sophorolipids showed spermicidal, antibacterial, and anti-HIV activities. Further, it has shown anti-inflammatory, antimicrobial, anticancer, and immunomodulatory activities against chronic inflammatory conditions and septic shock (Morya et al. 2013; Jahan et al. 2020). Most lactobacilli biosurfactants are usually surlactin type with high potential toward impeding pathogens adherence (Satpute et al. 2016). Some microorganisms are known to produce cell-free biosurfactants which have biomedical applications. *L. brevis* CV8LAC produced a cell-free biosurfactant which prevents the adhesion of *C. albicans* on medical-grade silicone elastomeric disks (Ceresa et al. 2015). These researchers recently demonstrated that sophorolpids could reduce biofilm formation *S. aureus* by 75% on pre-coated silicon discs (Ceresa et al. 2020). Moreover, sophorolipid inhibited the *C. albincans* biofilm effectively (Ceresa et al. 2020). Cyclic lipopeptides, fengycin and surfactin produced by *Bacillus subtilis* BBG111 induced systemic resistance against *Rhizoctonia solani* infection in rice (*Oryza sativa L.*) but it has no effect against *Magnaporthe oryzae* (Chandler et al. 2015).

12.5 Applications of Surfactin in Pharmaceutical Industry

Surfactin is an amphiphilic cyclic lipopeptide biomolecule. It consists of hydrophobic as well as hydrophilic moieties (Banat et al. 2010). Surfactin derived from *Bacillus subtilis* KLP2015 inhibited the growth of *Escherichia coli NCTC 10418*, *Klebsiella pnemoniae*, *Salmonella typhimurium* NCTC 74, and *Staphylococcus aureus* ATCC 6538. This surfactin had a strongly dislodged biofilm formed by *S. aureus* ATCC 6538 (Meena et al. 2020).

Cytotoxicity of surfactin toward cancer cells has been aimed in many studies (Gudiña et al. 2016). The cytotoxic effect of surfactin on K562 leukemia cells and human Bel-7402 hepatoma cells was investigated (Cao et al. 2009). Surfactin isolated from *Bacillus natto* TK-1 had cytotoxic effect on those cells as revealed by MTT assay. The IC_{50}–48 h values of K562 and Bel-7402 were 19.1 and 30.2 μg/mL, respectively, which indicated cancer chemo-preventive potential of surfactin (Cao et al. 2009). Cytotoxic potential of surfactin was studied in 47D and MDA-MB-231 cancer cells (Duarte et al. 2014). Cell viability was decreased when surfactin concentration and exposure duration was increased. The cell viability in MDA-MB-231 decreased sharply (reached up to 75% in 24 h) when surfactin concentration was 0.5 g/L; however, 0.5 g/L of surfactin resulted in 40% reduction in T47D cell viability after 24 h exposure (Duarte et al. 2014).

Surfactin-induced apoptosis pathways in MCF7 cells were investigated (Cao et al. 2010). They demonstrated reactive oxygen species (ROS)-driven apoptosis in MCF7 cells (Carrillo et al. 2003). Hydrophobic interactions help surfactin to penetrate into the plasma membrane. This increased the surface pressure which disturbs the order of hydrocarbon chain and eventually alters the thickness of the membrane. After this, conformational changes due to heptapeptide facilitate the mechanism of interaction (Maget-Dana and Ptak 1995). Seydlová et al. (2011) indicated that surfactin's integration into membranes induced specific dehydration of phospholipid polar heads. Therefore, destabilization of membrane integrity takes place which triggers the cascade of cellular events (Carrillo et al. 2003). Seydlová and Svobodová (2008) demonstrated that a low concentration of surfactin can be miscible with phospholipids, and penetrates into the membrane. Surfactin could exhibit strong detergent action at high concentration and hence could disrupt the membrane (Heerklotz and Seelig 2007).

Surfactin can rupture mycoplasmic wall completely through the osmotic influx of medium (Vollenbroich et al. 1997; Grau et al. 1999). Therefore, it may be used to control mycoplasma-related infections in manufacturing therapeutics and care products. The antiviral properties of surfactins against enveloped viruses (e.g., retrovirus and herpes virus) have also been reported. Surfactin can directly act on the lipid envelope of these viruses (Vollenbroich et al. 1997; Kracht et al. 1999). The nonpolar part of surfactin can penetrate into proteins and increase their stability by forming a noncovalent complex (Santos et al. 2018). It has potential application in oral delivery of insulin. Surfactin can pass through the gastrointestinal tract (Zhang et al. 2016). The potential capability of surfactin in boosting insulin uptake of intestine was investigated in mice model. Lipopeptide did not exert any noteworthy influence on bioavailability of hormones (Zhang et al. 2016).

Anti-biofilm property of surfactin is an important aspect. Surfactin, at low concentrations, disrupted the biofilm formed by various pathogens (Gomes and Nitschke 2012). Surfactin could be a potential antibiofilm agent in the medical industry; especially to eradicate bacterial biofilm from surfaces of surgical devices and biomaterials to prevent microbial colonization (Moryl et al. 2015). Presently, nanotechnology has been used to attenuate the toxicity of surfactin (Santos et al. 2018). To load doxorubicin (DOX), lipopeptide molecules were self-assembled into nanoparticles (NP) in treating multidrug-resistant problems in cancer treatment (Huang et al. 2018). Moreover, surfactin was applied in biological synthesis of NP (Singh et al. 2011). Surfactin-stabilized silver nanoparticles in diabetic mice reduced the wound surface (Krishnan et al. (2018).

12.6 Concluding Remarks

Although the biosurfactant market is expanding, the use of biosurfactants is being confined within a few specialized field of applications. Due to low toxicity, bioavailability, exceptional physicochemical characteristics, and renewable-resource origin, biosurfactants are a suitable alternative to their chemical counterparts. Several microbial surfactants exhibited antibacterial, antifungal, antiviral, anti-biofilm, and antiadhesive properties, which indicates their great potential application in the biomedical and healthcare industries. Biosurfactant molecules may be useful in blocking the growth of carcinoma cells. They damage the cell membranes and their mechanism of action associated with cellular lysis. To date, their major applications are limited to bioremediation. Due to less investment and lack of large-scale industrial production, wider application has not been achieved (Naughton et al. 2019).

In addition to an increasing demand for the biomedical applications of biosurfactants, their interactions with various components in microemulsions need to be explored further. Their use as adjuvant in microemulsion formulations are yet to be addressed adequately. The lack of adequate medical research about the use of biosurfactants is a challenge in the area of drug delivery. Few biosurfactants have fulfilled the criterions of drug regulatory systems. These successful outcomes will clear the path for the booming use of biosurfactant molecules in the pharmaceutical industry.

Acknowledgments S. S. Giri acknowledges the 'Korea Research Fellowship Program' (KRF: 2016H1D3A1909005) of National Research Foundation of Korea, Ministry of Science and ICT.

References

Arutchelvi J, Sangeetha J, Philip J, Doble M (2014) Self-assembly of surfactin in aqueous solution: role of divalent counterions. Colloids Surf B Biointerfaces 116:396–402

Augustin M, Hippolyte MT (2012) Screening of biosurfactants properties of cell-free supernatants of cultures of *Lactobacillus spp.* isolated from a local fermented milk (Pendidam) of Ngaoundere (Cameroon). Int J Eng Res Appl 2:974–985

Aziz N, Fatma T, Varma A, Prasad R (2014) Biogenic synthesis of silver nanoparticles using *Scenedesmus abundans* and evaluation of their antibacterial activity. J Nanoparticles. Article ID 689419. https://doi.org/10.1155/2014/689419

Aziz N, Faraz M, Pandey R, Sakir M, Fatma T, Varma A, Barman I, Prasad R (2015) Facile algae-derived route to biogenic silver nanoparticles: synthesis, antibacterial and photocatalytic properties. Langmuir 31:11605–11612. https://doi.org/10.1021/acs.langmuir.5b03081

Aziz N, Pandey R, Barman I, Prasad R (2016) Leveraging the attributes of *Mucor hiemalis*-derived silver nanoparticles for a synergistic broad-spectrum antimicrobial platform. Front Microbiol 7:1984. https://doi.org/10.3389/fmicb.2016.01984

Banat A, Franzetti I, Gandolfi G, Bestetti MG, Martinotti L, Fracchia TJ, Smyth R, Marchant R (2010) Microbial biosurfactants production, applications and future potential. Appl Microbiol Biotechnol 87:427–444

Benincasa M, Abalos A, Oliveira I, Manresa A (2004) Chemical structure, surface properties and biological activities of the biosurfactant produced by *Pseudomonas aeruginosa* LBI from soapstock. Antonie Van Leeuwenhoek 85(1):1–8

Bezza FA, Tichapondwa SM, Chirwa EMN (2020) Synthesis of biosurfactant stabilized silver nanoparticles, characterization and their potential application for bactericidal purposes. J Hazard Mater 393:122319

Bielinska AU, Janczak KW, Landers JJ et al (2007) Mucosal immunization with a novel nanoemulsion-based recombinant anthrax protective antigen vaccine protects against *Bacillus anthracis* spore challenge. Infect Immun 75:4020–4029

Bucci AR, Marcelino L, Mendes RK, Etchegaray A (2018) The antimicrobial and antiadhesion activities of micellar solutions of surfactin, CTAB and CPCl with terpinen-4-ol: applications to control oral pathogens. World J Microbiol Biotechnol 34:86

Cameotra SS, Makkar RS, Kaur J, Mehta SK (2010) Synthesis of biosurfactants and their advantages to microorganisms and mankind. In: Sen R (ed) Biosurfactants. Advances in experimental medicine and biology, vol 672. Springer, New York, pp 261–280

Cao XH, Liao ZY, Wang CL, Yang WY, Lu MF (2009) Evaluation of a lipopeptide biosurfactant from *Bacillus natto* TK-1 as a potential source of anti-adhesive, antimicrobial and antitumor activities. Braz J Microbiol 40:373–379

Cao XH, Wang AH, Wang CL et al (2010) Surfactin induces apoptosis in human breast cancer MCF-7 cells through a ROS/JNK-mediated mitochondrial/caspase pathway. Chem Biol Interact 183:357–362

Carrillo C, Teruel JA, Aranda FJ, Ortiz A (2003) Molecular mechanism of membrane permeabilization by the peptide antibiotic surfactin. Biochim Biophys Acta Biomembr 1611:91–97

Ceresa C, Tessarolo F, Caola I, Nollo G et al (2015) Inhibition of *Candida albicans* adhesion on medical-grade silicone by a *Lactobacillus*-derived biosurfactant. J Appl Microbiol 118:1116–1125

Ceresa C, Fracchia L, Williams M, Banat IM, Díaz De Rienzo MA (2020) The effect of sophorolipids against microbial biofilms on medical-grade silicone. J Biotechnol 309:34–43

Chandler S, Van Hese N, Coutte F, Jacques P, Höfte M, Vleesschauwer DD (2015) Role of cyclic lipopeptides produced by *Bacillus subtilis* in mounting induced immunity in rice (*Oryza sativa* L.). Physiol Mol Plant Pathol 91:20–30

Chen G (2004) Rhamnolipid biosurfactant behavior in solutions. J Biomater Sci Polym Ed 15 (2):229–235

Chen J et al (2006) Sophorolipid produced from the new yeast strain Wickerhamiella domercqiae induces apoptosis in H7402 human liver cancer cells. Appl Microbiol Biotechnol 72:52–59

Chen J, Zhifei LU, An Z, Ji P, Liu X (2020) Antibacterial activities of sophorolipids and nisin and their combination against foodborne pathogen *Staphylococcus aureus*. Eur J Lipid Sci Technol 122:1900333

Cheng A, Sun L, Xu X, Lou H (2009) The inhibitory effect of a macrocyclic bisbibenzyl riccardin D on the biofilms of Candida albicans. Biol Pharm Bull 32:1417–1421

Chiewpattanakul P et al (2010) Bioproduction and anticancer activity of biosurfactant produced by the dematiaceous fungus *Exophiala dermatitidis* SK80. J Microbiol Biotechnol 20:1664–1671

Craig VSJ, Ninham BW, Pashley RM (1993) Effect of electrolytes on bubble coalescence. Nature 364:317–319

Csizmazia E et al (2012) Ibuprofen penetration enhance by sucrose ester examined by ATR–FTIR in vivo. Pharm Dev Technol 17:125–128

Da Silva IA, Bezerra KGO, Durval IJB, Farias CBB et al (2020) Evaluation of the emulsifying and antioxidant capacity of the biosurfactant produced by candida bombicola URM 3718. Chem Eng Trans 79:67–72

Dalili D, Amini M, Faramarzi MA, Fazeli MR, Khoshayand MR, Samadi N (2015) Isolation and structural characterization of Coryxin, a novel cyclic lipopeptide from *Corynebacterium xerosis* NS5 having emulsifying and anti-biofilm activity. Colloids Surf B Biointerfaces 135:425–432

Das P, Mukherjee S, Sen R (2008) Antimicrobial potential of a lipopeptide biosurfactant derived from a marine Bacillus circulans. J Appl Microbiol 104:1675–1684

de Oliveira DWF, Franca IWL, Felix AKN et al (2013) Kinetic study of biosurfactant production by Bacillus subtilis Lami005 grown in clarified cashew apple juice. Colloids Surf B Biointerfaces 101:34–43

Duarte C, Gudiña EJ, Lima CF, Rodrigues LR (2014) Effects of biosurfactants on the viability and proliferation of human breast cancer cells. AMB Express 4:40

Elshikh M, Moya-Ramírez I, Moens H et al (2017) Rhamnolipids and lactonic sophorolipids: natural antimicrobial surfactants for oral hygiene. J Appl Microbiol 123(5):1111–1123

Feng W, Swift S, Singhal N (2013) Effects of surfactants on cell surface tension parameters and hydrophobicity of *pseudomonas putida* 852 and *Rhodococcus erythropolis* 3586. Colloids Surf B Biointerfaces 105:43–50

Fu SL et al (2008) Sophorolipids and their derivatives are lethal against human pancreatic cancer cells. J Surg Res 148:77–82

Gangwar M et al (2012) Recent advances in various emerging vesicular systems: an overview. Asian Pac J Trop Biomed 2:S1176–S1188

Gaur VK, Regar RK, Dhiman N, Gautam K, Srivastava JK et al (2019) Biosynthesis and characterization of sophorolipid biosurfactant by candida spp.: application as food emulsifier and antibacterial agent. Bioresour Technol 285:121314

Gaur VK, Tripathi V, Gupta P, Dhiman N, Regar RK, Gautam K, Srivastava JK, Patnaik S et al (2020) Rhamnolipids from *Planococcus spp.* and their mechanism of action against pathogenic bacteria. Bioresour Technol 307:123206

Giri SS, Ryu EC, Sukumaran V, Park SC (2019) Antioxidant, antibacterial, and anti-adhesive activities of biosurfactants isolated from *Bacillus* strains. Microb Pathog 132:66–72

Gomes MZV, Nitschke M (2012) Evaluation of rhamnolipid and surfactin to reduce the adhesion and remove biofilms of individual and mixed cultures of food pathogenic bacteria. Food Control 25:441–447

González-Jaramillo LM, Aranda FJ, Teruel JA, Villegas-Escobar V, Ortiz A (2017) Antimycotic activity of fengycin C biosurfactant and its interaction with phosphatidylcholine model membranes. Colloids Surf B Biointerfaces 156:114–122

Goussous SA, Casford MTL, Murphy AC et al (2017) Structure of the fundamental lipopeptide surfactin at the air/water interface investigated by sum frequency generation spectroscopy. J Phys Chem B 121(19):5072–5077

Grau A, Fernández JCG, Peypoux F, Ortiz A (1999) A study on the interactions of surfactin with phospholipid vesicles. Biochim Biophys Acta Biomembr 1418:307–319

Gudiña EJ, Rodrigues LR (2019) Microbial surfactants: alternative to vegetable oil surfactants. In: Balan V (ed) Microbial lipid production. Methods in molecular biology, vol 1995. Humana Press, New York, NY

Gudiña E, Rocha V, Teixeira J, Rodrigues L (2010) Antimicrobial and antiadhesive properties of a biosurfactant isolated from Lactobacillus paracasei ssp. paracasei A20. Lett Appl Microbiol 50:419–424

Gudiña EJ, Fernandes EC, Teixeira JA, Rodrigues LR (2015) Antimicrobial and anti-adhesive activities of cell-bound biosurfactant from *Lactobacillus agilis* CCUG31450. RSC Adv 5:90960–90968

Gudiña EJ, Teixeira JA, Rodrigues LR (2016) Biosurfactants produced by marine microorganisms with therapeutic applications. Mar Drugs 14:38. https://doi.org/10.3390/md14020038

Gupta S, Raghuwanshi N, Varshney R, Banat IM et al (2017) Accelerated in vivo wound healing evaluation of microbial glycolipid containing ointment as a transdermal substitute. Biomed Pharmacother 94:1186–1196

Haque F, Sajid M, Cameotra SS, Battacharyya MS (2017) Anti-biofilm activity of a sophorolipid-amphotericin B niosomal formulation against Candida albicans. Biofouling 33:768–779

Heeres AS, Picone CS, van der Wielen LA, Cunha RL, Cuellar MC (2014) Microbial advanced biofuels production: overcoming emulsification challenges for large-scale operation. Trends Biotechnol 32(4):221–229

Heerklotz H, Seelig J (2007) Leakage and lysis of lipid membranes induced by the lipopeptide surfactin. Eur Biophys J 36:305–314

Helvaci SS, Peker S, Ozdemir G (2004) Effect of electrolytes on the surface behavior of rhamnolipids R1 and R2. Colloids Surf B Biointerfaces 35:225–233

Hong HA, Duc LH, Cutting SM (2005) The use of bacterial spore formers as probiotics. FEMS Microbiol Rev 29:813–835

Huang W, Lang Y, Hakeem A, Lei Y et al (2018) Surfactin-based nanoparticles loaded with doxorubicin to overcome multidrug resistance in cancers. Int J Nanomedicine 13:1723–1736

Hung HC, Shreve GS (2001) Effect of the hydrocarbon phase on interfacial and thermodynamic properties of two anionic glycolipid biosurfactants in hydrocarbon/water systems. J Phys Chem B 105(50):12596–12600

Isoda H et al (1997) Microbial extracellular glycolipid induction of differentiation and inhibition of protein kinase C activity of human promyelocytic leukaemia cell line HL60. Biosci Biotechnol Biochem 61:609–614

Israelachvili J (1994) The science and applications of emulsions—an overview. Colloids Surf A Physicochem Eng Asp 91:1–8

Jahan R, Bodratti AM, Tsianou M, Alexandridis P (2020) Biosurfactants, natural alternatives to synthetic surfactants: physicochemical properties and applications. Adv Colloid Interface Sci 275:102061

Jana S, Mondal S, Kulkarni SS (2017) Chemical synthesis of biosurfactant succinoyl trehalose lipids. Org Lett 19:1784–1787

Janek T, Rodrigues LR, Czyznikowska Z (2018) Study of metal-lipopeptide complexes and their self-assembly behavior, micelle formation, interaction with bovine serum albumin and biological properties. J Mol Liq 268:743–753

Jimoh AA, Lin J (2019) Biosurfactant: a new frontier for greener technology and environmental sustainability. Ecotoxicol Environ Saf 184:109607

Kaizu K, Alexandridis P (2016) Effect of surfactant phase behavior on emulsification. J Colloid Interface Sci 466:138–149

Karlapudi AP, Venkateswarulu TC, Krupanidhi S, Rohini KK (2020) Evaluation of anti-cancer, anti-microbial and anti-biofilm potential of biosurfactant extracted from an Acinetobacter indicus M6 strain. J King Saud Univ Sci 32:223–227

Khan R, Irichhaiya R (2016) Niosomes: a potential tool for novel drug delivery. J Pharm Investig 46:195–204
Kitamoto D, Morita T, Fukuoka T, Konishi M, Imura T (2009) Self-assembling properties of glycolipid biosurfactants and their potential applications. Curr Opin Colloid Interface Sci 14 (5):315–328
Kolev VL, Danov KD, Kralchevsky PA, Broze G, Mehreteab A (2002) Comparison of the van der Waals and Frumkin adsorption isotherms for sodium dodecyl sulfate at various salt concentrations. Langmuir 18(23):9106–9109
Kracht M, Rokos H, Ozel M et al (1999) Antiviral and haemolytic activities of surfactin isoforms and their methyl ester derivatives. J Antibiot 52:613–619
Krishnan N, Velramar B, Pandiyan R, Velu RK (2018) Anti-pseudomonal and anti-endotoxic effects of surfactin-stabilized biogenic silver nanocubes ameliorated wound repair in streptozotocin-induced diabetic mice. Artif Cells Nanomed Biotechnol 46:488–499
Lahkar J, Goswami D, Deka S, Ahmed G (2018) Novel approaches for application of biosurfactant produced by *Pseudomonas aeruginosa* for biocontrol of *Colletotrichum capsici* responsible for anthracnose disease in chilli. Eur J Plant Pathol 150(1):57–71
Liu C, Xu Q, Yu S, Cheng J, Yuan Y (2020) Bio-removal of tetracycline antibiotics under the consortium with probiotics *Bacillus clausii* T and *Bacillus amyloliquefaciens* producing biosurfactants. Sci Total Environ 710:136329. https://doi.org/10.1016/j.scitotenv.2019.136329
Long X, He N, He Y, Jiang J et al (2017) Biosurfactant surfactin with pH-regulated emulsification activity for efficient oil separation when used as emulsifier. Bioresour Technol 241:200–206
Lovaglio RB, dos Santos FJ, Jafelicci M, Contiero J (2011) Rhamnolipid emulsifying activity and emulsion stability: pH rules. Colloids Surf B Biointerfaces 85(2):301–305
Lydon HL, Baccile N, Callaghan B, Marchant R et al (2017) Adjuvant antibiotic activity of acidic sophorolipids with potential for facilitating wound healing. Antimicrob Agents Chemother 61: e02547–e02516
Maget-Dana R, Ptak M (1995) Interactions of surfactin with membrane models. Biophys J 68:1937–1943
Manko D, Zdziennicka A, Janczuk B (2014) Thermodynamic properties of rhamnolipid micellization and adsorption. Colloids Surf B Biointerfaces 119:22–29
Marangon CA, Martins VCA, Ling MH, Melo CC, Plepis AMG, Meyer RL, Nitschke M (2020) Combination of rhamnolipid and chitosan in nanoparticles boosts their antimicrobial efficacy. ACS Appl Mater Interfaces 12(5):5488–5499
Marchant R, Banat IM (2012) Biosurfactants: a sustainable replacement for chemical surfactants? Biotechnol Lett 34:1597–1605
Martins PC, Martins VG (2018) Biosurfactant production from industrial wastes with potential remove of insoluble paint. Int Biodeter Biodegr 127:10–16
Mayer FL, Wilson D, Hube B (2013) Candida albicans pathogenicity mechanism. Virulence 4:119–128
Meena KR, Sharma A, Kanwar SS (2020) Antitumoral and antimicrobial activity of Surfactin extracted from *Bacillus subtilis* KLP2015. Int J Pept Res Ther 26:423–433
Morya VK, Ahn C, Jeon S, Kim EK (2013) Medicinal and cosmetic potentials of sophorolipids. Mini Rev Med Chem 13(12):1761–1768
Moryl M, Spetana M, Dziobek K, Paraszkiewicz K et al (2015) Antimicrobial, antiadhesive and antibiofilm potential of lipopeptides synthesised by *Bacillus subtilis*, on uropathogenic bacteria. Acta Biochim Pol 62:725–732
Naughton P, Marchant R, Naughton V, Banat I (2019) Microbial biosurfactants: current trends and applications in agricultural and biomedical industries. J Appl Microbiol 127:12–28
Neu T, Marshall KC (1990) Bacterial polymers: physicochemical aspects of their interactions at interfaces. J Biomater Appl 5:107–133
Nitschke M, Costa S, Contiero J (2011) Rhamnolipids and PHAs: recent reports on pseudomonas-derived molecules of increasing industrial interest. Process Biochem 46(3):621–630

Ohadi M, Forootanfar H, Rahimi HR et al (2017) Antioxidant potential and wound healing activity of biosurfactant produced by Acinetobacter junii B6. Curr Pharm Biotechnol 18:900–908
Ohadi M, Forootanfar H, Dehghannoudeh G et al (2020) Antimicrobial, anti-biofilm, and anti-proliferative activities of lipopeptide biosurfactant produced by *Acinetobacter junii* B6. Microb Pathog 138:103806
Onaizi SA, Nasser MS, Al-Lagtah NMA (2016) Benchmarking the self-assembly of surfactin biosurfactant at the liquid-air interface to those of synthetic surfactants. J Surfactant Deterg 19:645–652
Palanisamy P (2008) Biosurfactant mediated synthesis of NiO nanorods. Mater Lett 62:743–746
Perfumo A, Banat IM, Marchant R (2018) Going green and cold: biosurfactants from low-temperature environments to biotechnology applications. Trends Biotechnol 36:277–289
Prasad R, Pandey R, Varma A, Barman I (2017) Polymer based nanoparticles for drug delivery systems and cancer therapeutics. In: Kharkwal H, Janaswamy S (eds) Natural polymers for drug delivery. CAB International, Oxfordshire, pp 53–70
Preetha A et al (2005) Surface activity, lipid profiles and their implications in cervical cancer. J Cancer Res Ther 1:180–186
Raza ZA, Khalid ZM, Khan MS, Banat IM et al (2010) Surface properties and sub-surface aggregate assimilation of rhamnolipid surfactants in different aqueous systems. Biotechnol Lett 32(6):811–816
Razafindralambo H, Dufour S, Paquot M, Deleu M (2009) Thermodynamic studies of the binding interactions of surfactin analogues to lipid vesicles application of isothermal titration calorimetry. J Therm Anal Calorim 95(3):817–821
Rodrigues L, Banat IM, Teixeira J, Oliveira R (2006) Biosurfactants: potential applications in medicine. J Antimicrob Chemother 57:609–618
Salman JAS, Alimer AD (2014) *Lactobacillus rhamnosus* against some bacteria causing utiiniraqi women. Int J Curr Res 6:5368–5374
Sambanthamoorthy K, Feng X, Patel R, Patel S et al (2014) Antimicrobial and antibiofilm potential of biosurfactants isolated from lactobacilli against multi-drug-resistant pathogens. BMC Microbiol 14:197
Santos DKF, Rufino RD, Luna JM et al (2016) Biosurfactants: multifunctional biomolecules of the 21st century. Int J Mol Sci 17(3):401. https://doi.org/10.3390/ijms17030401
Santos VSV, Silveira E, Pereira BB (2018) Toxicity and applications of surfactin for health and environmental biotechnology. J Toxicol Environ Health Pt B 21(6–8):382–399
Saravanakumari P, Mani K (2010) Structural characterization of a novel xylolipid biosurfactant from *Lactococcus lactis* and analysis of antibacterial activity against multi-drug resistant pathogens. Bioresour Technol 101:8851–8854
Satpute SK, Kulkarni GR, Banpurkar AG et al (2016) Biosurfactant/s from lactobacilli species: properties, challenges and potential biomedical applications. J Basic Microbiol 56:1140–1158
Sen R (2010) Surfactin: biosynthesis, genetics and potential applications. In: Sen R (ed) Biosurfactants. Springer, New York, pp 316–323
Seydlová G, Svobodová J (2008) Review of surfactin chemical properties and the potential biomedical applications. Cent Eur J Med 3:123–133
Seydlová G, Čabala R, Svobodová J (2011) Surfactin—novel solutions for global issues. In: biomedical engineering, trends, research and technologies, 1st edn. InTechOpen, London. https://doi.org/10.5772/13015
Shao L, Song X, Ma X, Li H, Qu Y (2012) Bioactivities of Sophorolipid with different structures against human Esophageal Cancer cells. J Surg Res 173:286–291
Sharma D, Saharan BS (2016) Functional characterization of biomedical potential of biosurfactant produced by *Lactobacillus helveticus*. Biotechnol Rep 11:27–35
Shu Q, Niu Y, Zhao W, Chen Q (2019) Antibacterial activity and mannosylerythritol lipids against vegetative cells and spores of *Bacillus cereus* (2019). Food Control 106:106711

Singh BR, Dwivedi S, Al-Khedhairy AA, Musarrat J (2011) Synthesis of stable cadmium sulfide nanoparticles using surfactin produced by *Bacillus amyloliquifaciens* strain KSU-109. Colloids Surf B Biointerfaces 85:207–213

Ueno Y, Hirashima N, Inoh Y, Furuno T, Nakanishi M (2007) Characterization of biosurfactant-containing liposomes and their efficiency for gene transfection. Biol Pharm Bull 30(1):169–172

Varvaresou A, Iakovou K (2015) Biosurfactants in cosmetics and biopharmaceuticals. Lett Appl Microbiol 61:214–223

Vecino X, Rodríguez-López L, Ferreira D, Cruz JM, Moldes AB, Rodrigues LR (2018) Bioactivity of glycolipopeptide cell-bound biosurfactants against skin pathogens. Int J Biol Macromol 109:971–979

Vollenbroich D, Ozel M, Vater J, Kamp RM, Pauli G (1997) Mechanism of inactivation of enveloped viruses by the biosurfactant surfactin from Bacillus subtilis. Biologicals 25:289–297

Waghmode S, Swami S, Sarkar D, Suryavanshi M et al (2020) Exploring the pharmacological potentials of biosurfactant derived from *Planococcus maritimus* SAMP MCC 3013. Curr Microbiol 77:452–459

Zhang L, Gao J, Zhao X, Qi G (2016) A natural lipopeptide of surfactin for oral delivery of insulin. Drug Deliv 23:2084–2093

Zhao X et al (2000) Treatment of mouse melanoma cells with phorbol 12-myristate 13-acetate counteracts mannosylerythritol lipid-induced growth arrest and apoptosis. Cytotechnology 33:123–130

Zouari R, Hamden K, Feki AE et al (2016) Protective and curative effects of Bacillus subtilis SPB1 biosurfactant on high-fat-high-fructose diet induced hyperlipidemia, hypertriglyceridemia and deterioration of liver function in rats. Biomed Pharmacother 84:323–329

Antibacterial Biosurfactants 13

Debojyoti Mukherjee, Bipin Rooj, and Ujjwal Mandal

Abstract

When we say surfactant the first thing that comes in our mind is detergent. Obviously surfactants are detergents but they are more than that! Surfactants are very important for their versatile applications, as detergents and lubricants and in drug manufacturing, food processing, bioremediation, crude oil degradation, cosmetic production, removal of heavy metals, etc. The list can be really long but here we are only interested in the antibacterial activity of biosurfactants. Biosurfactants are formed by the microorganisms when the latter lives in a competitive environment. As a result, biosurfactants possess antibacterial properties. Different strains of microorganisms can produce various types of biosurfactants. Multiple drug resistance (MDR) against standard drugs is a challenge nowadays. Antibacterial biosurfactants, owing to their large variety can be a remedy to this problem. The sources of biosurfactants are very cheap, because, generally, they are produced from industrial waste, food waste, and cheap raw materials. Due to industrial application, medicinal importance, and economic viability, biosurfactants find a niche in our daily life. Lots of researches are ongoing to explore their full potential. This chapter provides a very basic discussion on biosurfactants and recent developments in this field.

Keywords

Biosurfactants · Antibacterial activity · Microorganisms · Glycolipids · Fatty acids

D. Mukherjee · B. Rooj · U. Mandal (✉)
Department of Chemistry, The University of Burdwan, Bardhaman, West Bengal, India
e-mail: umandal@chem.buruniv.ac.in

Inamuddin et al. (eds.), *Microbial Biosurfactants*, Environmental and Microbial Biotechnology, https://doi.org/10.1007/978-981-15-6607-3_13

13.1 Introduction

The history of biosurfactants (BS) is almost 70 years old. It started in 1949 when Jarvis and Johnson (1949), for the first time, showed that crystalline glycolipids isolated from *Pseudomonas aeruginosa* possess antibiotic activity against tuberculosis in mice. Back in 1949 there was very little knowledge about biosurfactants and their antibacterial activity. In 1968 Arima et al. (1968), from Tokyo University, Japan, first purified and characterized "surfactin." They were able to isolate it as white needle crystals and named it "surfactin." In 1971 another research group from Japan filed a US patent for a new antibiotic named macromomycin obtained from a new strain of Streptomyces designated as *Streptomyces macronomyceticus* (Umezawa et al. 1971). The antibiotic macromomycin was nothing but a biosurfactant. Many other groups later reported on different biosurfactants obtained from various microorganisms and among them some recent works are referenced here (Felix et al. 2019; Gaur et al. 2019; Mohd Hafez Mohd et al. 2020; Sood et al. 2020; Ashitha et al. 2020; Ohadi et al. 2020; Waghmode et al. 2020; Hussain et al. 2020; Cheffi et al. 2020; Meena et al. 2020; Ceresa et al. 2020; Lima et al. 2020). Depending on the nature of sources, surfactants can be classified into two categories (1). One is chemical surfactant and the other is biosurfactant (BS). Petroleum resources, mainly, are used for the production of chemical surfactants. Biosurfactants are naturally occurring, and different types of living cells are used to produce them. Our increased concern for the environment compelled us to use nature-derived products wherever possible and for that reason nowadays people are more inclined to use BS compared to chemical surfactants. The main reasons for preferring BS are as follows: reduced toxicity, higher biodegradability, environmental friendliness (Giri et al. 2019). They are often considered as alternatives to synthetically produced surfactants not only due to their physicochemical properties but also because of the trend towards renewable resources and environmentally friendly compounds. In addition, BS can function well under very adverse conditions like high salinity, elevated temperatures, and wide pH ranges, where chemical surfactants fail to work (Bezza and Chirwa 2015). Biosurfactants are very benign to the nature and work very well at extreme conditions. In the long run they may replace synthetic surfactants (Kaur et al. 2017).

Hydrocarbons (C_xH_y) are a rich source of carbon and energy. Microorganisms use hydrocarbons as their food source (Soussi et al. 2019). It is necessary to carry these hydrocarbons inside cells. Microorganisms produce various types of compounds to solubilize them for transport inside cells. These compounds are known as biosurfactants. There are also some other types of microorganisms which can modify the cell wall by producing nonionic surfactants enabling the passage of the hydrocarbons or any other water-insoluble substrates. Sometimes biosurfactants are classified into two categories: low molecular weight and high molecular weight (Ron and Rosenberg 2001). Biosurfactants that show their action through dissolution are known as emulsifiers. Generally high molecular weight BS fall in this class. Low molecular weight biosurfactants show their activity by modulating the surface tension, interfacial tension, and other surface properties.

Biosurfactants may also be divided into different classes depending on the chemical structures of their hydrophilic and hydrophobic groups. These classes are broadly glycolipids, lipopeptides, phospholipids, fatty acids, and polymeric structures (Desai and Banat 1997). Glycolipids, lipopeptides, and phospholipids are examples of low molecular weight biosurfactants. For glycolipids the carbohydrate part is often formed by monosaccharides like rhamnose or disaccharides, such as sophorose or trehalose (Zajic and Seffens 1983). Biosurfactants may be derived from a broad range of bacterial and fungal species. But there are some typical producers of distinct surfactant groups. For instance, many rhamnolipids are formed by Pseudomonads, while sophorolipids can often be found in yeasts. On the other hand, Bacillus species typically act as producers of many lipopeptides and lipoproteins (Desai and Banat 1997). Different types of microorganisms produce a variety of biosurfactants and as a result they show different physiological roles. These surfactants have huge potential for its antibacterial, antifungal, and cosmetic applications. Biosurfactants show their activity mainly through changing surface tension and or interfacial tension. The antiadhesive property of biosurfactants is another reason for which they show antibacterial activity as microorganisms find it very difficult to stick to the surfaces and tissues (Abdelli et al. 2019). Strong detergency effect in general can explain the antibacterial property. Some other mechanisms are interaction with membrane phospholipids or the alteration of the electrical conductance of membranes through which antibacterial properties are shown. Biosurfactants are used for the extraction of heavy metals and find applications in the production of antimicrobial and antibiofilm compounds (Sharma and Saharan 2016). They are also used as natural food preservatives. According to WHO, one death out of ten is due to contaminated food. Food-borne diseases are responsible for this. People nowadays prefer natural additives over artificial ones. Rhamnolipids show antibacterial activity and have been used to preserve foods. From a long list of microorganisms and their different strains we can get even a longer list of biosurfactants. Hence it is necessary to find some way to verify whether a biosurfactant possesses antibacterial property or not. Minimum inhibitory concentration (MIC) is the lowest amount required to inhibit the growth of a microorganism after 24 h. Low value of MIC indicates better antibacterial property. The MIC test for a biosurfactant is performed against the microorganisms we are interested in (Abdelli et al. 2019). Microorganisms which can diminish the surface tension by 40 mN/m are considered as good biosurfactant producers and this fact is used to find antibacterial biosurfactants (Shete et al. 2006; De Jesus et al. 2013; Gaur et al. 2019).

Major functions played by biosurfactants include surface property modification (Rosenberg and Ron 1999). Through solubilization, biosurfactants increase the bioavailability of hydrophobic substrates. Hydrophilic and hydrophobic groups of biosurfactants help them to aggregate at interface of fluids with different polarity and decrease their interfacial tension (Banat 1995; Karanth et al. 1999). All these properties make biosurfactants potential antibacterial agents.

13.2 Glycolipids

Among all the biosurfactants glycolipids are the most important and most studied. These exist as a combination of one or more carbohydrates with one or more fatty acids. Here hydroxy fatty acids are connected by ester or ether group. Here the lipid backbones are connected with sugar having α or β configuration. Though there are different types of glycolipids, the most important glycolipids are sophorolipids, rhamnolipids, and trehalose lipids (Rikalovic et al. 2015).

13.2.1 Rhamnolipids

Rhamnolipids are surface-active glycolipids. They are a crystalline acid. They normally originate from *Pseudomonas aeruginosa* (Abdel-Mawgoud et al. 2010; Shekhar et al. 2015). Rhamnolipids are divided into mono- and dirhamnolipids (Figs. 13.1 and 13.2). There are many pseudomonas species reported which produce rhamnolipids. Also there are some bacteria which produce rhamnolipids. In the structure of rhamnolipids there are two parts, one part contains one rhamnose molecule and other part contains two rhamnose molecules. Another part is the aglycon part which contains β-hydroxy fatty acid chain. The fatty acids are linked through ester bond formation. In these glycolipids, rhamnose molecules are connected with β-hydroxy carboxylic acid. But β-hydroxy decanoic acid is the most studied one. The first rhamnolipid contained two molecules of rhamnose and two molecules of β-hydroxy decanoic acid (Fig. 13.3) (Edwards and Hayashi 1965). These types of lipids are found to be an excellent emulsifying agent (Hisatsuka et al. 1971). In the past decades vast research has been done on rhamnolipids and they've been found to have many applications in different fields. Among the biosurfactants,

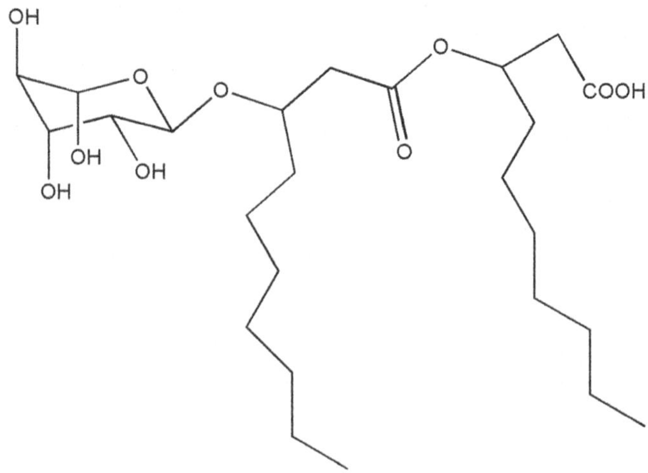

Fig. 13.1 Structure of mono rhamnolipids

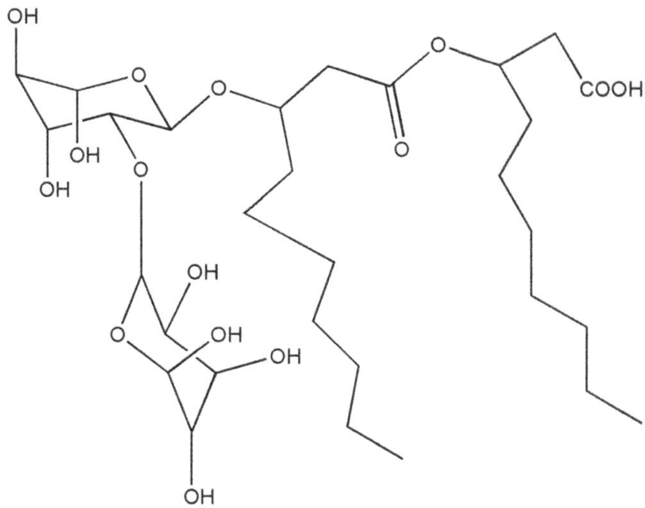

Fig. 13.2 Structure of dirhamnolipids

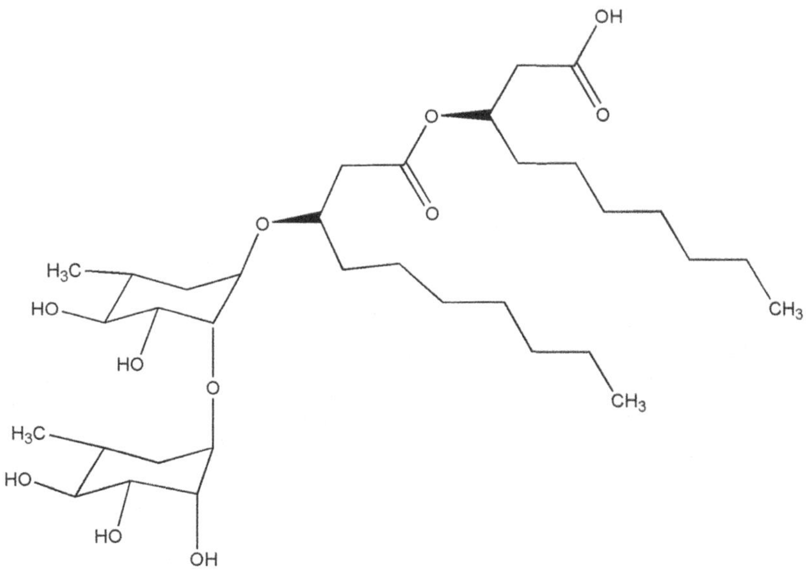

Fig. 13.3 Chemical structure of the first identified rhamnolipids

rhamnolipids have some noticeable activity, which finds a broad range of industrial applications. The major applications of rhamnolipids are bioremediation, pharmaceuticals, therapeutics, cosmetics, detergents, cleaners, agriculture, antimicrobial, and many more areas. Hisatsuka et al. (1971) measured the surfactant properties of the firstly characterized lipid and got its surface tension value

40 dyn/cm. It was found that the lipid forms very stable emulsion more efficiently than the two commercial surfactants Noigen EA 141 and Tween 20.

13.2.2 Sophorolipids

Sophorolipids, important class of biosurfactants, fall under the category of glycolipids. *Torulopsis magnoliae* is the source from where Gorin et al. in 1961 first isolated sophorolipids (Gorin et al. 1961). They are produced by several microorganisms, but among these microorganisms *Candida bombicola* ATCC 2214 is the most significant (Shah et al. 2007). They have both polar and nonpolar characteristics and are produced in 40 different types and associated isomers (Lang and Philp 1998). Here a disaccharide sophorose is linked to the hydroxyl function of a carboxylic acid chain through a glycosidic linkage (Li et al. 2015). Here one fatty acid is bound through one or two acetate group(s) with the disaccharide moiety but at that time the site of attachment was not determined. Later a modified sophorolipid was isolated by Jones (1967) and Tulloch et al. (1967) in which both the hydroxyl and carboxylic acid groups are attached to the disaccharide and the sites of attachment were determined (Copper et al. 1980). Addition of a secondary substrate can influence their structure and yield (Tulloch et al. 1967; Jones 1967). Here we draw the lactone configuration of a sophorolipid (Fig. 13.4).

13.2.3 Trehalose Lipids

Trehalose lipids are nonreducing disaccharides. 6,6′–dimycolate is the most reported trehalose lipid. Suzuki et al. in 1969 isolated trehalose lipids from *Arthrobacter Paraffineus* KY4303 grown on n-paraffins (Martin et al. 1991). Each lipid molecule contains trehalose and mycolic acid. Mycolic acid is β-hydroxy α-branched fatty acid. Arthrobacter, Nocardia, Rhodococcus, Gordonia are the mycolate group microorganisms that are responsible for the production of different trehalose-containing glycolipids (Barenholz and Thompson 1980). Glycolipids containing

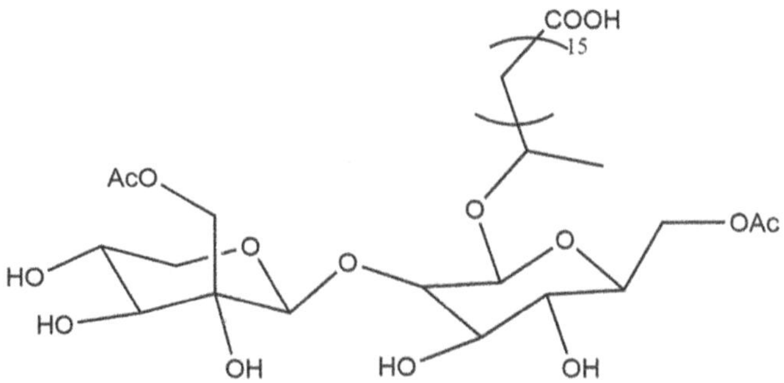

Fig. 13.4 Structure of sophorolipids in lactone configuration

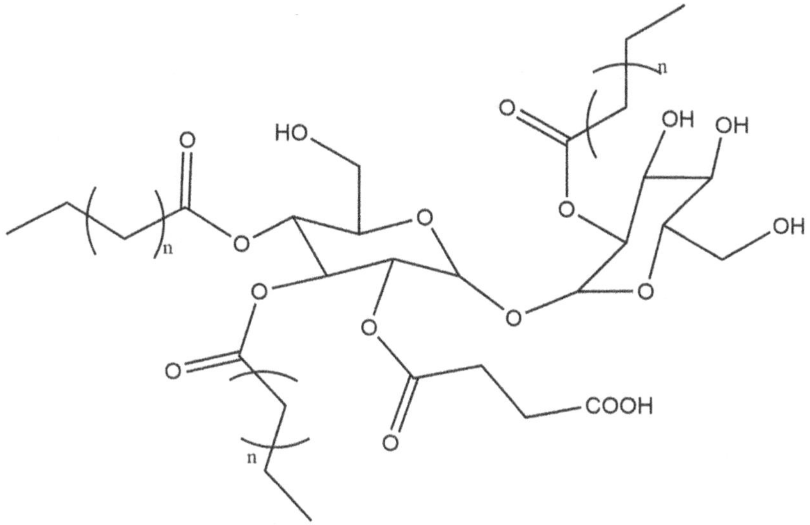

Fig. 13.5 Structure of trehalolipids $n = 6$–11

trehalose lipids have also been isolated from different sources (Rapp et al. 1979). The most studied trehalose lipids are cord factors (Das et al. 1969). Mycolic acid has the general formula R^1-CH(OH)-CHR^2-COOH. Here R^1 vary from 18–23 carbons and R^2 from 7 to 12 carbon longs. Trehalose lipids from various organisms vary in size and structure for mycolic acid (Asselineau and Asselineau 1978). Here we see a general trehalose lipid structure in Fig. 13.5.

13.3 Lipopeptides

Lipopeptides are the most exoteric biosurfactants. These are the amino acid-containing biosurfactants. In 1949 the first lipopeptide polymyxin A was isolated from *Bacillus polymyxa* (Jones 1949). But lipopeptides produced from bacillus subtilis is the most useful biosurfactants reported. It has diverse applications in different fields, like pharmaceutical industry, food industry, biotech industry, etc. Lipopeptides, isolated from different *Bacillus* strains, have excellent antibacterial activity (Nishikiori et al. 1986; Peypoux et al. 1984; Grangemard et al. 1999). They have diverse structure due to variation of length, configuration, number, and composition of lipids and amino acids (Tally et al. 1999). Lipopeptides obtained from *Bacillus* are divided into three types surfactin, fengycin, and iturin (Raaijmakers et al. 2010).

13.4 Phospholipids

From the term phospholipid it is clear that the lipid contains phosphorus. Phospholipids are amphiphilic molecules. They have an ionic part and a nonionic part in their structure. On the basis of alcohol contained they are categorized into two types: glycerophospholipids and sphingomyelins. In glycerophospholipids the backbone is glycerol. The chemical structure of different glycerophospholipids depends upon different factors, like head group, length of hydrophobic side chain, number of aliphatic chain, nature of linkage between the aliphatic moiety, and glycerol backbone. Phosphatidylcholine, phosphatidylserine, phosphatidic acid, etc., are the different glycerophospholipids which are different due to their different head groups. Dioleoyl and distearoyl PC are different glycerophospholipids which are different due to saturation of aliphatic groups (Lecithin 1995).

The next type of phospholipids are sphingomyelins. Johann Ludwig Wilhelm Thudichum in 1884 first reported sphingomyelins. Here a phosphatidylcholine is attached to an acyl chain of fatty acids.

All sphingomyelins obtained from natural sources have a D-erythro configuration (Paltauf and Hermetter 1990). Here the backbone is sphingosine. Again on the basis of sources, phospholipids are of two types, natural and synthetic. Natural phospholipids are generally found in vegetable oils and animal tissues. Soybean, sunflower, and corns are the main sources of vegetable oil. Egg yolk and bovine brain are the main sources for animal tissues (Eibl 1980). Naturally occurring phospholipids are cheap (Shapiro 1962). Another type of phospholipids are synthetic phospholipids. To get proper structure and configuration, chemists try to synthesize phospholipids in the laboratory (Garedew et al. 2004). There are two types of synthetic methods, semisynthesis and total synthesis. Head group, tail group, or both are changed in the semisynthesis method. Formation of ether or ester bond to glycerol backbone falls under the total synthesis method. Here we present a general sphingomyelins phospholipid structure (Fig. 13.6).

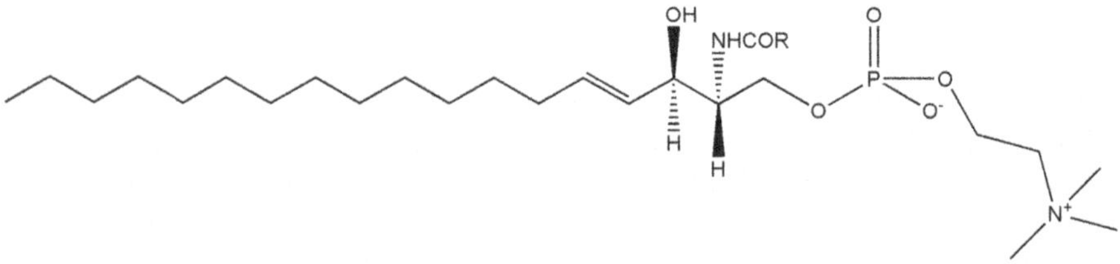

Fig. 13.6 Structure of general sphingomyelins phospholipids

13.5 Antibacterial Activity

Recently, microbial infection has become a severe clinical threat. There are many bacteria which are very harmful in our daily life. They attack our foods and spread diseases. Some examples are,

1. *Salmonella* is one type of bacteria which can taint any type of food, but mostly affects eggs, tomatoes, salad, greens, and processed foods.
2. *E. coli* generally affect ground beef, raw juice, and milk. They also taint fruit and vegetables.
3. Botulism bacteria are known to damage canned foods.
4. Hepatitis A is causes by improper food handling.

There are many more bacteria beside these which can trouble our daily life as they cater various types of diseases. Researchers have adopted different methods to study antibacterial activity. But here we will discuss some main methods. The methods are used to check whether the testing samples are antibacterial active or not and to determine the MIC value. A high-throughput system was proposed by Pitner et al. (2000). Diffusion is a very important and old method to study antimicrobial activity. There are many types of diffusion methods; one of them is agar disk diffusion. This method was developed in 1940 (Balouiri et al. 2016). This is a well-known method for routine antibacterial activity test. In this process agar plates are vaccinated with the tested microorganisms. Then a tested microorganism containing filter paper disc with known concentration is placed on the agar surface. Generally diffusion of antimicrobial agent onto the agar surface takes place and suppresses the growth of microorganisms, and the diameters of reduced growth zones are measured. But this method is not properly applicable to measure MIC. But an imprecise MIC can be calculated for some microorganisms (Nijs et al. 2003). Although not very accurate, this process has many benefits, like cheap and capable to test huge number of samples. The second diffusion method is antimicrobial gradient method. This method works based on the combined principle of dilution and diffusion methods. For MIC determination of antibiotics, this method has been used. It is a simple method, but costly if many samples are tested (Goodall and Levi 1946). Agar plug, agar well, poisoned food, cross streak, etc., are few other diffusion methods. Besides this diffusion method, another testing method is Thin Layer Chromatography (TLC) Bioautography method. Goodall and Levi (1946) combined paper chromatography with bioautography to detect different penicillins (Goodall and Levi 1946). Later Fisher and Lautner introduced TLC in the same field (Fischer and Lautner 1961). Generally three bioautography techniques are used, i.e., agar diffusion, direct bioautography, and agar overlay assay. In the agar diffusion process at first the tested microorganisms are vaccinated to the agar plate. Then the transfer of antimicrobial agent through diffusion process to the agar plate takes place (Choma and Grzelak 2011). Direct bioautography method is better among the three. In this method in a microbial suspension the TLC plate is dipped. Tetrazolium salts are applied for viewing the microbial growth. These salts are switched to the respective intensely colored formazan (Marston 2011). The next TLC method, agar overlay bioassay, is a hybrid method of the agar diffusion and direct bioautography methods. This method is used for the separation of complex mixture; it is straightforward and cheap (Al-Bakri and Afifi 2007; Liang et al. 2012; Monteiro et al. 2012; Kuhn et al. 2003). In this method to cover TLC plate it has been treated with a molten agar medium. Here the plates are kept at reasonable temperature before incubation. After incubation, a mark is made with tetrazolium dye. The next method is the dilution method, which is the best one for the determination of MIC. To check antimicrobial activity, broth dilution method is a good method among the dilution methods. Here we can monitor antibacterial activity more accurately. Among many instructions available for antibacterial activity test the most standard and uniform guidelines are given by CLSI. There are two types of broth dilution: macro and micro. Some drawbacks of this method are time consuming, manual handling, and risk of errors during sample preparation for each test. For detecting MIC end point in the analysis of antibacterial activity, some dye reagents have been used. They act as growth indicators (Meletiadis et al. 2001; Gomez-Lopez et al. 2005). Inoculum size, growth medium, incubation time, etc., can influence MIC, so this method has been standardized by CLSI (Rodriguez-Tudela et al. 2003). One more dilution method is agar dilution method. This method is somewhat rapid and simple, and no sophisticated equipment is required. Here known concentration of test material is fused into the agar end and bacteria are put into the surface. Many sets can be made by varying the concentration of the test substances. This method have several limitations, like some fungi may be very slow growing, separation from agar by hydrophobic extracts, etc. But the most problematic work is incorporating stably essential oil in aqueous situation. To overcome this problem the most common method is the use of surfactants (Hammer et al. 1999).

13.6 Polymeric Surfactants

Extracellular polymeric substances (EPS) are secreted from some bacteria when they stick to a surface, where the bacteria develop and cover the surface like a film, often called as biofilm. Polymeric substances generally contain polysaccharides, DNA, proteins, and lipids (Falk 2019). Extracellular polymeric substances are biopolymers formed from different genera of bacteria. There are three major paths by which polymeric biosurfactants are produced. These are (1) secretion by microorganisms when they interacts with the environment, (2) formation of compounds due to substrate metabolism, and (3) during the lysis of microorganisms. EPS comes with a variety of chemical structures and as a result they show various unique properties. EPS find wide application in wound healing, food preservation, bio remediation of heavy metals, and hence waste water treatment, oil cleaning, etc. *Acinetobacter* strains are most well known to produce different types of polymeric surfactants. A few examples of the polymeric surfactants along with the key microorganisms that produce them are Emulsan: *Acinetobacter calcoaceticus*, Yasan: *Yarrowia lipolytica*, Alasan: *Acinetobacter radioresistens*, Biodispersan: *Acinetobacter calcoaceticus* RAG-1, Liposan: *Acinetobacter radioresistens*, and many more (Fenibo et al. 2019). In a current article Dhagat and Jujjavarapu (2020) reported the synthesis of a bioemulsifier (emulsan) capable of working at high temperatures. The synthesis of emulsan in *Acinetobacter calcoaceticus* was started with adding glucose as the carbon source. Glucose is taken into the bacterial cell membrane and through a series of enzyme-catalyzed reactions, UDP-*N*-acetylglucosamine is produced (Dhagat and Jujjavarapu (2020).

Polymeric surfactants find wide application in the petroleum industry. They reduce the viscosity of oil products and facilitate their pipeline transport (Zhao et al. 2016). Polymeric surfactants are well known for their antibacterial and antifungal properties (Hayder 2015). Bio emulsifier produced from *Acinetobacter* sp. BE-254 has the potential to be used as a cleaning agent (Kim et al. 1996) as it forms stable emulsions with different hydrocarbons, organic solvents, and waste oils.

13.7 Fatty Acids

Fatty acids are organic carboxylic acids having long carbon chain which may have double bonds and which are found in almost every living creature on earth. They may be in two isomeric forms (with respect to double bonds) one is cis and another one is trans. In nature most of them are of cis form and generally trans form is rare. Fatty acids exist in the biosystems mostly in bounded form of glycerol, fats, or lipids. Animal origin generally consists of a large extent of saturated fatty acids, whereas those originating from plants are mostly unsaturated fatty acids. Fatty acids have a carboxylic head with a long hydrocarbon chain, while in biological systems these are made of 10 to 28 even-numbered carbon atoms. All fatty acids have one carboxylic end (hydrophilic end) and a methyl group end (hydrophobic end). On the basis of carbon atoms fatty acids are of two types. Fatty acid having carbon atoms less than 6 are fatty acids of small chain and carbon atoms greater than 18 are called fatty acids of longer chain. In most of the cases free fatty acids do not occur because of their chemical affinity to react with different protein molecules and with others like glycerol, sugars, or phosphate head groups resulting in the formation of lipids, a biological potent compound, in chloroplast, mitochondria. Lipid is a vital component of almost all cell structure. Cell membranes, which are constituted by phospholipids are important sources of molecular signaling, composed of fatty acids, and also act as energy stores. These biomolecules, with fatty acid as a major component, can function as intracellular referees or as extracellular indicators, and are important in interspecies connection and plant defense. As mentioned earlier, free fatty acids have higher affinity towards other biomolecules, but release of fatty acids are possible from membrane by the activity of enzymes which are highly specific towards particular fatty acids. Free fatty acids have potent applications in the biological field. Lipase is such a lipolytic enzyme that can break fatty acid attached to lipids and this enzymatic activity depends on various factors like the position of the head group, carbon chain length, and also in the position of unsaturation.

13.7.1 Bio-Sources of Fatty Acids

Plants and animals are sources of different free fatty acids. The seeds and fruits of many plants contain fatty acids in the form of lipids acting as storage of energy for the utilization of various physiological activities. There are lots of saturated fatty acids starting from the smallest one named butyric acid with the molecular formula $CH_3(CH_2)_2COOH$ to the largest one named cerotic acid with the molecular formula $CH_3(CH_2)_{24}COOH$. Some major sources of butyric acid are different plant oils, butter, and animal fat. Cerotic acid is a long chain fatty acid having molecular formula $C_{26}H_{52}O_2$ mostly found in beeswax. The other examples are caproic acid, caprylic acid, capric acid, lauric acid, and myristic acid. Palmitic acid having 16 carbon atoms is the most abundant fatty acids found in plant leaf lipids majorly though it can be seen in seed of plant to some extent. Another fatty acid, stearic acid (saturated fatty acid containing 18 carbon atoms) is abundant in seed fats. On the other hand, fatty acids with unsaturation contain C=C double bonds in the hydrocarbon chain which gives rise to two isomeric form, cis and trans. The degree of unsaturation generally varies between one and two. With respect to the abundance of palmitic, linoleic are major fatty acids and, in particular, α-linolenic acids. The tri-unsaturated linolenic acid is get-at-able. The membranes of chloroplasts contain exceptionally high (about 90% in some lamellae) percentages of α-linolenic acid. Trans fatty acids are not obtained in nature.

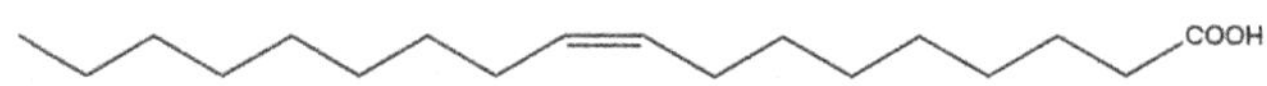

Oieicacid (cis)

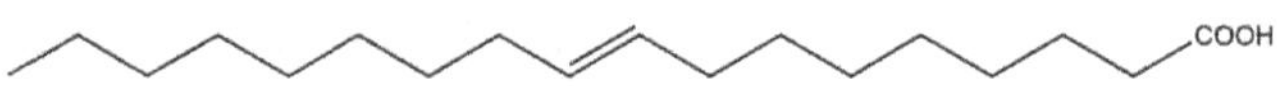

Elaidicacid (trans)

The major source of trans fatty acids are meat and milk of livestock. These are produced inside their body through fermentation process in the presence of different microorganisms. This is also available in milk derivatives. Some other examples of fatty acids with monounsaturation are myristoleic acid, palmitoleic acid, sapienic acid, oleic acid, elaidic acid, vaccenic acid. Fatty acids with double unsaturation are linoleic acid, linoelaidic acid. Fatty acids with tri-unsaturation are α-linolenic acid, etc. Tetra-, penta-, hexa-unsaturated fatty acids are also available.

13.7.2 Role of Fatty Acids as Antimicrobials

The antibacterial role of free fatty acids is well established since several years. Fatty acids are occasionally added as antibacterial food additives to cease the growth of unhygienic microorganisms. The antibacterial activity of fatty acids is equally comparable with antimicrobial polypeptides in vitro. Cerdeiras et al. identified 11-O-(6′-O-acetyl-β-D-glucopyranosyl)-stearic acid as the main antibacterial component of aerial parts of Ibicellalutea. This fatty acid derivative exhibits antibacterial activity towards various Gram-positive and Gram-negative bacteria. Again fatty acids can have a role in different systems of multicellular organisms like mollusks, mammalians, plants, seaweeds, and amphibians. Fatty acids can perform a major role in human defense systems among which lauric acid, myristic acid, palmitic acid, sapienic acid, and cis-8-octadecenoic acid are the most abundant in the skin and mucosal surfaces. They are good enough to stop bacterial activity on the skin. The most effective fatty acids found in human skin are long chain monounsaturated fatty acid exudates. The lack of these and other fatty acids make the skin more sensitive to colonization by the pathogen, *Staphylococcus aureus*. Fatty acids that are available on the epidermis and dermis layer of the skin always developed environment which retard the process of growth of certain bacteria by controlling the pH of the surface within acidic range. It has also the capability to kill the bacteria and inhibit the activity of bacteria. Most important is that free fatty acids can disrupt cell-to-cell signaling, which is mandatory for bacteria to lead to infection which is also known as bacterial virulence factors. The saturated and unsaturated fatty acids can also stop bacterial adhesion and biofilm formation. Saturated fatty acids having carbon number greater than five can stop the swarming behavior of the urinary tract pathogens, proteus mirabilis. Some fatty acids can also down the activity of certain toxins, enzymes, and hemolysins that can suppress the activity of drugs. Fatty acids can be produced from coconut oil by the enzymatic activity of *Candida rugosalipase* and are used against Gram-negative and Gram-positive bacteria. Jae-Suk Choi et al. (2013) studied the antibacterial efficacy of various saturated and unsaturated fatty acids against various oral pathogens responsible for dental caries, stomatitis, gingivitis, and periodontitis. They conclude that the ε-3 and the ε-6 poly unsaturated fatty acids are strong antimicrobial against *Porphyromona gingivalis* KCTC 381.

13.7.3 Structural Effect on the Antibacterial Activity of Fatty Acids

There is always an inherent relationship between the structure and antibacterial activity of carboxylic acids with a long carbon chain. Fatty acids having less carbon atoms (C6 or less) are effective toward Gram-negative bacteria at higher concentrations and this antibacterial activity is influenced by the pH of the medium (Knapp and Melly 1986; Bergsson et al. 1998). Fatty acids having greater than C8 cannot affect the Gram-negative bacteria. Whereas long chain fatty acids are susceptible towards Gram-positive bacteria (Galbraith et al. 1971) at low concentrations and their response is independent of pH. For fatty acids having greater than 12 carbon atoms the number and location of double bonds are more important than fatty acids having less than 12 carbon atoms. The most effective saturated, mono-, and polyunsaturated fatty acids are C12, C(16:1), C(18:2) (Benkendorff et al. 2005). Another aspect is that Gram-positive bacteria get more influenced by the fatty acids than

Gram-negative bacteria. Fatty acids exert a greater effect on yeast with 10–12 carbon atoms (Cerantola et al. 2007).

13.7.4 Mechanism

The scenario of how fatty acids act against microorganisms is not transparent. The fatty acids mainly target the cell membrane and stop some of the vital physiological activities that happen inside the bacterial cell. Fatty acids have a hydrophobic tail (nonpolar part) and a hydrophilic head (polar part). This property makes them very susceptible to microorganisms. It enhances the capability of fatty acids to stick to the cell membrane of microorganisms and creates porous cell membrane. This hydrophobic interaction leads to solubilization of the cell membrane and releases lots of proteins and lipid molecules. Other ways that can lead to the death or inhibit the growth of microorganisms are the following:

1. Cell lysis.
2. Suppression of enzyme activity.
3. Deprivation of nutrient uptake.
4. Production of toxic peroxidation and auto oxidation products,
5. Disruption of electron transport chain.
6. Interruption of oxidative phosphorylation.

13.7.5 Cell Lysis

When a cell loses its integrity by breakdown of cell membrane, it is known as cell lysis. Due to the favorable structure of free unsaturated fatty acids they can easily be encapsulated inside the cell membrane of microorganisms, leading to a lot of porosity on the cell membrane of different sizes resulting in greater cell membrane permeability and increased fluidity of the cell (Wang et al. 1992). This placing of unsaturated medium and long chain fatty acids results in the leakage of cellular granules to outside of the cell. This can inhibit the growth and sometimes lysis of the cell (Galbraith and Miller 1973).

13.7.6 Suppression of Enzyme Activity

Free fatty acids are good inhibitors of various enzymes. Fatty acids with unsaturation can exhibit greater enzyme inhibitory effect than those without unsaturation. Fatty acids can inhibit the activity of enzymes that are vital for membrane formation, cell division, and also for molecular signaling. The incorporated fatty acids can affect the synthesis of fatty acids by the microorganisms itself and thus alter the composition of the cell membrane, which alters the fluidity and permeability of the cell membrane that results in cell death (Sado Kamdem et al. 2009).

13.7.7 Deprivation of Nutrient Consumption by Cell

The process of taking nutrients like amino acids by the cell of bacteria can be interrupted by saturated and unsaturated fatty acids (Galbraith and Miller 1973). Therefore the bacteria starve for nutrients to survive. But the exact mechanism of this inhibition process by fatty acids is not clear enough. It may be that the fatty acids directly bind to the transporter proteins that are attached to the cell membrane or fall short of the energy that is essential for active transport.

13.7.8 Peroxidation and Autooxidation

Another concept for the antibacterial action of fatty acids is the peroxidation and autooxidation of fatty acids. The physiological activity of bacteria gets hampered by hydrogen peroxide and oxygen species that are formed during peroxidation of fatty acids (Schonfeld and Wojtczak 2008; Wang et al. 1992; Knapp and Melly 1986; Hazell and Graham 1990). On the other hand, unsaturated fatty acids undergo self-oxidation that generates oxylipins and aldehydes of low molecular weight that have inherent antibacterial properties. The exact mechanism by which the fatty acid shows its antibacterial property by either stopping the bacterial growth or causing death will depend on the structure and concentration of the fatty acid, nature of the bacteria, binding sides, pH, and temperature. There may be involvement of multiple mechanisms depending on various factors (Galbraith and Miller 1973; Kabara and Varble 1977; Coulondre and Miller 1977; Greenway and Dyke 1979; Wang et al. 1992; Sun et al. 2004).

13.7.9 Disruption of Electron Transport Chain

Electron transport chain is a system of complexes that transport electron from electron donors to electron acceptors via redox reactions and also attached with transport of proton (H+) and ultimately produces ATP which is essential to drive all the physiological process for bacteria. In the case of Gram-positive and Gram-negative bacteria, the electron transport chains are situated on the inner side of the cell membrane and thus energy production takes place there. The various electron transporters residing in the cell membrane supply electrons from the carrier to the receptor as far as two electrons are attached to the receptor, resulting in the production of oxygen and two protons which finally gets converted into water. The proton produced inside the cell is exported to the outside which creates a higher gathering of electrons inside the cell. It develops an electrochemical gradient of proton and finally a potential difference across the cell membrane. ATP synthase produce ATP under this cell membrane potential (Mitchell 1961). Now the medium and long chain saturated and unsaturated fatty acids which are attached to the cell membrane can disrupt this process by binding with them directly or displacing them from the cell membrane entirely (Galbraith and Miller 1973). The saturated and unsaturated fatty acids can attach with the electron transport chain directly but only unsaturated fatty acids can displace them from the cell membrane due to their better capability to increase membrane fluidity (Greenway and Dyke 1979). The reason behind this greater activity of the cis isomer of fatty acid with unsaturation is its curved structure; it cannot bind to cell membrane tightly. Thus incorporation of medium and fatty acids with double bonds and a greater number of carbon atoms imposes defects in cell membrane structure that increases the fluidity of the cell membrane (Stulnig et al. 2001). On the other hand, trans isomer of medium and long chain fatty acids due to their linear structure bind to the cell membrane tightly and reduce membrane fluidity and disrupt the electron transport chain within the cell membrane (Sheu and Freese 1971). In all aspects, the process of electron transfer by the electron transport chain is impaired and thus the production of ATP gets reduced. These causes death of bacteria cells.

13.7.10 Interruption of Oxidative Phosphorylation

The energy that is produced within the cell is utilized by ATP synthase (Sheu and Freese 1971; Greenway and Dyke 1979; Beck et al. 2007) to convert ADP to ATP; this is known as oxidative phosphorylation, which takes place within the cytosol of bacteria cells. Free fatty acids interrupt this process in many ways. One of which is saturated or unsaturated fatty acid can directly bind to the ATP synthase itself and diminish its activity. In other ways fatty acids can alter the electrochemical gradient across the cell membrane, which changes the driving force for proton transportation and the activity of ATP synthase. For these reasons energy produced by the different electron transporters results heating effect rather than be used by ATP synthase which converts it into useful form of energy.

13.8 Conclusion

Biosurfactants have a wide range of applications such as in food, cosmetics, pharmaceuticals, oil industry, and other different fields. They are produced from microorganisms. Different strains of microorganisms are responsible for producing different types of biosurfactants. Here in this chapter we have discussed various types of biosurfactants and their antibacterial activity. Biosurfactants are mostly biodegradable, nontoxic, and very efficient antibacterial agents. So people are more inclined to use them. The sources for biosurfactants are cheaper than chemical surfactants. Several biosurfactants have been reported so far but there are lots more to see. Research in the field of biosurfactants will be highly rewarded in the near future.

Acknowledgments This work is supported by DST-SERB research grant (File no. EEQ/2018/000964). DM thanks DST-SERB for his fellowship. The authors would like to thank the Department of Chemistry, The University of Burdwan.

References

Abdelli F, Jardak M, Elloumi J (2019) Antibacterial, anti-adherent and cytotoxic activities of surfactin(s) from a lipolytic strain *Bacillus safensis* F4. Biodegradation 30:287–300

Abdel-Mawgoud AM, Lepine F, Deziel E (2010) Rhamnolipids: diversity of structures, microbial origins and roles. Appl Microbiol Biotechnol 86:1323–1336

Al-Bakri AG, Afifi FU (2007) Evaluation of antimicrobial activity of selected plant extracts by rapid XTT colorimetry and bacterial enumeration. J Microbiol Methods 68:19–25

Arima K, Kakinuma A, Tamura G (1968) Surfactin, a crystalline peptide lipid surfactant produced by *Bacillus subtilis*: isolation, characterization and its inhibition of fibrin clot formation. Biochem Biophys Res Commun 31:488–494

Ashitha A, Radhakrishnan EK, Mathew J (2020) Characterization of biosurfactant produced by the endophyte Burkholderia sp. WYAT7 and evaluation of its antibacterial and antibiofilm potentials. J Biotechnol 313:1–10

Asselineau C, Asselineau J (1978) Trehalose containing glycolipids. Prog Chem Fats Other Lipids 16:59–99

Balouiri M, Sadiki M, Ibnsouda SK (2016) Methods for *in vitro* evaluating antimicrobial activity: a review. J Pharmaceut Anal 6:71–79

Banat IM (1995) Biosurfactants production and possible uses in microbial enhanced oil recovery and oil pollution remediation: a review. Bioresour Technol 51:1–12

Barenholz Y, Thompson T (1980) Sphingomyelins in bilayers and biological membranes. Biochim Biophys Acta 604:129–158

Beck MR, Levin DT, Angelone BL (2007) Metacognitive errors in change detection: lab and life converge. Conscious Cogn 16:58–62

Benkendorff K, Davis AR, Rogers CN, Bremner JB (2005) Free fatty acids and sterols in the benthic spawn of aquatic molluscs and their associated antimicrobial properties. J Exp Mar Biol Ecol 316:29–44

Bergsson G, Arnfinnsson J, Karlsson SM, Steingrimsson O, Thormar H (1998) In vitro inactivation of chlamydia trachomatis by fatty acids and monoglycerides. Antimicrob Agents Chemother 42 (9):2290–2294

Bezza FA, Chirwa EMN (2015) Production and applications of lipopeptide biosurfactant for bioremediation and oil recovery by Bacillus subtilis CN2. Biochem Eng J 101:168–178

Cerantola V, Vionnet C, Aebischer OF, Jenny T, Khudsen J, Conzelmann A (2007) Yeast sphingolipids do not need to contain very long chain fatty acids. Biochem J 401:205–216

Ceresa C, Hutton S, Lajarin-Cuesta M (2020) Production of Mannosylerythritol lipids (MELs) to be used as antimicrobial agents against *S. aureus* ATCC 6538. Curr Microbiol 77(8):1373–1380. https://doi.org/10.1007/s00284-020-01927-2

Cheffi M, Hentati D, Chebbi A (2020) Isolation and characterization of a newly naphthalene-degrading *Halomonas pacifica,* strain Cnaph3: biodegradation and biosurfactant production studies. 3 Biotech 10:89. https://doi.org/10.1007/s13205-020-2085-x

Choi JS, Park NH, Hwang SY, Sohn JH, Kwak I, Cho KK, Choi IS (2013) The antibacterial activity of various saturated and unsaturated fatty acids against several oral pathogens. J Environ Biol 34 (4):673–676

Choma LM, Grzelak EM (2011) Bioautography detection in thin-layer chromatography. J Chromatogr A 1218:2684–2691

Copper TG, Lam C, Turoscy V (1980) Structural analysis of the dur loci in S. cerevisiae: two domains of a single multifunctional gene. Genetics 94(3):555–580

Coulondre C, Miller JH (1977) Genetic studies of the lac repressor. IV Mutagenic specificity in the lacI gene of Escherichia coli. J Mol Biol 117:577–606

Das BC, Gero SD, Lederer E (1969) An acetylated N-terminus of adenovirus type 2 hexon protein. Biochem Biophys Res Commun 29:211–215

De Jesus C-SA, Hernandez-Sanchez H, Jaramillo-Flores ME (2013) Biological activity of glycolipids produced by microorganisms: new trends and possible therapeutic alternatives. Microbiol Res 168(1):22–32

Desai JD, Banat IM (1997) Microbial production of surfactants and their commercial potential. Microbiol Mol Biol Rev 61:47–64

Dhagat S, Jujjavarapu SE (2020) Isolation of a novel thermophilic bacterium capable of producing high-yield bioemulsifier and its kinetic modelling aspects along with proposed metabolic pathway. Braz J Microbiol 51:135–143

Edwards JR, Hayashi JA (1965) Structure of a Rhamnolipid from Pseudomonas aeruginosa. Arch Biochem Biophys 111:415–421

Eibl H (1980) Synthesis of glycerophospholipids. Chem Phys Lipids 26(4):405–429

Falk NA (2019) Surfactants as antimicrobials: a brief overview of microbial interfacial chemistry and surfactant antimicrobial activity. J Surfactant Deterg 22:1119–1127

Felix AKN, Martins JJ, Almeida JGL, Giro MEA, Cavalcante KF, Melo VMM, de Santiago AR (2019) Purification and characterization of a biosurfactant produced by Bacillus subtilis in cashew apple juice and its application in the remediation of oil-contaminated soil. Colloids Surf B Biointerfaces 175:256–263

Fenibo EO, Douglas SI, Stanley HO (2019) A review on microbial surfactants: production, classifications, properties and characterization. J Adv Microbiol 18(3):1–22

Fischer R, Lautner H (1961) On the paper chromatographic detection of penicillin preparations. Arch Pharm 294:1–7

Galbraith H, Miller TB (1973) Physicochemical effects of long chain fatty acids on bacterial cells and their protoplasts. J Appl Bacteriol 36:647–658

Galbraith H, Miller TB, Paton AM, Thompson JK (1971) Antibacterial activity of long chain fatty acids and the reversal with calcium, magnesium, ergocalciferol and cholesterol. J Appl Bacteriol 34(4):803–813

Garedew A, Schnmolz E, Lamprecht I (2004) Modern phytomedicine: turning medical plants into drugs. Thermochim Acta 415:99–106

Gaur VK, Bajaj A, Regar RK, Kamthan M, Jha RR, Srivastava JK, Manickam N (2019) Rhamnolipid from a Lysinibacillus sphaericus strain IITR51 and its potential application for dissolution of hydrophobic pesticides. Bioresour Technol 272:19–25

Giri SS, Ryua EC, Sukumaran V, Park SC (2019) Antioxidant, antibacterial, and anti-adhesive activities of biosurfactants isolated from Bacillus strains. Microb Pathog 132:66–72

Gomez-Lopez A, Aberkane A, Petrikkou E, Mellado E, Rodriguez-Tudela JL, Cuenca-Estrella M (2005) Analysis of the influence of tween concentration, inoculum size, assay medium, and reading time on susceptibility testing of Aspergillus spp. J Clin Microbiol 43:1251–1255

Goodall RR, Levi AA (1946) A microchromatographic method for the detection and approximate determination of the different Penicillins in a mixture. Nature 158:675–676

Gorin PAJ, Spencer JFT, Tullock AP (1961) Biosurfactants, production, properties, applications. Can J Chem 39:846–855

Grangemard I, Bonmatin JM, Bernillon J, Das BC, Peypoux F, Lichenysins G (1999) A novel family of lipopeptide biosurfactants from Bacillus licheniformis IM 1307: production, isolation and structural evaluation by NMR and mass spectrometry. J Antibiot 52:363–373

Greenway DLA, Dyke KGH (1979) Mechanism of the inhibitory action of linoleic acid on the growth of *Staphylococcus aureus*. J Gen Microbiol 115:233–245

Hammer KA, Carson CF, Riley TV (1999) Influence of organic matter, cations and surfactants on the antimicrobial activity of Melaleuca alternifolia (tea tree) oil in vitro. J Appl Microbiol 86:446–452

Hayder HN (2015) Production, characterization and antimicrobial activity of a bioemulsifier produced by *Acinetobacter baumanii* AC5 utilizing edible oils. Iraq J Biotechnol 14:55–70

Hazell SL, Graham DY (1990) Unsaturated fatty acids and viability of helicobacter (campylobacter) pylori. J Clin Microbiol 28(5):1060–1061

Hisatsuka K, Nakahara T, Sano N, Yamada K (1971) Formation of Rhamnolipid by *Pseudomonas aeruginosa* and its function in hydrocarbon fermentation. Agric Biol Chem 35:686–692

Hussain T, Haris M, Shakeel A (2020) Bio-nematicidal activities by culture filtrate of *Bacillus subtilis HussainT-AMU*: new promising biosurfactant bioagent for the management of root galling caused by *Meloidogyne incognita*. Vegetos 33:229. https://doi.org/10.1007/s42535-020-00099-5

Jarvis FG, Johnson MJ (1949) A glyco-lipide produced by *Pseudomonas Aeruginosa*. J Am Chem Soc 71(12):4124–4126

Jones TS (1949) Chemical evidence for the multiplicity of the antibiotics produced by Bacillus polymyxa. Ann N Y Acad Sci 51:909–916

Jones DF (1967) Novel macrocyclic glycolipids from Torulopsis gropengiesseri. J Chem Soc C 6:479–484

Kabara JJ, Varble R (1977) Antibacterial free fatty acids: activities, mechanisms of action and biotechnological potential. Lipids 12:753–759

Karanth NGK, Deo PG, Veenanadig NK (1999) Microbial production of biosurfactants and their importance. Curr Sci 77:116–126

Kaur K, Sangwan S, Kaur H (2017) Biosurfactant production by yeasts isolated from hydrocarbon polluted environments. Environ Monit Assess 189(12):603–612

Kim SH, Lee JD, Kim BC, Lee TH (1996) Purification and characterization of bioemulsifier produced by *Acinetobacter* sp. BE-254. J Microbiol Biotechnol 6:184–188

Knapp HR, Melly MA (1986) Bactericidal effects of polyunsaturated fatty acids. J Infect Dis 154:84–94

Kuhn DM, Balkis MJ, Chandra J, Mukherjee PK, Ghannoum MA (2003) Uses and limitations of the XTT assay in studies of Candida growth and metabolism. J Clin Microbiol 41:506–508

Lang S, Philp JC (1998) Surface-active lipids in Rhodococci. Anton Leeuw Int J Gen Mol Biol 74:59–70

Lecithin WA (1995) Kirk-othmer encyclopedia of chemical technology. Wiley-Interscience, New York

Li J, Wang X, Zhang T, Wang C, Huang Z, Luo X, Deng Y (2015) A review on phospholipids and their main applications in drug delivery systems. Asian J Pharmaceut Sci 10:81–98

Liang H, Xing Y, Chen J et al (2012) Antimicrobial activities of endophytic fungi isolated from Ophiopogon japonicus (Liliaceae). BMC Complement Altern Med 12:238

Lima TA, Etchegaray A, Machini MT (2020) Design, synthesis and valued properties of surfactin oversimplified analogues. Amino Acids 52:25–33

Marston A (2011) Thin-layer chromatography with biological detection in phytochemistry. J Chromatogr A 1218:2676–2683

Martin M, Bosch P, Parra JL, Espuny MJ, Virgili A (1991) Structure and bioconversion of trehalose lipids. Carbohydr Res 220:93–100

Meena KR, Sharma A, Kanwar SS (2020) Antitumoral and antimicrobial activity of Surfactin extracted from Bacillus subtilis KLP2015. Int J Pept Res Ther 26:423–433

Meletiadis J, Meis JFGM, Mouton JW, Verweij PE (2001) Analysis of growth characteristics of filamentous fungi in different nutrient media. J Clin Microbiol 39:478–484

Mitchell P (1961) Coupling of phosphorylation to electron and hydrogen transfer by a chemi-osmotic type of mechanism. Nature 191:144–148

Mohd Hafez Mohd I, Nur Hashimah S, Najeeb KNA, Fahd AA, Mohd Sahaid K (2020) Evaluation of antibacterial potential of biosurfactant produced by surfactin-producing *Bacillus*isolated from selected Malaysian fermented foods. Food Biotechnol 34(1):1–24

Monteiro MC, de la Cruz M, Cantizani J (2012) A new approach to drug discovery: high-throughput screening of microbial natural extracts against Aspergillus fumigatus using resazurin. J Biomol Screen 17:524–529

Nijs A, Cartuyvels R, Mewis A, Peeters V, Rummens JL, Magerman K (2003) Comparison and evaluation of Osiris and Sirscan 2000 antimicrobial susceptibility Systems in the Clinical Microbiology Laboratory. J Clin Microbiol 41:3627–3630

Nishikiori T, Naganawa H, Muraoka Y, Aoyagi T, Umezawa H (1986) Plipastatins: new inhibitors of phospholipase A2, produced by Bacillus cereus BMG302-fF67. III Structural elucidation of plipastatins. J Antibiot (Tokyo) 39:755–761

Ohadi M, Forootanfar H, Dehghannoudeh G, Eslaminejad T, Ameri A, Shakibaie M, Adeli-Sardou M (2020) Antimicrobial, anti-biofilm, and anti-proliferative activities of lipopeptide biosurfactant produced by Acinetobacter junii B6. Microb Pathog 138:103806–103816

Paltauf F, Hermetter A (1990) Phospholipids natural, semisynthetic, synthetic [M]. In: Phospholipids. Springer, Berlin, pp 1–12

Peypoux F, Pommier MT, Das BC, Besson F, Delcambe L, Michel G (1984) Structures of bacillomycin D and bacillomycin L peptidolipid antibiotics from Bacillus subtilis. J Antibiot (Tokyo) 37:1600–1604

Pitner JB, Timmins MR, Kashdan M, Nagar M, Stitt DT (2000) High-throughput assay system for the discovery of antibacterial drugs. In: AAPS proceedings. BD Biosciences, San Diego, CA

Raaijmakers JM, De Bruijn I, Nybroe O, Ongena M (2010) Natural functions of lipopeptides from *Bacillus* and *Pseudomonas*: more than surfactants and antibiotics. FEMS Microbiol Rev 34:1037. https://doi.org/10.1111/j.1574-6976.2010.00221.x

Rapp P, Beck CH, Wagner F (1979) Formation of exopolysaccharides by Rhodococcus erythropolis and partial characterization of a heteropolysaccharide of high molecular weight. Eur J Appl Microbiol Biotechnol 7:67–78

Rikalovic MG, Vrvic MM, Karadzic IM (2015) Rhamnolipid biosurfactant from Pseudomonas aeruginosa – from discovery to application in contemporary technology. J Serb Chem Soc 80 (3):279–304

Rodriguez-Tudela JL, Chryssanthou E, Petrikkou E (2003) Interlaboratory evaluation of hematocytometer method of inoculum preparation for testing antifungal susceptibilities of filamentous fungi. J Clin Microbiol 41:5236–5237

Ron EZ, Rosenberg E (2001) Natural roles of biosurfactants. Environ Microbiol 3(4):229–236

Rosenberg E, Ron EZ (1999) High and low molecular mass microbial surfactants. Appl Microbiol Biotechnol 52:154–162

Sado Kamdem SL, Vannini L, Guerzoni E (2009) Effect of alpha-linolenic, capric and lauric acid on the fatty acid biosynthesis in Staphylococcus aureus. Int J Food Microbiol 129(3):288–294

Schonfeld P, Wojtczak L (2008) Fatty acids as modulators of the cellular production of reactive oxygen species. Free Radic Biol Med 45:231–241

Shah V, Badia D, Ratsep P (2007) Sophorose lipids having enhanced antibacterial activity. Antimicrob Agents Chemother 51:397–400

Shapiro D (1962) Flowers H studies on sphingolipids. VII Synthesis and configuration of natural sphingomyelins. J Am Chem Soc 84:1047–1050

Sharma D, Saharan BS (2016) Functional characterization of biomedical potential of biosurfactant produced by lactobacillus helveticus. Biotechnol Rep 11:27–35

Shekhar S, Sundaramanickam A, Balasubramanian T (2015) Biosurfactant producing microbes and their potential applications: a review. Crit Rev Environ Sci Technol 45(14):1522–1554

Shete AM, Wadhawa G, Banat IM, Chopade BA (2006) Mapping of patents on bioemulsifier and biosurfactant: a review. J Sci Ind Res 65:91–115

Sheu CW, Freese E (1971) Effects of fatty acids on growth and envelope proteins of Bacillus subtilis. J Bacteriol 111:516–524

Sood U, Singh DN, Hira P, Lee JK, Kalia VC, Lal R, Shakarad M (2020) Rapid and solitary production of mono-rhamnolipid biosurfactant and biofilm inhibiting pyocyanin by a taxonomic outlier Pseudomonas aeruginosa strain CR1. J Biotechnol 307:98–106

Soussi S, Essid R, Hardouin J (2019) Utilization of grape seed flour for antimicrobial Lipopeptide production by *Bacillus amyloliquefaciens* C5 strain. Appl Biochem Biotechnol 187:1460–1474

Stulnig TM, Huber J, Leitinger N, Imre EM, Angelisova P, Nowotny P, Waldhausl W (2001) Polyunsaturated Eicosapentaenoic acid displaces proteins from membrane rafts by altering raft lipid composition. J Biol Chem 276:37335–37340

Sun T, Xie W, Xu P (2004) Superoxide anion scavenging activity of graft chitosan derivatives. Carbohydr Polym 58:379–382

Tally FP, Zeckel M, Wasilewski MM, Carini C, Berman CL, Drusano GL (1999) Daptomycin: a novel agent for gram-positive infections. Expert Opin Investig Drugs 8:1223–1238

Tulloch AP, Hill A, Spencer JFT (1967) A new type of macrocyclic lactone from *Torulopsis apicola*. Chem Commun 7:584–586

Umezawa H, Takeuch T, Hamada M, Ishizuka M, Chimura H, Maeda K (1971) Antibiotic macromomycin and process for making same. Patents US3595954A

Waghmode S, Swami S, Sarkar D (2020) Exploring the pharmacological potentials of biosurfactant derived from *Planococcus maritimus* SAMP MCC 3013. Curr Microbiol 77:452–459

Wang RF, Cao WW, Johnson MG (1992) 16S rRNA-based probes and polymerase chain reaction method to detect Listeria monocytogenes cells added to foods. Appl Environ Microbiol 58 (9):2827–2831

Zajic JE, Seffens W (1983) Biosurfactants. CRC Crit Rev Biotechnol 1(2):87–107

Zhao YH, Chen LY, Tian ZJ, Sun Y (2016) Characterization and application of a novel bioemulsifier in crude oil degradation by *Acinetobacter beijerinckii* ZRS. J Basic Microbiol 56:184–195

Microbial Biosurfactants as Cleaning and Washing Agents

14

Sameer Nadaf, Vijay M. Kumbar, Suresh Killedar, Anand I. Torvi, Joy H. Hoskeri, and Arun K. Shettar

Abstract

Surfactants derived from microbes belong to the diverse group of surface-active metabolites which are secreted during their growth on hydrophobic substrates. The use of chemical surfactants as a detergent in different industries such as leather, petroleum, paper, dairy, cosmeceuticals, and pharmaceuticals is limited due to their hazardous effects on the aqueous and territorial ecosystem. This found the basis for the use of biosurfactants as a detergent for industrial and household applications. In recent years, the use of biosurfactants in the cleaning of storage tanks in petroleum industries, cleaning of membranes during ultrafiltration, and remediation of leather dust from the leather industry is increased. Different companies are manufacturing biosurfactant-based dish-washing agents. Some patents are also claiming the role of biosurfactants in hair and skin cosmetics. This chapter describes the chemical nature of biosurfactants, media composition required for microbial growth, genetic regulation and biosynthesis of surfactants, and the application of biosurfactants in different fields as cleansing agents.

S. Nadaf · S. Killedar
Sant Gajanan Maharaj College of Pharmacy, Mahagoan Site, Chinchewadi, Maharashtra, India

V. M. Kumbar
Central Research Laboratory, Maratha Mandal's Nathajirao G. Halgekar Institute of Dental Sciences and Research Centre, Belgavi, Karnataka, India

A. I. Torvi
Department of Chemistry, Karnatak University, Dharwad, Karnataka, India

J. H. Hoskeri
Department of Bioinformatics and Biotechnology, Karnataka State Akkamahadevi Women's University, Vijayapura, Karnataka, India

A. K. Shettar (✉)
Department of Biotechnology and Microbiology, Karnatak University, Dharwad, Karnataka, India

Inamuddin et al. (eds.), *Microbial Biosurfactants*, Environmental and Microbial Biotechnology, https://doi.org/10.1007/978-981-15-6607-3_14

Keywords

Surfactants · Microbes · Genes · Cleaning and washing agent

14.1 Introduction

For many years, surfactant molecules of amphiphilic characteristics are well recognized for decreasing the surface tension in two different phases and termed as surface-active agents. Being amphiphilic, i.e., presence of both hydrophilic head and non-polar tail, makes to adsorb in interfaces between water and oil systems or solid and gases. Surfactants are utilized for diverse applications, for example, detergent, emulsifiers, dispersant, wetting agents, cleaning agents, and foaming agents (Marchant and Banat 2012). Surfactants are a diverse group of chemical molecules, comprised of alkyl chains with 8–22 carbons which form the molecular clusters in solution and named as micelles (Nakama 2017).

Nowadays, chemical surfactants have occupied the market enormously, and this is ascribable to their high use in many of the household detergents and cleaning products (Kogawa et al. 2017). However, to improve the efficacy of chemical detergents as cleaning agents, some additives such as phosphates are used. After using these surfactants, their products are discarded as effluents into the environment mainly into the surface waters and sludge disposal on lands (Ying 2006). Excessive discharge of phosphates into the surface water system promotes the growth of algae due to their nutrient phenomenon and leads to eutrophication of waters (Kogawa et al. 2017; Yang et al. 2008). Phosphates from the discharge of surfactants form the white foam on the water surface which restricts the entry of oxygen and light in water. Different surfactants affect the environment differently. Positively charged and uncharged surfactants exhibit relatively major sorption for soil and oil sediments compared to anionic surfactants (Ying 2006). Hence, it was essential to find some alternatives to combat these circumstances. This caused the existence and exploration of biosurfactants.

Biosurfactants are amphiphilic in characters and generally refers to surfactants of microbial origin. When microbes like bacteria, fungi, and actinomycetes are cultured on media containing hydrophilic and hydrophobic substrates, they lead to the production of surfactants as extracellular products (Havasi 2011). Surfactants producing microorganisms mainly belong to *Bacillus*, *Agrobacterium*, *Streptomyces*, *Pseudomonas*, and *Thiobacillus* (Liu et al. 2015).

Their amphiphilic characters allow them to accumulate at interphase existing between two liquids or solid/gas interphase. Nowadays, microbial surfactants have gained huge popularity because of their unique characteristics, viz. specificity, less toxicity, and ease of preparation (Vijayakumar and Saravanan 2015). Due to this, biosurfactants are finding their applicability in different industries such as agriculture, pharmaceutical, cosmeceuticals, petroleum, petrochemicals, food processing, mining, metallurgy, agrochemicals, and fertilizers (Volkering et al. 1998).

14.2 Chemical Nature of Biosurfactants

Unlike chemical surfactants, biosurfactants consist of diversity in their chemical composition and microbial origin. Based on the chemical structure of hydrophobic moieties, biosurfactants are categorized as (1) glycolipids, (2) fatty acids, neutral lipids, and phospholipids, (3) lipopeptide type, and (4) polymeric surfactants (Morita et al. 2016). The major types of microbial surfactants with their properties are reported in Table 14.1.

14.3 Microbial Production of Biosurfactants

Usually microbial surfactants produced through culture conditions consist of water-immiscible substrate. They reduce the interfacial tension existing at the interphase and improve the availability of potential biodegradable substrate for uptake (Huang and Tang 2007). Certain microorganisms produce the non-ionic biosurfactants to change the structure of their cell wall (Vijayakumar and Saravanan 2015). The rate of production of biosurfactants is affected by different factors.

14.3.1 Carbon Source

Carbon sources are used mainly to promote the cell growth and formation of biosurfactants. Specifically, microorganism produces surfactants when they are cultured in rich carbon source like hydrocarbon-polluted soil or produced using proven agriculture waste as a carbon source (Fenibo et al. 2019). The carbon consumption rate of microorganisms during cultivation decides the biomass and product formation (Singh et al. 2017). The carbon source is a prime rate-limiting factor for biosurfactants production which are classified into various categories: carbohydrates, hydrocarbons, and vegetable oils and fats (Tan and Li 2018; Nurfarahin et al. 2018).

14.3.1.1 Carbohydrates

Glucose considered as frequently used major carbon source for microbial production of biosurfactants and microorganisms can metabolize it with ease by glycolysis pathway in the production of energy and also results into higher yield product. It acts as a precursor for the production of both hydrophilic and hydrophobic moieties (Singh et al. 2017). Different researchers revealed the use of sugars for the production of biosurfactants (Table 14.2). *Pseudomonas* and *Vibrionaceae* strains have been used previously for biosurfactant production (Persson and Molin 1987).

14.3.1.2 Vegetable Oil and Fats

The cost for production of biosurfactants depends on the type of carbon source used. This demands on the application of cheap substrates in the production. Vegetable oils are made up of saturated or unsaturated fatty acids chain with 16–18 carbon

Table 14.1 Types of biosurfactants and their properties

Type	Composition	Microorganism	Surface tension (mN/m)	CMC (mg/L)	EI_{24} (%)	References
Glycolipids						
Rhamnolipids	One or two L-rhamnose interconnected by one or two β-hydroxy fatty acids	*Pseudomonas aeruginosa* HAK01	28.1	120 ppm	–	Khademolhosseini et al. (2019)
		Pseudomonas aeruginosa	27.9[a]	60	80.3	Zhao et al. (2018)
			33.1[b]	80	62.3	
Sophorolipids	Sophorose linked to fatty acids (16 or 18 C atoms)	*Candida bombicola*	28.56	–	68.75	Elshafie et al. (2015)
		Rhodotorula babjevae YS3	32.6	130	100	Sen et al. (2017)
Trehalolipids	Trehalose linked to fatty acids	*Rhodococcus* sp.; PML026	29	250	26–50	White et al. (2013)
		Fusarium fujikuroi	20	~30	–	Reis et al. (2018)
Mannosylerythritol lipids	Mannose linked to fatty acids and erythritol linked to other structure	*Pseudozyma aphidis* ZJUDM34	30–35	20		Niu et al. (2019)
		Pseudozyma tsukubaensis	28.5	2.2×10^{-5} M	–	Yamamoto et al. (2013)
Lipopeptides and lipoproteins						
Serrawettin	Cyclodepsipeptide	*Serratia marcescens*	28.8–32.2	–	–	Matsuyama et al. (2011)
Rubiwettin	Cyclic peptide	*Serratia rubidaea*	25.5	–	–	
Surfactin	Heptapeptide	*Bacillus subtilis*	27			Nayarisseri et al. (2018)
Viscosin	Hexapeptide (antibiotic)	*Pseudomonas fluorescens*	26.5	0.15 mg/mL		Neu et al. (1990)

Subtilisin	Protease (275-residue globular protein)	*Bacillus subtilis*				Kamal et al. (1995)
Gramicidins	Ionophoric antibiotics	*B. brevis* *Aneurinibacillus migulanus*	34			Alenezi et al. (2017)
Polymyxins	Lipopeptide antibiotics	*Paenibacillus polymyxa*				Shaheen et al. (2011)
Polymeric surfactants						
Emulsan	Acylated polysaccharide	*Acinetobacter calcoaceticus*				Rubinovitz et al. (1982)
Liposan	Heteropolysaccharide	*Candida lipolytica*				Cirigliano and Carman (1985)
Biodispersan	Polysaccharide	*Acinetobacter calcoaceticus*				Rosenberg et al. (1988)

[a]Aerobic production
[b]Non-aerobic production

Table 14.2 Carbohydrates as a carbon source for biosurfactant production

Source	Concentration (g/L)	Microorganism	Biosurfactant	References
Glucose	24.2	*Pseudomonas aeruginosa*	7.5 mg/mL	Rashedi et al. (2006)
	10–70	*Bacillus* sp.	1–2.46 g/L	Heryani and Putra (2017)
	30	*Fusarium fujikuroi*	–	Reis et al. (2018)
	40	*Bacillus subtilis* SPB1 strain	4.92 g/L	Ghribi and Ellouze-Chaabouni (2011)
Sucrose	20	*Bacillus subtilis* strain ANSKLAB03	0.324 g/ 100 mL	Nayarisseri et al. (2018)
	20	*Pseudomonas putida* MTCC 2467	1.3 g/L	Kanna et al. (2014)

Apart from glucose and sucrose, other sugars such as maltose, lactose, starch, and xylose as a carbon source (Hu et al. 2015)

Table 14.3 Vegetable oils, fats, and industrial waste products for biosurfactant production

Substrate	Microorganism	Biosurfactant	Surface tension (mN/m)	References
Used cooking oil	*Bacillus* sp. HIP3	9.5 g/L	38.15	Md Badrul Hisham et al. (2019)
Vegetable oil refinery wastes	*Pseudomonas aeruginosa* EBN-8	8.50	28.5	Raza et al. (2007)
Waste cooking oil	*Pseudomonas* SWP-4	13.93	24.1	Lan et al. (2015)
Sunflower acid oil	*Pseudomonas aeruginosa*	4.9	30.12	Jadhav et al. (2019)
Sunflower oil	*Serratia marcescens*		29.75	Ferraz et al. (2002)
Soybean oil	*Candida antarctica*	13.86	32	Accorsini et al. (2012)
Vegetable fat waste	*C. glabrata*		24	de Gusmão et al. (2010)
Animal fat and corn steep liquor	*C. lipolytica UCP0988*	1–2.2	28	Santos et al. (2013)

atoms. High uses of oils in homes and restaurants are drastically increased in recent years and have found to produce the waste product at a higher rate (Table 14.3). Disposal of these wastes into the environment causes the pollution of soil and water, and their disposal is a severe issue (Md Badrul Hisham et al. 2019). This motivated many researchers to use different sources of oils and fats as a substrate for biosurfactant production. Different oils namely soybean, olive, castor, sunflower, and coconut fat are reported to produce biosurfactants.

Waste from different industries such as corn steep liquor (Gudiña et al. 2015), sugarcane molasses (Takahashi et al. 2011), banana peel (Chooklin et al. 2014), orange peel (Kumar et al. 2016), soybean meal and rice husk (Massi et al. 2014), potato process effluents (Thompson et al. 2000), potato peels (Sharma et al. 2015), and processing have also been utilized as a potential carbon source for biosurfactant production.

14.3.2 Nitrogen Source

Like carbon source, even nitrogen is considered as essential requirement for the growth of microorganisms. Nitrogen used is either organic or inorganic, and both can impact the biosurfactant production (Desai et al. 1994). Yeast extract, meat extract, tryptone or peptone, beef extract, and urea are widely used as organic nitrogen sources, whereas ammonium chloride, ammonium sulfate, potassium nitrate, etc. are used as inorganic nitrogen sources (Ghribi and Ellouze-Chaabouni 2011). Different nitrogen sources used as a component of fermentation medium are reported in Table 14.4.

14.3.3 Carbon to Nitrogen Ratio (C/N)

The balance of carbon to nitrogen ratio (C/N) is an essential factor that needs to be monitored and optimized for escalating the biosurfactant production. The optimized C/N ratio significantly impacts the biosurfactant yield. Higher proportion of C/N ratios inhibits the growth of bacteria, favoring metabolism of cell in the production of metabolites (Nurfarahin et al. 2018). The yield of biosurfactant in the case of *Bacillus* sp. with glucose as a carbon source was around 1–2.46 g/L. The maximum yield was noted when the C/N ratio was kept at 12.4 (Heryani and Putra 2017).

Table 14.4 Nitrogen sources used for biosurfactant production

Nitrogen source	Microorganism	Biosurfactant (mg/L)	References
Urea	*B. subtilis*	720	Ghribi and Ellouze-Chaabouni (2011)
Ammonium nitrate	*Pseudomonas fluorescens*	3.3 g/L	Abouseoud et al. (2008)
Ammonium sulfate and yeast extract	*Yarrowia lipolytica* IMUFRJ 50682		Fontes et al. (2010)
Yeast extract	Torulopsis bombicola ATCC22214	18.0	Cooper and Paddock (1984)
Peptone	*Candida* sp. SY 16	37	Kim et al. (2006)
Ammonium nitrate	*P. aeruginosa* RS29	0.80	Saikia et al. (2012)
Ammonium nitrate	*C. lipolytica UCP 0988*	8	Rufino et al. (2014)
Urea	*Virgibacillus salaries*		Elazzazy et al. (2015)

Abouseoud et al. screened *Pseudomonas fluorescens* Migula 1895-DSMZ for biosurfactant production. The higher yield was found with the use of olive oil, and ammonium nitrate was used as carbon and nitrogen sources at the ratio of 10 (Stoimenova et al. 2009). In another study, isolated *Bacillus subtilis* produced a high-yield biosurfactant when cultured in medium containing crystal sugar and ammonium nitrate at the ratio of 3 (Fonseca et al. 2007).

14.3.4 Minerals

Calcium, iron, potassium, an magnesium are added as a nutrient to the production medium to facilitate the growth of microorganisms. Cameotra and Makkar et al. studied the effect of metal cations in the production of biosurfactants in the case of *Bacillus subtilis*. It has already been reported that the presence of metal cations has improved the biosurfactant production by twofold. The maximum yield of biosurfactant was obtained when Mg^{2+} and Ca^{2+} were used at a concentration of 2.43 mM and 0.36 mM, respectively (Makkar and Cameotra 2002). Lu et al. studied the effect of Mn^{2+} on the production of surfactin by *Bacillus subtilis* ATCC 21332. Media enriched with Mn^{2+} increased surfactin production by 6.2-fold compared to media devoid of Mn^{2+}. Mn^{2+} enhanced the glutamate synthase activity, changed nitrogen utilization, and increased the absorption of nitrogen and thereby amino acid availability for surfactin synthesis (Huang et al. 2015). Fermentation medium supplemented with Fe^{2+} (4.0 mM) improved the surfactant yield from *B. subtilis* ATCC 21332 by tenfold compared to those without Fe^{2+} provision (Wei et al. 2004). Use of metal ions causes in the formation of novel surfactin variants (Bartal et al. 2018). Use of chemicals like potassium dihydrogen phosphate and dipotassium hydrogen phosphate in fermentation medium maintains the required pH throughout the process (Nurfarahin et al. 2018).

14.3.5 Amino Acids

Role of different amino acids such as aspartic acid, asparagine, and glutamic acid in the fermentation process and thereby biosurfactant production are well established. *B. subtilis* MTCC 2423 separately cultivated in numerous amino acids, namely aspartic acid, asparagine, glutamic acid, valine, and lysine increased the final yield of biosurfactant by about 60% (Makkar and Cameotra 2002). In another interesting study, cultivation of *B. subtilis* TD7 shows that the presence of different amino acids and fatty acids increases the proportion of surfactin variants. In this, culture medium supplemented with arginine, glutamine, or valine increased the production of surfactin variants with even β-hydroxy fatty acids, whereas the addition of cysteine, histidine, isoleucine, leucine, methionine, serine, or threonine enhanced the proportion of surfactin variants composed of odd β-hydroxy fatty acids (Liu et al. 2012).

14.3.6 Vitamins and Growth Promoters

Although useful, vitamins are rarely used during biosurfactant production. This is attributed to the fact that the majority of carbon sources such as vegetable oils, molasses, cashew apples, etc. contain some amount of vitamins (Makkar et al. 2011). Sidkey and Al Hadry reported the effect of vitamins for biosurfactant production in *Bacillus cereus* B_7. None of the tested vitamins increased the biosurfactant yield (Sidkey and Al Hadry 2014). Qazi et al. used vitamin B2 in culture medium prepare for on the biosurfactant production by Pseudomonas putida SOL-10 (Qazi et al. 2013).

14.4 Genetic Regulation and Biosynthesis of Surfactants

Biosynthesis of surfactants involves the formation of different hydrophilic and hydrophobic moieties using carbon, nitrogen sources, as well as micronutrients followed by their joining to produce biosurfactants (Marchant and Banat 2012). However, this is a complex process governed by a particular metabolic pathway, certain enzymes, and genetic makeup of the producer organisms as well (Das et al. 2008). Among all biosurfactants, rhamnolipid produced by *Pseudomonas aeruginosa* and lipopeptide biosurfactants (surfactin) produced by *Bacillus* and *Pseudomonas* spp. are primarily studied for molecular genetics. Lipopeptide biosurfactants (LPBSs) consist of peptide groups and fatty acids. The cyclic structure is formed by binding between two different functional groups like C-terminal peptide residue with β-hydroxy fatty acid, hydroxyl group of the peptide residue, or β-amino acid (Roongsawang et al. 2010). An enzyme called nonribosomal peptide synthetases (NRPSs) causes the formation of this linkage (Koglin and Walsh 2009).

Surfactin and lichenysin are LPBSs mainly produced by *B. subtilis* and *B. licheniformis*, respectively. NRPSs involved in their synthesis are surfactin and lichenysin synthetases; both are identical in structure. Surfactin synthetase comprised of three protein subunits (*SrfA*, *ComA*, and *SrfC*) and four open reading frames (ORFs) (*srfA-A*, *srfA-B*, *srfA-C*, and *srfA-Te*) that can be further divided into functional domains. *Lichenysin synthetases* also composed of four ORFs, namely *licA*, *licB*, *licC*, and *lic-Te* (Das et al. 2008; Roongsawang et al. 2010; Konz et al. 1999). Other LPBSs such as fengycin, bacillomycin, iturin, and mycosubtilin are also produced non-ribosomally by a multienzyme peptide synthetase complex (Tsuge et al. 2001; Wu et al. 2007).

Rhamnolipids are the type of glycolipids composed of either one or two (L)-rhamnose molecules, glycosidically linked to one or two β-hydroxy fatty acids. *P. aeruginosa* is the largest producer of rhamnolipid and its synthesis is catalyzed by a different rhamnosyltransferase. The biosynthesis of rhamnolipids is cross-linked with the production of different species of polysaccharide species (Müller and Hausmann 2011). Rhamnolipid biosynthesis involves the conversion of glucose to D-glucose-1-phosphate by gluconeogenesis and Entner–Doudoroff pathways. Substrate D-glucose-1-phosphate is then producing L-rhamnose (Chong and Li

2017). Three quorum sensing systems, namely *las*, *rhl*, and *pqs*, are present in *P. aeruginosa*, which regulate rhamnolipid production. LasI and RhlI synthases produce their signal molecules, homoserine lactones $3OC_{12}$-HSL and C_4-HSL. These signal molecules bind and modulate LasR and RhlR, respectively, and control the biosynthesis of rhamnolipids (Dobler et al. 2016). The three key enzymes that govern the rhamnolipid biosynthesis are RhlA, RhlB, and RhlC. Synthesis of precursor 3-(3-hydroxyalkanoyloxy) alkanoic acid (HAA) that constitutes the hydrophobic component of rhamnolipid is catalyzed by RhlA, whereas RhlB and RhlC catalyze the reaction between L-rhamnose and HAA or mono-rhamnolipid (Déziel et al. 2003). Different genes clusters involved in production of different biosurfactants are summarized in Table 14.5.

14.5 Application of Biosurfactants as Cleansing and Washing Agents

Washing and cleaning agents are the chemical substances which are used for removing dirt, including dust, stains, bad smell, and clutter. These are available as liquids, powders, sprays, or granules (Nitsch et al. 2003). Chemical cleaning agents are mainly grouped based on composition of acid/base, surfactant, enzyme, chelating agent, and oxidants (Li and Elimelech 2004). Modern cleansing agents are either surfactants or sequestering agents. Uses of these chemical surfactants are limited due to their different environmental impact. They are known for several properties like more reactive, low toxicity, easily biodegradable, and produced from renewable sources (Mahanti et al. 2017). Biosurfactants are used as cleaning agents in different fields.

14.5.1 Dairy Industry

The reduction of the membrane performance during the ultrafiltration process involved in dairy industries is very common, and it happens due to membrane fouling. Chemical agents such as sodium dodecyl sulfate (SDS) and polysorbate (Tween) have been used to remove fouling and sustenance of permeability; however, their use is limited due to their damaging effect on the membrane and hazardous impact on the environment (Madaeni et al. 2010; Kim et al. 2015). Zhang et al. (Kim et al. 2015) used rhamnolipid for the cleaning of ultrafiltration membranes. Rhamnolipid (below pH 9) removed the foulant from the polysulfone, polyacrylonitrile, and polysulfone-g-polyethylene glycol membranes and restored the water flux to about 94% of the initial level, whereas SDS and Tween 20–treated membranes showed lower flux recovery (50–70%). Rahimpour et al. (Aghajani et al. 2018) also explored the rhamnolipid which novel cleaning agent used in ultrafiltration membrane fouled by whey and compared with chemical cleaners such as sodium hydroxide (NaOH), SDS and Tween 20. Rhamnolipid along with NaOH as cleaning agents showed 100% of flux recovery (Aghajani et al. 2018).

Table 14.5 Genetic regulation of microbial surfactants

Biosurfactant	Microorganism	Enzyme	ORFs/ gene cluster/ modules/ proteins	References
Surfactin	*B. subtilis*	*Surfactin synthetase*	*SrfA–A, SrfA–B, SrfA–C and SrfT–e*	Das et al. (2008) and Roongsawang et al. (2010)
Lichenysin	*B. licheniformis*	*Lichenysin synthetase*	*licA*, *licB*, *licC*, and *lic-Te*	Das et al. (2008) and Roongsawang et al. (2010)
Iturin	*Bacillus subtilis*	*Malonyl coenzyme A transacylase*	*ituD*, *ituA*, *ituB*, and *ituC*	Tsuge et al. (2001)
Fengycin	*Bacillus* spp.	*Fengycin synthetase*	*FenC, FenD, FenE, FenA*, and *FenB*	Wu et al. (2007)
Locillomycins	*B. subtilis* 916	*Peptide synthetases*	*Loc*	Luo et al. (2014)
Mycosubtilin	*Bacillus subtilis* ATCC6633	*Mycosubtilin synthetase*	*fenF*, *mycA*, *mycB*, *and mycC*	Duitman et al. (1999)
Bacillomycin D	*B. subtilis* 916	*Peptide synthetases*	*bamD*, *bamA*, *bamB*, *and bamC*	Moyne et al. (2004)
Fusaricidin	*Paenibacillus polymyxa* PKB1	*Fusaricidin synthetase*	*fusA*	Li and Jensen (2008)
Syringomycin	*Pseudomonas syringae* B301D	*Syringomycin synthetase*	*SyrB1, SyrE, (SyrB2, SyrC, SyrP*	Scholz-Schroeder et al. (2001)
Syringopeptin	*Pseudomonas syringae* B301D	*Syringopeptin synthetase*	*SypA, SypB, SypC*	Roongsawang et al. (2010) and Scholz-Schroeder et al. (2001)

(continued)

Table 14.5 (continued)

Biosurfactant	Microorganism	Enzyme	ORFs/ gene cluster/ modules/ proteins	References
Arthrofactin	*Pseudomonas* sp. MIS38	*Arthrofactin synthetase*	*arfA*, *arfB*, and *arfC*	Roongsawang et al. (2003)
Orfamide	*Pseudomonas* sp. CMR12a	*Orfamide synthetase*	*ofaA*, *ofaB*, and *ofaC*	Olorunleke et al. (2017)
Sessilins	*Pseudomonas* sp. CMR12a	*Sessilins synthetase*	*sesA*, *sesB*, and *sesC*	
Tolaasin	*Pseudomonas tolaasii* NCPPB 1116	*Peptide synthetase*	*TL1*, *TL2*, and *TL*	Rainey et al. (1993)
Viscosin	*Pseudomonas fluorescens* PfA7B	*Viscosin synthetase*	*Vsp1*, *Vsp2*, and *Vsp3*	Braun et al. (2001)
Massetolide	*P. fluorescens* SS101	*Massetolide synthetase*	*massA*, *massB*, and *massC*	de Bruijn et al. (2008)
Syringafactin A to F	*Pseudomonas syringae pv. tomato* DC3000	*Syringafactin synthetase*	*syfR*, *syfA*, *syfB*, *syfC*, and *syfD*	Berti et al. (2007)
Entolysin	*Pseudomonas entomophila*	*Entolysin synthetase*	*EtlA*, *EtlB*, *EtlC*	Vallet-Gely et al. (2010)
Rhamnolipids	*Burkholderia thailandensis P. aeruginosa*	*Rhamnosyl transferase*	*rhlA*, *rhlB*, *and rhlC*	Dubeau et al. (2009)
Sophorolipid	*Starmerella bombicola*	*Glucosyltransferase Cytochrome P450 monooxygenase*	*adh*, *ugtB1*, *ugtA1*, *ugtA1*, and *cyp52m1*	Van Bogaert et al. (2013)
Amphisin	*Pseudomonas* sp. *strain* DSS73	*Amphisin synthetase and sensor kinase*	*amsY* and *gacS*	Nielsen et al. (2005)
Emulsan	*Acinetobacter lwoffii RAG-1*	*Emulsan synthetase*	*Wee A–Week*, *wza*, *wzb*, *wzc*, *wzx*, *wzy*	Nakar and Gutnick (2001)

(continued)

Table 14.5 (continued)

Biosurfactant	Microorganism	Enzyme	ORFs/ gene cluster/ modules/ proteins	References
Alasan	*Acinetobacter radioresistens* KA53	*Alasan synthetase*	*AlnA, AlnB, and AlnC*	Toren et al. (2002)
Serrawettin	*Surfactantfaciens* sp. *nov. YD25T*	*Peptide synthetase*	*Pig, swrA*	Su et al. (2016)
Mannosylerythritol lipids	*Ustilago maydis*	*Glycosyl transferase*	*Emt1, Mac1, Mac2, and Mat1*	Hewald et al. (2006)

ORFs open reading frames

14.5.2 Textile Detergent

Nowadays, textile end products are needed to satisfy the different quality attributes. Hence, they are processed through different pretreatment that involves the removal of natural and synthetic fiber admixtures and lubrication of fabrics using oils as well as waxes. Removal of these lubricants is a critical step because of repeated emulsification and redeposition of oil on its fiber surface. Traditional treatments with detergents have a hazardous effect surrounding atmosphere. This urges the use of biosurfactants (Kesting et al. 1996). Campos-Takaki et al. investigated the ability of surfactants isolated from *Cunninghamella echinulata* and used as a textile detergent to clean cotton fabric used to remove traces of hydrophobic residues from the automobile industry (Andrade et al. 2018). Biosurfactant obtained from *Ochrobactrum intermedium* strain MZV101 has shown good stability at pH 9–13 and shown strong oil removal (Zarinviarsagh et al. 2017). Lipopeptide biosurfactant produced by *Bacillus subtilis* SPB1 together with commercial detergents improved their oil stain removal and tea stain removing ability (Bouassida et al. 2018). Similarly, two other *Bacillus subtilis* strains (DM-03 and DM-04) produced cyclic lipopeptide biosurfactants and exhibited thermal stability and surface-active property (Mukherjee 2007). Rhamnolipid-based washing powder produced by Bafghi and Fazaelipoor effectively removed the edible oil, chocolate, and albumen stains from cotton clothes (Khaje and Fazaelipoor 2012). Sophorolipids synthesized using *C. bombicola* ATCC22214 also exhibited stain removal capability (Joshi-Navare et al. 2013).

14.5.3 Petroleum Industry

The petroleum industry is one of the continuously growing fields and involves the purification, transport, as well as storage of its crude products in storage tanks. Being crud in nature, the contaminants from the oil settle at lower surface and deposit on the sidewalls of the storage tanks. These viscous deposits later solidify and cannot be detached even with pumping. Periodical cleaning with cleaning agents is time-consuming and labor-intensive, and final disposals are hazardous to the environment (Matsui et al. 2012). Biosurfactants are effectual in the washing of storage tanks due to their ability to form emulsion that diminishes the viscosity of sludges as well as solid deposits and thereby facilitates pumping of waste (de Cássia et al. 2014). Notably, biosurfactants produced by *Pseudomonas aeruginosa* SH 29 effectively cleaned the oil storage tanks and observed the complete recovery of oil from the bottom and sidewalls of the tanks after 15 min of treatment (Diab and Din 2013). Matsui et al. investigated the potential of biosurfactant produced by *Gordonia* sp. strain JE-1058 and reported that biosurfactant at concentrations of 1–10 g/L effectively removed the oil tank bottom sludge (Matsui et al. 2012).

Petroleum industries generate waste products during extraction, refining, and transportation, and oil spillage in the marine environment and groundwater poses a serious issue which disturbs the marine ecosystems drastically and causes the low penetration of light which results in the death of aquatic organisms (Fenibo et al. 2019; de Cássia et al. 2014). Biosurfactants effectively reduce interfacial tension, disperse the oil particles, and degrade them into non-toxic debris (Patel et al. 2019). Further, in the microorganisms, they induce the changes in the surface of cells to convert into more hydrophobic and subsequently increase the pinocytosis index of hydrocarbons (Fenibo et al. 2019; Patel et al. 2019). Among all biosurfactants, the rhamnolipid, sophorolipid, and surfactin have been explored for this purpose. Biosurfactant produced from *Gordonia* sp. strain JE-1058 effectively clean up oil spills from contaminated seawater and sea sand (Saeki et al. 2009). Another study of Feng et al. revealed the use of lipopeptide surfactant produced from *Bacillus subtilis* HSO121 for oil spill remediation (Feng et al. 2019).

14.5.4 Leather Industry

Pretreatment processes on leather, for example, shaving and buffering generate the waste products comprised of numerous contaminants mainly chromium that can hazardously impact the aquatic and terrestrial ecosystem. Many studies showed that already microorganisms were effectively used in the bioremediation and recycling of waste materials (Greenwell et al. 2016). Raman et al. investigated the usefulness of biosurfactant produced from *Bacillus subtilis* SA-6 for bioremediation of leather dust. Results showed that biosurfactant increased the chromium concentration (190.81 ± 20.18 mg/L) in cell-free supernatant and reduced the surface tension (30.13 ± 0.15 mN/m) during fermentation (Greenwell et al. 2016).

14.5.5 Food Industry

Several classes of biosurfactants are being used as antimicrobial agents to avoid contamination in food processing industries. Effective cleaning is essential to reduce or eliminate microorganisms found on food contact surfaces (Sharma 2016). Biosurfactants (rhamnolipids) have explored in food processing as a cleaning agent (Meng and Zhang 2012).

14.5.6 Household Detergent and Dish Wash

The use of detergents and dishwashing products in the household is inevitable. Biosurfactants are useful as a component of household detergents for the advantage of their unique properties such as easily degradability, cost-effectiveness, and environmental compatibility. Sophorolipids produced from *C. bombicola* ATCC 22214 were found to be useful in hard surface cleaning and automatic dishwashing rinse aid formulations. At a concentration of 36 mg/L, the contact angle of sophorolipids was similar to that of the reference surfactant. Sophorolipid exhibited comparable surface activity with that of alkyl polyglucosides (Develter and Lauryssen 2010). Glycoside surfactants effectively increased the wetting characteristics of conventional low foaming non-ionic surfactant-based compositions (Furuta et al. 2004).

14.5.7 Cosmetic Industry

Cosmetic industries make use of biosurfactants (glycolipids and lipopeptides) particularly in dermatological preparations due to their cleansing, foaming, skin hydrating properties, and antimicrobial potential (Varvaresou and Iakovou 2015). Cosmetic formulation comprised of lipopeptide biosurfactant exhibited an excellent washability and low skin irritation and provided high skin comfort (Yoneda 2006). Moreover, Allef et al. prepared different hair and skin cleaning formulations comprised of biosurfactants, for example, shower gels, shampoos, conditioners, body cleansers, or skin cleansers (Allef et al. 2014).

14.6 Conclusion

Biosurfactants are the naturally produced surfactants using different microbial strains and have occupied the market due to their superior characteristics compared to chemical surfactants. Biosurfactants can be produced on a larger scale using abundantly available cheap raw material and waste products of different industries. In addition to this, biodegradable, non-toxic nature, and environmental compatibility of biosurfactants make it feasible for different industries to use biosurfactants for cleansing and washing applications.

References

Abouseoud M, Maachi R, Amrane A, Boudergua S, Nabi A (2008) Evaluation of different carbon and nitrogen sources in production of biosurfactant by *Pseudomonas fluorescens*. Desalination 223(1–3):143–151

Accorsini FR, Mutton MJ, Lemos EG, Benincasa M (2012) Biosurfactants production by yeasts using soybean oil and glycerol as low cost substrate. Braz J Microbiol 43(1):116–125

Aghajani M, Rahimpour A, Amani H, Taherzadeh MJ (2018) Rhamnolipid as new bio-agent for cleaning of ultrafiltration membrane fouled by whey. Eng Life Sci 18:272–280

Alenezi FN, Rekik I, Chenari Bouket A, Luptakova L, Weitz HJ, Rateb ME, Jaspars M, Woodward S, Belbahri L (2017) Increased biological activity of *Aneurinibacillus migulanus* strains correlates with the production of new gramicidin secondary metabolites. Front Microbiol 8:517

Allef P, Hartung C, Schilling M (2014) Aqueous hair and skin cleaning compositions comprising biosurfactants. Patent US 20140349902 A1

Andrade RFS, Silva TAL, Ribeaux DR, Rodriguez DM, Souza AF, Lima MAL et al (2018) Promising biosurfactant produced by *Cunninghamella echinulata* UCP 1299 using renewable resources and its application in cotton fabric cleaning process. Adv Mater Sci Eng 2018:1624573. https://doi.org/10.1155/2018/1624573

Bartal A, Vigneshwari A, Bóka B, Vörös M, Takács I, Kredics L, Manczinger L, Varga M, Vágvölgyi C, Szekeres A (2018) Effects of different cultivation parameters on the production of Surfactin variants by a *Bacillus subtilis* strain. Molecules 23(10):2675

Berti AD, Greve NJ, Christensen QH, Thomas MG (2007) Identification of a biosynthetic gene cluster and the six associated lipopeptides involved in swarming motility of *Pseudomonas syringae* pv. Tomato DC3000. J Bacteriol 189(17):6312–6323

Bouassida M, Fourati N, Ghazala I, Ellouze-Chaabouni S, Ghribi D (2018) Potential application of Bacillus subtilis SPB1 biosurfactants in laundry detergent formulations: compatibility study with detergent ingredients and washing performance. Eng Life Sci 18:70–77

Braun PG, Hildebrand PD, Ells TC, Kobayashi DY (2001) Evidence and characterization of a gene cluster required for the production of viscosin, a lipopeptide biosurfactant, by a strain of *Pseudomonas fluorescens*. Can J Microbiol 47(4):294–301

Chong H, Li Q (2017) Microbial production of rhamnolipids: opportunities, challenges and strategies. Microb Cell Fact 16(1):137

Chooklin CS, Maneerat S, Saimmai A (2014) Utilization of banana peel as a novel substrate for biosurfactant production by *Halobacteriaceae archaeon* AS65. Appl Biochem Biotechnol 173 (2):624–645

Cirigliano MC, Carman GM (1985) Purification and characterization of Liposan, a bioemulsifier from *Candida lipolytica*. Appl Environ Microbiol 50(4):846–850

Cooper DG, Paddock DA (1984) Production of a biosurfactant from *Torulopsis bombicola*. Appl Environ Microbiol 47:173–176

Das P, Mukherjee S, Sen R (2008) Genetic regulations of the biosynthesis of microbial surfactants: an overview. Biotechnol Genet Eng Rev 25:165–185

de Bruijn I, de Kock MJ, de Waard P, van Beek TA, Raaijmakers JM (2008) Massetolide A biosynthesis in *Pseudomonas fluorescens*. J Bacteriol 190(8):2777–2789

de Cássia FSSR, Almeida DG, Rufino RD, Luna JM, Santos VA, Sarubbo LA (2014) Applications of biosurfactants in the petroleum industry and the remediation of oil spills. Int J Mol Sci 15 (7):12523–12542

de Gusmão CAB, Rufino RD, Sarubbo LA (2010) Laboratory production and characterization of a new biosurfactant from *Candida glabrata* UCP1002 cultivated in vegetable fat waste applied to the removal of hydrophobic contaminant. World J Microbiol Biotechnol 26:1683–1692

Desai AJ, Patel RM, Desai JD (1994) Advances in production of biosurfactant and their commercial applications. J Sci Ind Res 53:619–629

Develter DWG, Lauryssen LML (2010) Properties and industrial applications of sophorolipids. Eur J Lipid Sci Technol 112(6):628–638. https://doi.org/10.1002/ejlt.200900153
Déziel E, Lépine F, Milot S, Villemur R (2003) rhlA is required for the production of a novel biosurfactant promoting swarming motility in *Pseudomonas aeruginosa*: 3-(3-hydroxyalkanoyloxy)alkanoic acids (HAAs), the precursors of rhamnolipids. Microbiology 149:2005–2013
Diab A, Din GE (2013) Application of the biosurfactants produced by *Bacillus* sp. (SH 20 and SH 26) and *P. aeruginosa* SH 29 isolated from the rhizosphere soil of an Egyptian salt marsh plant for the cleaning of oil-contaminated vessels and enhancing the biodegradation. Afr J Environ Sci Technol 7:671–679
Dobler L, Vilela LF, Almeida RV, Neves BC (2016) Rhamnolipids in perspective: gene regulatory pathways, metabolic engineering, production and technological forecasting. N Biotechnol 33 (1):123–135
Dubeau D, Déziel E, Woods DE, Lépine F (2009) *Burkholderia thailandensis* harbors two identical rhl gene clusters responsible for the biosynthesis of rhamnolipids. BMC Microbiol 9:263
Duitman EH, Hamoen LW, Rembold M, Venema G, Seitz H, Saenger W et al (1999) The mycosubtilin synthetase of *Bacillus subtilis* ATCC6633: a multifunctional hybrid between a peptide synthetase, an amino transferase, and a fatty acid synthase. Proc Natl Acad Sci U S A 96 (23):13294–13299
Elazzazy AM, Abdelmoneim TS, Almaghrabi OA (2015) Isolation and characterization of biosurfactant production under extreme environmental conditions by alkali-halo-thermophilic bacteria from Saudi Arabia. Saudi J Biol Sci 22(4):466–475
Elshafie AE, Joshi SJ, Al-Wahaibi YM, Al-Bemani AS, Al-Bahry SN, Al-Maqbali D, Banat IM (2015) Sophorolipids production by *Candida bombicola* ATCC 22214 and its potential application in microbial enhanced oil recovery. Front Microbiol 6:1324
Feng JQ, Gang HZ, Li DS, Liu JF, Yang SZ, Mu BZ (2019) Characterization of biosurfactant lipopeptide and its performance evaluation for oil-spill remediation. RSC Adv 9:9629–9632
Fenibo EO, Ijoma GN, Selvarajan R, Chikere CB (2019) Microbial surfactants: the next generation multifunctional biomolecules for applications in the petroleum industry and its associated environmental remediation. Microorganisms 7(11):581
Ferraz C, De Araújo AA, Pastore GM (2002) The influence of vegetable oils on biosurfactant production by Serratia marcescens. Appl Biochem Biotechnol 98–100:841–847
Fonseca RR, Silva AJ, De França FP, Cardoso VL, Sérvulo EF (2007) Optimizing carbon/nitrogen ratio for biosurfactant production by a Bacillus subtilis strain. Appl Biochem Biotechnol 137–140(1–12):471–486. https://doi.org/10.1007/s12010-007-9073-z
Fontes GC, Amaral PF, Nele M, Coelho MA (2010) Factorial design to optimize biosurfactant production by *Yarrowia lipolytica*. J Biomed Biotechnol 2010:821306
Furuta T, Hirata Y, Igarashi K (2004) Low-foaming detergent compositions. US Patent 20040171512 A1
Ghribi D, Ellouze-Chaabouni S (2011) Enhancement of *Bacillus subtilis* lipopeptide biosurfactants production through optimization of medium composition and adequate control of aeration. Biotechnol Res Int 2011:653654. https://doi.org/10.4061/2011/653654
Greenwell M, Sarker M, Rahman P (2016) Biosurfactant production and biodegradation of leather dust from tannery. Open Biotechnol J 10(Suppl-2, M4):312–325
Gudiña EJ, Fernandes EC, Rodrigues AI, Teixeira JA, Rodrigues LR (2015) Biosurfactant production by Bacillus subtilis using corn steep liquor as culture medium. Front Microbiol 6:59
Havasi R (2011) Microbial biosurfactants: from an environmental application point of view. J Bioremed Biodegr 2:104e
Heryani H, Putra MD (2017) Kinetic study and modeling of biosurfactant production using *Bacillus* sp. Electron J Biotechnol 27:49–54
Hewald S, Linne U, Scherer M, Marahiel MA, Kämper J, Bölker M (2006) Identification of a gene cluster for biosynthesis of mannosylerythritol lipids in the basidiomycetous fungus *Ustilago maydis*. Appl Environ Microbiol 72(8):5469–5477

Hu X, Wang C, Wang P (2015) Optimization and characterization of biosurfactant production from marine *Vibrio* sp. strain 3B-2. Front Microbiol 6:976

Huang W-C, Tang I-C (2007) Bacterial and yeast cultures—process characteristics, products, and applications. In: Bioprocessing for value-added products from renewable research, pp 185–223

Huang X, Liu J, Wang Y, Liu J, Lu L (2015) The positive effects of Mn^{2+} on nitrogen use and surfactin production by *Bacillus subtilis* ATCC 21332. Biotechnol Biotechnol Equip 29 (2):381–389

Jadhav JV, Anbu P, Yadav S, Pratap AP, Kale SB (2019) Sunflower acid oil-based production of rhamnolipid using *Pseudomonas aeruginosa* and its application in liquid detergents. J Surfactant Deterg 21(1):113–125. https://doi.org/10.1002/jsde.12255

Joshi-Navare K, Khanvilkar P, Prabhune A (2013) Jatropha oil derived sophorolipids: production and characterization as laundry detergent additive. Biochem Res Int 2013:169797

Kamal M, Höög JO, Kaiser R, Shafqat J, Razzaki T, Zaidi ZH, Jörnvall H (1995) Isolation, characterization and structure of subtilisin from a thermostable Bacillus subtilis isolate. FEBS Lett 374(3):363–366

Kanna R, Gummadi SN, Kumar GS (2014) Production and characterization of biosurfactant by Pseudomonas putida MTCC 2467. Aust J Biol Sci 14:436–445

Kesting W, Tummuscheit M, Schacht H, Schollmeyer E (1996) Ecological washing of textiles with microbial surfactants. In: Jacobasch HJ (ed) Interfaces, surfactants and colloids in engineering. Progress in colloid & polymer science, vol 101. Steinkopff, Dresden

Khademolhosseini R, Jafari A, Mousavi SM, Hajfarajollah H, Noghabi KA, Manteghian M (2019) Physicochemical characterization and optimization of glycolipid biosurfactant production by a native strain of Pseudomonas aeruginosa HAK01 and its performance evaluation for the MEOR process. RSC Adv 9(14):7932–7947

Khaje BM, Fazaelipoor MH (2012) Application of rhamnolipid in the formulation of a detergent. J Surfactant Deterg 15:679–684

Kim HS, Jeon JW, Kim BH, Ahn CY, Oh HM, Yoon BD (2006) Extracellular production of a glycolipid biosurfactant, mannosylerythritol lipid, by *Candida* sp. SY16 using fed-batch fermentation. Appl Microbiol Biotechnol 70(4):391–396

Kim LH, Jung Y, Kim SJ, Kim CM, Yu HW, Park HD et al (2015) Use of rhamnolipid biosurfactant for membrane biofouling prevention and cleaning. Biofouling 31(2):211–220

Kogawa AC, Cernic BG, do Couto LGD, Salgado HRN (2017) Synthetic detergents: 100 years of history. Saudi Pharm J 25(6):934–938

Koglin A, Walsh CT (2009) Review structural insights into nonribosomal peptide enzymatic assembly lines. Nat Prod Rep 26(8):987–1000

Konz D, Doekel S, Marahiel MA (1999) Molecular and biochemical characterization of the protein template controlling biosynthesis of the lipopeptide lichenysin. J Bacteriol 181(1):133–140

Kumar AP, Janardhan A, Viswanath B, Monika K, Jung JY, Narasimha G (2016) Evaluation of orange peel for biosurfactant production by *Bacillus licheniformis* and their ability to degrade naphthalene and crude oil. 3 Biotech 6(1):43

Lan G, Fan Q, Liu Y, Chen C, Li G, Liu Y, Yin X (2015) Rhamnolipid production from waste cooking oil using *Pseudomonas* SWP-4. Biochem Eng J 101:44–54

Li Q, Elimelech M (2004) Organic fouling and chemical cleaning of nanofiltration membranes: measurements and mechanisms. Environ Sci Technol 38:4683–4693

Li J, Jensen SE (2008) Nonribosomal biosynthesis of fusaricidins by *Paenibacillus polymyxa* PKB1 involves direct activation of a D-amino acid. Chem Biol 15(2):118–127

Liu JF, Yang J, Yang SZ, Ye RQ, Mu BZ (2012) Effects of different amino acids in culture media on surfactin variants produced by *Bacillus subtilis* TD7. Appl Biochem Biotechnol 166 (8):2091–2100

Liu JF, Mbadinga SM, Yang SZ, Gu JD, Mu BZ (2015) Chemical structure, property and potential applications of biosurfactants produced by Bacillus subtilis in petroleum recovery and spill mitigation. Int J Mol Sci 16(3):4814–4837

Luo C, Liu X, Zhou H, Wang X, Chen Z (2014) Nonribosomal peptide synthase gene clusters for lipopeptide biosynthesis in *Bacillus subtilis* 916 and their phenotypic functions. Appl Environ Microbiol 81(1):422–431

Madaeni SS, Rostami E, Rahimpour A (2010) Surfactant cleaning of ultrafiltration membranes fouled by whey. Int J Dairy Technol 63:273–283

Mahanti P, Kumar S, Patra JK (2017) Biosurfactants: an agent to keep environment clean. In: Patra J, Vishnuprasad C, Das G (eds) Microbial biotechnology. Springer, Singapore

Makkar RS, Cameotra SS (2002) Effects of various nutritional supplements on biosurfactant production by a strain of *Bacillus subtilis* at 45°C. J Surfactant Deterg 5:11–17

Makkar RS, Cameotra SS, Banat IM (2011) Advances in utilization of renewable substrates for biosurfactant production. AMB Expr 1:5

Marchant R, Banat IM (2012) Microbial biosurfactants: challenges and opportunities for future exploitation. Trends Biotechnol 30(11):558–565. https://doi.org/10.1016/j.tibtech.2012.07.003

Massi J, Neto DC, Rezende MI (2014) Production of biossurfactants using soybean meal andrice husk with *Bacillus amyloliquefaciens* mo-04b bysolid state fermentation (ssf). BMC Proc 8 (Suppl 4):223

Matsui T, Namihira T, Mitsuta T, Saeki H (2012) Removal of oil tank bottom sludge by novel biosurfactant, JE1058BS. J Jpn Pet Inst 55:138–141

Matsuyama T, Tanikawa T, Nakagawa Y (2011) Serrawettins and other surfactants produced by Serratia. In: Soberón-Chávez G (ed) Biosurfactants. Microbiology monographs, vol 20. Springer, Berlin

Md Badrul Hisham NH, Ibrahim MF, Ramli N, Abd-Aziz S (2019) Production of biosurfactant produced from used cooking oil by *Bacillus* sp. HIP3 for heavy metals removal. Molecules 24 (14):2617

Meng Q, Zhang G (2012) Application of rhamnolipid as biological cleaning agent. Chinese Patent CN 102399644

Morita T, Fukuoka T, Imura T, Kitamoto D (2016) Glycolipid biosurfactants. In: Reference module in chemistry, molecular sciences and chemical engineering. https://doi.org/10.1016/b978-0-12-409547-2.11565-3

Moyne AL, Cleveland TE, Tuzun S (2004) Molecular characterization and analysis of the operon encoding the antifungal lipopeptide bacillomycin D. FEMS Microbiol Lett 234(1):43–49

Mukherjee AK (2007) Potential application of cyclic lipopeptide biosurfactants produced by *Bacillus subtilis* strains in laundry detergent formulations. Lett Appl Microbiol 45(3):330–335

Müller MM, Hausmann R (2011) Regulatory and metabolic network of rhamnolipid biosynthesis: traditional and advanced engineering towards biotechnological production. Appl Microbiol Biotechnol 91(2):251–264

Nakama Y (2017) Surfactants. In: Cosmetic science and technology. Elsevier, Amsterdam, pp 231–244. https://doi.org/10.1016/b978-0-12-802005-0.00015-x

Nakar D, Gutnick DL (2001) Analysis of the wee gene cluster responsible for the biosynthesis of the polymeric bioemulsifier from the oil-degrading strain *Acinetobacter lwoffii* RAG-1. Microbiology 147(Pt 7):1937–1946

Nayarisseri A, Singh P, Singh SK (2018) Screening, isolation and characterization of biosurfactant producing *Bacillus subtilis* strain ANSKLAB03. Bioinformation 14(6):304–314

Neu TR, Härtner T, Poralla K (1990) Surface active properties of viscosin: a peptidolipid antibiotic. Appl Microbiol Biotechnol 32:518–520. https://doi.org/10.1007/BF00173720

Nielsen TH, Nybroe O, Koch B, Hansen M, Sørensen J (2005) Genes involved in cyclic lipopeptide production are important for seed and straw colonization by *Pseudomonas* sp. strain DSS73. Appl Environ Microbiol 71(7):4112–4116

Nitsch C, Heitland H-J, Marsen H, Schlüssler H-J (2003) Cleansing agents. In: Ullmann's encyclopedia of industrial chemistry. https://doi.org/10.1002/14356007.a07_137

Niu Y, Wu J, Wang W, Chen Q (2019) Production and characterization of a new glycolipid, mannosylerythritol lipid, from waste cooking oil biotransformation by *Pseudozyma aphidis* ZJUDM34. Food Sci Nutr 7(3):937–948

Nurfarahin AH, Mohamed MS, Phang LY (2018) Culture medium development for microbial-derived surfactants production—an overview. Molecules 23(5):1049
Olorunleke FE, Kieu NP, De Waele E, Timmerman M, Ongena M, Höfte M (2017) Coregulation of the cyclic lipopeptides orfamide and sessilin in the biocontrol strain Pseudomonas sp. CMR12a. Microbiologyopen 6(5):e00499. https://doi.org/10.1002/mbo3.499
Patel S, Homaei A, Patil S, Daverey A (2019) Microbial biosurfactants for oil spill remediation: pitfalls and potentials. Appl Microbiol Biotechnol 103(1):27–37
Persson A, Molin G (1987) Capacity for biosurfactant production of environmental Pseudomonas and Vibrionaceae growing on carbohydrates. Appl Microbiol Biotechnol 26:439–442
Qazi MA, Malik ZA, Qureshi GD, Hameed A, Ahmed S (2013) Yeast extract as the most preferable substrate for optimized biosurfactant production by rhlB gene positive *Pseudomonas putida* SOL-10 isolate. Bioremediat Biodegrad 4:1–10
Rainey PB, Brodey CL, Johnstone K (1993) Identification of a gene cluster encoding three high-molecular-weight proteins, which is required for synthesis of tolaasin by the mushroom pathogen *Pseudomonas tolaasii*. Mol Microbiol 8(4):643–652
Rashedi H, Jamshidi E, Mazaheri AM, Bonakdarpour B (2006) Biosurfactant production with glucose as a carbon source. Chem Biochem Eng Q 20(1):99–106
Raza ZA, Rehman A, Khan MS, Khalid ZM (2007) Improved production of biosurfactant by a Pseudomonas aeruginosa mutant using vegetable oil refinery wastes. Biodegradation 18 (1):115–121
Reis CBLD, Morandini LMB, Bevilacqua CB et al (2018) First report of the production of a potent biosurfactant with α,β-trehalose by *Fusarium fujikuroi* under optimized conditions of submerged fermentation. Braz J Microbiol 49(Suppl 1):185–192
Roongsawang N, Hase KI, Haruki M, Imanaka T, Morikawa M, Kanaya S (2003) Cloning and characterization of the gene cluster encoding arthrofactin synthetase from *Pseudomonas* sp. MIS38. Chem Biol 10(9):869–880
Roongsawang N, Washio K, Morikawa M (2010) Diversity of nonribosomal peptide synthetases involved in the biosynthesis of lipopeptide biosurfactants. Int J Mol Sci 12(1):141–172
Rosenberg E, Rubinovitz C, Gottlieb A, Rosenhak S, Ron EZ (1988) Production of Biodispersan by *Acinetobacter calcoaceticus* A2. Appl Environ Microbiol 54(2):317–322
Rubinovitz C, Gutnick DL, Rosenberg E (1982) Emulsan production by Acinetobacter calcoaceticus in the presence of chloramphenicol. J Bacteriol 152(1):126–132
Rufino RD, de Luna JM, de Campos Takaki GM, Sarubbo LA (2014) Characterization and properties of the biosurfactant produced by *Candida lipolytica* UCP 0988. Electron J Biotechnol 17:34–38
Saeki H, Sasaki M, Komatsu K, Miura A, Matsuda H (2009) Oil spill remediation by using the remediation agent JE1058BS that contains a biosurfactant produced by *Gordonia* sp. strain JE-1058. Bioresour Technol 100(2):572–577
Saikia RR, Deka S, Deka M, Sarma H (2012) Optimization of environmental factors for improved production of rhamnolipid biosurfactant by *Pseudomonas aeruginosa* RS29 on glycerol. J Basic Microbiol 52(4):446–457
Santos DKF, Rufino RD, Luna JM, Santos VA, Salgueiro AA, Sarubbo LA (2013) Synthesis and evaluation of biosurfactant produced by *Candida lipolytica* using animal fat and corn steep liquor. J Petrol Sci Eng 105:43–50
Scholz-Schroeder BK, Soule JD, Lu SE, Grgurina I, Gross DC (2001) A physical map of the syringomycin and syringopeptin gene clusters localized to an approximately 145-kb DNA region of *Pseudomonas syringae* pv. syringae strain B301D. Mol Plant Microbe Interact 14 (12):1426–1435
Sen S, Borah SN, Bora A, Deka S (2017) Production, characterization, and antifungal activity of a biosurfactant produced by Rhodotorula babjevae YS3. Microb Cell Fact 16(1):95
Shaheen M, Li J, Ross AC, Vederas JC, Jensen SE (2011) *Paenibacillus polymyxa* PKB1 produces variants of polymyxin B-type antibiotics. Chem Biol 18(12):1640–1648

Sharma D (2016) Applications of biosurfactants in food. In: SpringerBriefs in food, health, and nutrition, pp 43–80

Sharma D, Ansari MJ, Gupta S, Al Ghamdi A, Pruthi P, Pruthi V (2015) Structural characterization and antimicrobial activity of a biosurfactant obtained from *Bacillus pumilus* DSVP18 grown on potato peels. Jundishapur J Microbiol 8(9):e21257

Sidkey NM, Al Hadry EA (2014) Biosurfactant production by *Bacillus cereus*, B7 from lubricant oil waste. Int J Sci Res 3(12):498–509

Singh V, Haque S, Niwas R, Srivastava A (2017) Strategies for fermentation medium optimization: an in-depth review. Front Microbiol 7:1–16

Stoimenova E, Vasileva-Tonkova E, Sotirova A, Galabova D, Lalchev Z (2009) Evaluation of different carbon sources for growth and biosurfactant production by *Pseudomonas fluorescens* isolated from wastewaters. Z Naturforsch C J Biosci 64(1–2):96–102

Su C, Xiang Z, Liu Y, Zhao X, Sun Y, Li Z et al (2016) Analysis of the genomic sequences and metabolites of *Serratia surfactantfaciens* sp. nov. YD25T that simultaneously produces prodigiosin and serrawettin W2. BMC Genomics 17(1):865

Takahashi M, Morita T, Wada K, Hirose N, Fukuoka T, Imura T, Kitamoto D (2011) Production of sophorolipid glycolipid biosurfactants from sugarcane molasses using *Starmerella bombicola* NBRC 10243. J Oleo Sci 60(5):267–273

Tan YN, Li Q (2018) Microbial production of rhamnolipids using sugars as carbon sources. Microb Cell Fact 17(1):89

Thompson DN, Fox SL, Bala GA (2000) Biosurfactants from potato process effluents. Appl Biochem Biotechnol 84–86:917–930

Toren A, Orr E, Paitan Y, Ron EZ, Rosenberg E (2002) The active component of the bioemulsifier alasan from *Acinetobacter radioresistens* KA53 is an OmpA-like protein. J Bacteriol 184 (1):165–170

Tsuge K, Akiyama T, Shoda M (2001) Cloning, sequencing and characterization of the iturin a operon. J Bacteriol 183:6265–6273

Vallet-Gely I, Novikov A, Augusto L, Liehl P, Bolbach G, Péchy-Tarr M et al (2010) Association of hemolytic activity of *Pseudomonas entomophila*, a versatile soil bacterium, with cyclic lipopeptide production. Appl Environ Microbiol 76(3):910–921

Van Bogaert IN, Holvoet K, Roelants SL, Li B, Lin YC, Van de Peer Y et al (2013) The biosynthetic gene cluster for sophorolipids: a biotechnological interesting biosurfactant produced by *Starmerella bombicola*. Mol Microbiol 88(3):501–509

Varvaresou A, Iakovou K (2015) Biosurfactants in cosmetics and biopharmaceuticals. Lett Appl Microbiol 61(3):214–223

Vijayakumar S, Saravanan V (2015) Biosurfactants—types, sources and applications. Res J Microbiol 10:181–192

Volkering F, Breure AM, Rulkens WH (1998) Microbiological aspects of surfactant use for biological soil remediation. Biodegradation 8:401–417

Wei YH, Wang LF, Chang JS (2004) Optimizing iron supplement strategies for enhanced surfactin production with *Bacillus subtilis*. Biotechnol Prog 20(3):979–983. https://doi.org/10.1021/bp030051a

White DA, Hird LC, Ali ST (2013) Production and characterization of a trehalolipid biosurfactant produced by the novel marine bacterium *Rhodococcus* sp., strain PML026. J Appl Microbiol 115(3):744–755

Wu CY, Chen CL, Lee YH, Cheng YC, Wu YC, Shu HY, Götz F, Liu ST (2007) Nonribosomal synthesis of fengycin on an enzyme complex formed by fengycin synthetases. J Biol Chem 282 (8):5608–5616

Yamamoto S, Fukuoka T, Imura T, Morita T, Yanagidani S, Kitamoto D, Kitagawa M (2013) Production of a novel mannosylerythritol lipid containing a hydroxy fatty acid from castor oil by *Pseudozyma tsukubaensis*. J Oleo Sci 62(6):381–389

Yang XE, Wu X, Hao HL, He ZL (2008) Mechanisms and assessment of water eutrophication. J Zhejiang Univ Sci B 9(3):197–209

Ying GG (2006) Fate, behavior and effects of surfactants and their degradation products in the environment. Environ Int 32(3):417–431

Yoneda T (2006) Cosmetic composition comprising A and A lipopeptide. US Patent 0222616

Zarinviarsagh M, Ebrahimipour G, Sadeghi H (2017) Lipase and biosurfactant from *Ochrobactrum intermedium* strain MZV101 isolated by washing powder for detergent application. Lipids Health Dis 16:177

Zhao F, Shi R, Ma F, Han S, Zhang Y (2018) Oxygen effects on rhamnolipids production by Pseudomonas aeruginosa. Microb Cell Fact 17(1):39

The manufacturer's authorised representative in the EU is Springer Nature Customer Service Centre GmbH, Europaplatz 3, 69115 Heidelberg, Germany. If you have any concerns regarding our products, please contact ProductSafety@springernature.com

Printed and bound by CPI Group (UK) Ltd, Croydon, CR0 4YY
30/09/2026
02225293-0009